Stephenson College

T22428

Object-Oriented Systems Analysis and Design using UML

STEPHENSON COLLEGE COALVILLE
LIBRARY

Date Recd 7/3/01

Requisition No. 445697

Accession No. T22428

Location 004.2 Ben

HE

D1394580

H E COLLECTION

H E COLLECTION

Object-Oriented Systems Analysis and Design using UML

Simon Bennett, Steve McRobb and Ray Farmer

THE McGRAW-HILL COMPANIES

London · Burr Ridge, IL · New York · St Louis · San Francisco · Auckland
Bogotá · Caracas · Lisbon · Madrid · Mexico · Milan
Montreal · New Delhi · Panama · Paris · San Juan · São Paulo
Singapore · Sydney · Tokyo · Toronto

Published by
McGraw-Hill Publishing Company
SHOPPENHANGERS ROAD, MAIDENHEAD, BERKSHIRE, SL6 2QL,
ENGLAND
Telephone +44 (0) 1628 502500
Fax: +44 (0) 1628 770224   Web site: http://www.mcgraw-hill.co.uk

---

A catalogue record for this book is available from the British Library

ISBN 007 709497 2

Library of Congress Cataloging-in-Publication Data

The LOC data for this book has been applied for and may be obtained from the
Library of Congress, Washington, D.C.

Further information on this and other McGraw-Hill titles is to be found at
http://www.mcgraw-hill.co.uk/
Authors' Website address: http://www.mcgraw-hill.co.uk/bennett/

Publishing Director: Alfred Waller
Publisher: David Hatter
Typeset by: the Authors
Produced by: Steven Gardiner Ltd
Cover by: Hybert Design

Copyright © 1999 Simon Bennett, Stephen McRobb, Raymond Farmer
All rights reserved. No part of this publication may be
reproduced, stored in a retrieval system, or transmitted,
in any form or by any means, electronic or otherwise without the prior
permission of McGraw-Hill International (UK) Limited.

45CUP3210

Printed in Great Britain at the University Press, Cambridge

To all our families and friends

Contents

Acknowledgments

Our thanks to the reviewers of the original proposal for the book and to everyone who has read part of the book and provided us with feedback. Particular thanks go to David Howe who has reviewed each chapter as it was produced and who provided helpful comments throughout. However, the authors accept all responsibility for any errors, either of omission or of commission, that have crept into the text.

The names of products and of companies mentioned in this book may be claimed as trademarks or registered trademarks of their respective companies. Where those names appear in this book and the authors or publisher were aware of a claim, the name has been printed in all capitals, with initial capitals or with the addition of ™ or ®.

Extracts from Gamma/Helm/Johnson/Vlissides, DESIGN PATTERNS. (pages 185, 186, 187). © 1995 by Addison-Wesley Publishing Company. Reprinted by permission of Addison Wesley Longman.

Definition of ethics reprinted by permission of Chambers Harrap Publishers Ltd.

Extracts from Checkland/Scholes, SOFT SYSTEM METHODOLOGY IN ACTION. © 1990 John Wiley and Sons Ltd. Reproduced with permission.

Extracts from *Computing* reproduced by permission of *Computing*.

Preface

Background to the Book

The Faculty of Computing Sciences and Engineering at De Montfort University began teaching object-oriented systems analysis and design to first-year students on undergraduate programmes in Business Information Systems and Multimedia Computing in 1997, having taught structured approaches for many years. Object-oriented approaches to analysis, design and programming are now taught on all our diploma, undergraduate and postgraduate programmes in Computer Science and Information Systems. We wanted a book to support our teaching, one that put the analysis and design activities in the context of the whole systems life cycle (whatever kind of life cycle that is), and that included generic analysis and design issues, such as fact finding. Most books on object-oriented approaches to analysis and design concentrate on object-orientation and on the notation. We have, so far, only found one general systems analysis and design text that uses an object-oriented approach. However, the author of that text has chosen to base it on a notation that is nowhere near the developing industry standard of the Unified Modelling Language (UML). An approach from Dave Hatter at McGraw-Hill coincided with our search for a suitable book that we could use with our students, and we ended up writing it ourselves.

Another concern with many textbooks that we have used is that they do not employ a consistent case study as a source of examples throughout the book. We try to teach by means of practical case studies so that students experience something that is close to the development of a real system rather than a series of disjointed exercises. We particularly like the approach taken by two former colleagues, Barbara Robinson and Mary Prior in their book *Systems Analysis Techniques* (1995). Throughout the book they use two case studies, one for examples and one for exercises. However, their book teaches a structured approach using the notation of Structured Systems Analysis and Design Method (SSADM). In our book we have taken a similar approach and have developed two case studies through the course of the text, although not to the same level of detail.

In the rest of this Introduction, we explain who we think the book is useful for and different paths that readers may want to take through the book, depending on their interest and previous experience.

Who Should Read this Book?

The three authors of this book have been teaching in a Department of Information Systems and believe that systems analysis and design are activities that should take place in the context of the organizations that will use the information systems that are the result. The examples we use are based on business organizations, but they could be any kind of organization in the public or voluntary (not for profit) sectors, and the approach we adopt is suitable for most kinds of information system including real-time systems. The book starts with three chapters that set the development of information systems in this context.

We expect most of our readers to be students undertaking a Diploma, Bachelor's or Master's course in a computing or information systems subject. It will also be of relevance to some students of other subjects, such as business studies, who want to understand how business information systems are developed without wanting to be programmers.

The book is also suitable for professionals in computing and information systems, many of whom began their professional career before the advent of object-oriented development techniques and who want to upgrade their skills by learning about object-oriented analysis and design. We have used the Unified Modelling Language (UML), which we believe is the emerging *de facto* standard notation for object-oriented development.

Case Studies

In our teaching we use case studies as the basis of tutorials and practical work. We also use the same case studies to provide examples in lectures and in student assessments. We believe that it is important that students see analysis and design as a coherent process that goes from initial fact-finding through to implementation, and not as a series of disjointed exercises. This book uses two practical case studies. The first of these, Agate Ltd, is an advertising company. Agate is used for examples in most of the chapters of the book that explain techniques and for some exercises. The second case study, FoodCo Ltd, is a grower of fruit and vegetables and a manufacturer of packaged foods. FoodCo is used for most of the exercises that are included in chapters for the reader to complete.

The two case studies are introduced in short chapters (A1 and B1) that can be found between Chapters 4 and 5. In these first two case study chapters, we provide background material about the two companies and explain some of their requirements for computerized information systems. Chapter A2 (between Chapters 10 and 11) brings together some of the models produced in the analysis chapters of the book as an excerpt from the analysis specification for Agate's new system. Chapter A3 (between Chapters 17 and 18) brings together some of the models produced in the design chapters of the book as an excerpt from the design specification for the new system. We do not provide full models for the FoodCo case study, but FoodCo forms the basis of most of the practical exercises at the end of chapters. A few partial solutions are provided where they are required as the basis for later exercises.

If you are using this book as a teacher, you are welcome to use these materials and to develop them further as the basis of practical exercises and assessments. Some exercises that you may want to use are provided in each chapter.

Exercises for Readers

Each chapter contains two kinds of exercises for readers. Firstly we provide Review Questions. The aim of these is to allow you to check that you have understood the material in the chapter or section of a chapter that you have just read. Most of these Review

Questions should only take a few minutes at most to complete. Solutions to some of these questions are to be found at the back of the book. The answers to many of them are to be found in the text. Some require you to apply the techniques you have learned, and the answers to these are not always provided.

At the end of each chapter are Case Study Work, Exercises and Projects. These are exercises that will take longer to complete. Some are suitable to be used as tutorial exercises or homework, some could be used as assignments, and some are longer projects that could be developed over a matter of weeks. We have provided answer pointers to some of these exercises to help you.

Solutions to more of these exercises are available to bona fide university and college teachers who adopt the book as the set text for their courses. Please contact McGraw-Hill for details of the support materials.

Structure of the Book

Although we have not formally divided the book into sections, there are four parts to the book, each of which has a different focus.

Part 1

Chapters 1 to 4 provide the background both to information systems analysis and design and to object-orientation. In the first three of these chapters we explain why analysis and design are important in the development of computerized systems and introduce fundamental concepts such as those of systems theory. Chapter 4 introduces some of the ideas of object-orientation that will be developed in the second part.

Part 2

The second part of the book begins with the first two case study chapters (A1 and B1) and includes Chapters 5 to 10. The focus of this part of the book is on the activities of systems analysis and the basic notation of the Unified Modelling Language (UML). In it we introduce use cases, class diagrams, sequence diagrams, collaboration diagrams, activity diagrams, statechart diagrams and the Object Constraint Language (OCL). This part concludes with Chapter A2, in which we provide an excerpt from the analysis specification for Agate Ltd., one of our case study companies. The purpose of this is to draw together in one place diagrams and text that are developed in different chapters to illustrate different techniques. We do not have the space to provide all the analysis documentation in the book.

Part 3

The third part of the book is about systems design. It includes Chapters 11 to 17 and concludes with the design specification for the Agate case study (Chapter A3). In this part we develop the use of most of the diagramming techniques introduced in Part 2. We do this by enhancing the analysis models that we have produced to take design decisions into account. We also introduce package notation. This part covers the transition to design, the distinction between system design and object design, system architecture, design patterns, and the design of objects, user interfaces and data storage. The design specification at the end of this part serves the same purpose as the one at the end of the analysis chapters.

Part 4

In the final part we cover the later stages of the systems development life cycle and what we have loosely called 'advanced issues'. Originally we intended to include the chapter on design patterns in this part of the book, as some would regard it as an advanced aspect of design. However, we found that we wanted to use patterns in the other design chapters and moved it forwards. This leaves us with only three 'advanced' chapters. The chapter on implementation introduces the last of the UML diagrams, component diagrams and deployment diagrams.

Pathways through the Book

Whatever the formal structure of the book, you the reader are welcome to work through it in whatever order you like.

We teach two undergraduate analysis and design modules: a first year module on analysis and design, with the emphasis on the lifecycle and on analysis; and a second year module with an emphasis on design and the use of methodologies. For this purpose, we would expect to cover the following chapters in each module.

Year 1—Chapters 1, 2, 3, 4, 5, 6, 7, 8, 9, 11, 12, 13. (statechart and activity diagrams are omitted from this so that students are not overloaded with different diagramming notations in the first year.)

Year 2—Chapters 11, 12, 13 (recap), 14, 10, 15, 16, 17, 18, 19, 20, 21.

You might well ask why we did not write the book in exactly this order. We have tried to group together techniques under the general headings of analysis and design, even though many of them are used throughout the life cycle. This does not mean that we necessarily advocate following a life cycle model that treats analysis and design as separate phases. In many projects, developers will follow an iterative life cycle in which the models of the system are progressively elaborated and in which analysis and design are interwoven. However, we do believe that analysis and design are separate activities, even if they are not separate stages in a project's life cycle. We also think that it is easier to learn analysis and design as separate activities rather than merged together. (This is like a chef being taught to cook desserts and main courses separately. Later this chef can either follow a structured approach to cooking in which a dessert is prepared in advance and then chilled and stored until required, and the main course is then cooked separately. Or, when she is experienced, she can take an iterative approach and progressively build the meal, switching from main course to dessert and back again.)

If you are familiar with the general aspects of information systems development and are reading this book in order to gain an understanding of how you can use UML in analysis and design, then you can either start at Chapter 4, if you are not familiar with object-oriented approaches, or at Chapter A1, the first of the case study chapters, and work your way through from there.

Transferable Skills

Some of the skills of the systems analyst and designer can be classified as transferable or professional skills. Most employers place a high value on these skills. Many colleges and universities provide special modules on these skills, embed them in other modules or provide self-study packages for students. We have included material on fact finding skills, particularly interviewing, within the text of the book. We have not included other skills explicitly, but there are opportunities for teachers to use the exercises to develop skills in problem solving, group work, report writing and oral presentation.

Web Site and Support Materials

This book is supported by materials that we have placed on the book's web site. The web site is accessible at www.mcgraw-hill.co.uk/bennett/. These materials consist mainly of links to other web pages that we think that you the reader will find interesting or that extend the material that we have been able to include in the book. They also include solutions to some of the exercises and copies of some of the figures from the book that can be used for teaching.

   We welcome feedback about the book. You can email us at umlisad@dmu.ac.uk or write to us care of McGraw-Hill at the address at the beginning of the book.

Latest Version of UML

This book is based on UML Version 1.1. During production of the book, an editorial Version 1.2 was released, and a draft Version 1.3 was available at the time of going to press. Information on how the Version 1.3 notation differs from the Version 1.1 notation will be available on the book's web site.

Simon Bennett, Steve McRobb, Ray Farmer
Leicester, UK
March 1999

1

Information Systems—What Are They?

Objectives
In this chapter you will learn:
- how to define an information system;
- some examples and types of information system;
- how to apply the basic concepts of systems theory;
- how information systems are related to organizations.

1.1 Introduction

Information systems have played an important part in human affairs since our most distant ancestors first became capable of organized collective action. For example, Palaeolithic cave paintings of hunters and animals that date back 30,000 years or more can be seen as a sort of information system, used to capture, store, organize and display information.

The application of information technology (IT) in modern times has brought immense changes to the scope and nature of information systems. So profound are these changes that some writers believe we are living through an information revolution, on a scale that is little short of a second industrial revolution. This has been a recurrent theme since Bell coined the term 'post-industrial society' (1973), and remains an implicit undercurrent in Zuboff's classic *In The Age Of The Smart Machine* (1988).

Not everyone agrees that a direct comparison with the industrial revolution is valid. For example, Webster (1995) argues that contemporary changes in society, while certainly significant, do not represent the kind of radical break with the past implied by a 'revolution'. The picture is also complicated by the fact that, in some countries, the introduction of computers is interwoven with industrialization. Still, it is clear that computers have had a dramatic and pervasive effect on our lives. At the time of writing, it seems that both the spread of computerized information systems and the pace of technological development will continue to accelerate for the foreseeable future. Recent technical developments like the Internet have made possible information systems that would not have been feasible even a few years ago.

Despite the antiquity of some of its subject material, the academic study of information systems is young, even by the standards of the computer age, and owes its importance today chiefly to the rise of the computer. Those who developed the earliest computer systems gave very little thought to the issues with which this book is primarily concerned. Wartime imperatives drove them to apply the new technology to military problems like code-breaking, naval gunnery calculations and other mathematical tasks. Their focus was on the technical difficulties of implementing hardware based on ideas from the very cutting edge of research in electronics and control logic. It was also necessary to invent efficient techniques for controlling these new machines. In due course these techniques became today's computer programming.

By the time that computer scientists had gained a degree of mastery over their hardware, the world was once more at peace. Businesses began to be aware of the potential that computers held for commercial activities, and attention gradually turned to wider questions. Those with which we are concerned are the following.

- How do we establish the business requirements for a new system (often much subtler and more complex than the role of the earliest machines)?
- What effects will the new system have on the organization?
- How do we ensure that the system we build will meet its requirements?

It is primarily with these questions that the field of information systems is concerned, and they are also the main subjects of this book.

Within the relatively new field of information systems, object-oriented analysis and design are even newer, although they are derived from object-oriented programming, which has been around a lot longer. Object-oriented analysis and design were first conceived of only about a decade ago, and today they are still very much a novelty. Only in the last few years has there been a real surge of interest in object-oriented systems development. This relative youth of object-oriented analysis and design might give cause for concern. Are they no more than a passing fashion? We believe they are no mere fad, but on the contrary, for many information system applications—although not all[1]—object-orientation is simply the best way yet found of carrying out the analysis, design and implementation.

So far we have skirted around a question that we must address right at the start. What is an information system? In this chapter we provide the beginnings of an answer (the rest must come through practical experience). We set the scene by describing some information systems, ranging from one that is completely up to date, to others that are considerably older (Section 1.2). We discuss what is meant by 'system' (Section 1.3), by 'information' and finally by 'information system' (Section 1.4). We also put the question in context by raising some business issues, in particular how the managers of an organization decide (or discover) the things that are important for the business to do, and we outline the role of information systems in doing these things successfully (Section 1.5). For business managers, this is the real bottom line. An information system that did not help to ensure the success and prosperity of the organization, its investors and its employees would simply not be worth the cost of its development.

1.2 Information Systems in Practice

We set the contemporary scene in Box 1.1 by describing a complex modern information system. Many readers will be familiar with systems that resemble this one, and some may have used them to make purchases over the Internet. This particular example is entirely fictitious, although it is loosely based on elements taken from a number of real companies.

---

[1] We return to this question in Chapter 4.

Box 1.1 An on-line retail system

McGregor plc is a chain of retail stores that sells kitchen appliances, mobile phones and electronic home entertainment equipment. The company has recently created an on-line shopping centre, accessed via the World Wide Web. After registering their name and address, shoppers can browse through various products, selecting items and placing them in a virtual trolley. At the end of the trip, a shopper can buy what is in their trolley, remove items or quit without making a purchase. Payment is made by submitting credit card details on-line, or by entering part of the card details and phoning to give the rest. Delivery times are usually within three working days for small items such as mobile phones, but up to three weeks for larger items such as cookers. Goods are despatched direct to the customer's home. Credit cards are debited only on the day of despatch, and between purchase and delivery, customers can use the web site to check on the progress of their order.

   This is how an on-line shopper interacts with the system, but beneath the surface a great deal more is going on. A whole network of hardware connects the shopper's home PC and modem, through a phone line, a telephone exchange, another phone line and modem, to a computer that acts as a web server. This is connected to other networks at McGregor's head office and shops. Many software applications are also busy processing information captured through the web pages, and feeding various support activities. Some of these are computerized and others are carried out by people.

- Marketing staff keep prices and product details up to date on the electronic product catalogue system. This can also be accessed by touch-screen PCs in the shops.
- Credit card details are stored electronically for relay to the card processing centre when the goods are despatched.
- Robot forklift trucks in the warehouse fetch items to the loading bay when they are due for despatch, and warehouse staff load them onto delivery trucks.
- Delivery drivers follow a schedule linked to an electronic map in the vehicle cab. This is updated by radio every few minutes, helping to avoid traffic jams.
- Out of stock items are re-ordered from the supplier by electronic data interchange (EDI). On delivery, warehouse, delivery and charging begin quite automatically.
- At each significant point in the sequence, an entry on a web page is automatically updated, allowing shoppers to discover what stage their order has reached.

There are many users besides the shopper, each with a different view of the overall system. A network manager monitors the number of hits on the web server and network traffic within McGregor, checking for hardware or software failure and breaches of network security (e.g. if hackers try to break in). Her concern is the efficient and secure flow of information; she is not interested in its content. A financial controller uses a linked accounting system to monitor sales transactions and cash flow. A despatch clerk regularly checks forklift schedules on her PC, and compares them to delivery truck schedules faxed daily from the courier company. She smoothes the flow of goods through the loading bay, and spends long hours on the phone sorting out delays. A market researcher uses a specialized statistical package on a portable PC to analyse on-line sales, assessing the success of the web pages' presentation styles. Registration allows individual customers to be tracked as they browse the site; using information about their preferences, the design can be finely tuned to attract high-spending customers.

Box 1.2 Fighter Command's 1940 information system

This vital communications and control system was used to monitor and control Spitfire and Hurricane fighter squadrons as they flew against Luftwaffe bombing raids during the Battle of Britain. It is quite certain that, without it, the RAF could not have succeeded in the defence of Britain in 1940. The main 'hardware' was a large map kept at Command headquarters at Bentley Priory. Coloured counters were used to mark the current position of various aircraft formations, and were moved around the map by hand as reports came in via radar, telephones, teleprinters and radio. Reports of enemy raids were collected both by radar and a network of observer posts throughout the country, while information about RAF deployment was sent in via Group and Sector control rooms. A special filter room correlated and checked all reports before passing them through to the main control room for display. Other displays included a blackboard showing at a glance the status of all current Luftwaffe raids, and a row of coloured lights flagged the readiness of RAF fighter squadrons. Carefully co-ordinated duplicates of these displays were based at other control rooms, including those at Sector level where the controller spoke directly by radio to the tense young pilots in their aeroplanes. Using this basic, but effective technology, human operators performed communication, filtering, processing and display tasks that would today be carried out by computers.

Some accounts of the Battle of Britain have claimed that it was won chiefly due to the courage of British pilots or to the RAF's use of radar. But these miss a most important point. The new radar technology was certainly important, but the British variety was still technically far inferior to its German equivalent, already in use for naval and anti-aircraft gunnery. Equally, there could have been no victory without the bravery of the RAF pilots, but was this really greater than that of their German opponents, who also faced extreme danger each time they flew over the English coast?

Important as these factors were, it was only through Fighter Command's information system that the British human and technical resources were deployed so effectively.

The staff described at McGregor all rely on computer systems for the information they need to do their jobs. But only recently have information systems begun to use information technology in such a sophisticated way. Our next example, described in Box 1.2, is almost exactly contemporary with the birth of the electronic digital computer, yet it contains little that we would recognize today as information technology . Checkland and Holwell (1998) describe it aptly as 'the information system that won the war.'

With a little imagination, it is easy to see how a modern system could use computers to fulfil similar functions to those described in Box 1.2, and indeed this is the situation in most airforces today. Box 1.3 looks even further back in time, introducing briefly an information system that dates back over a century, yet still operates effectively today with similar technology in many parts of the world.

The distinguishing characteristics of the pre-computer information systems described in Boxes 1.2 and 1.3 would change relatively little if they were to be re-implemented today using computers. But the system described in Box 1.1 would not be feasible at all without the use of modern information technology. Thus, while the computer is not a necessary precondition for the existence of many information systems, in some cases today computers have made possible information systems that could not exist without them.

Box 1.3 A railway signalling system

This example is particularly interesting because, according to most classifications of work and social status, an old-fashioned railway signalman would be regarded as essentially a manual worker of the early industrial age. Yet on closer examination he can be seen clearly to be an information worker, differing from many modern computer users only in that the technology available to him was much less sophisticated than would now be the case. According to Webster (1995) '. . .The railway signalman must have a stock of knowledge about tracks and timetables, about roles and routines; he needs to communicate with other signalmen down the line, with station personnel and engine drivers, is required to "know the block" of his own and other cabins, must keep a precise and comprehensive ledger of all the traffic that moves through his area. . .'. In other words, the signalman operates an information system that comprises his ledgers, the levers, signals and lamps and what he knows in his mind.

Nowadays many railway signalling systems are largely automated[2]. However, whether the work is done by networked computers linked electronically to electric motors that switch points and operate signal lights, or by a traditional signalman who pulls levers connected by steel cables to the points and signals, and keeps careful records in a hand-written ledger, the tasks that are actually carried out are little different. All that has changed is the technology that implements the system.

1.3 General Systems Theory

We now introduce some basic concepts of general systems theory. This is essentially a way of understanding a system in terms of those components and characteristics that are common to all systems. We use the term *system* here in a very specific sense, although one that has much broader application than just computer systems. In everyday speech, people may refer to the legal system, a tropical storm system, the system of parliamentary democracy, an eco-system, a system for winning at roulette, a computer system in someone's office, a system for shelving books in a library, a system-built house, and many more. Some of these certainly meet our definition of a system, and others do not. Probably the only thing that they all have in common is that they have some kind of organization. But in general systems theory, a system is rather more than just anything that shows a degree of organization.

1.3.1 Characteristics of a system

So what are systems? For our purposes, a system has the following characteristics.
- A system exists in an environment.
- A system is separated from its environment by some kind of boundary.

---

[2] This raises questions about the ethics and efficacy of the computer simulation of human judgement in safety-critical situations, but such arguments are beyond the scope of this book.

- Systems have inputs and outputs. They receive inputs from their environment, and send outputs into their environment.
- Systems have interfaces. An interface allows communication between two systems.
- A system may have sub-systems. A sub-system is also a system, and may have further sub-systems of its own.
- Systems that endure have a control mechanism.
- System control relies on feedback (and sometimes feed-forward). These comprise information about the system's operations or its environment, that is passed to the control mechanism.
- A system has some properties that are not directly dependant on the properties of its parts. These are called emergent properties, as they only emerge at the level of the system as a whole.

Figure 1.1 presents the relationships between these concepts schematically.

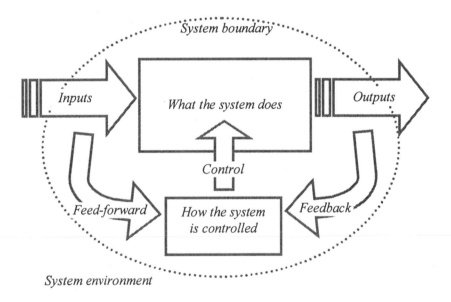

Figure 1.1 *Parts of a system, and their relationship to each other.*

Boundary and environment

An understanding of the related concepts of *boundary* and *environment* is essential to making sense of any system. In fact the first step in understanding a system is to choose the system that you wish to understand, and this largely means choosing its boundary.

For example, a cell biologist may think about a single human cell as a system. Her interest, and therefore the system she thinks about in detail, is bounded by the cell membrane. A biochemist might be interested in a slightly larger system, perhaps a chemical reaction that takes place in a particular group of cells. A specialist physician attempting to diagnose an illness may consider a patient's kidney as a system. The boundary of this system may coincide with the organ, or it may be wider still, depending on the particular illness. For a doctor in general practice, a person's whole body may be considered as a system bounded by its skin. These different boundaries are illustrated in Figure 1.2. What is readily apparent is that we can continue zooming out in this way, perhaps until we reach a cosmologist (whose interest might be bounded by the limits of the physical universe) or even a theologian (whose interest are even wider).

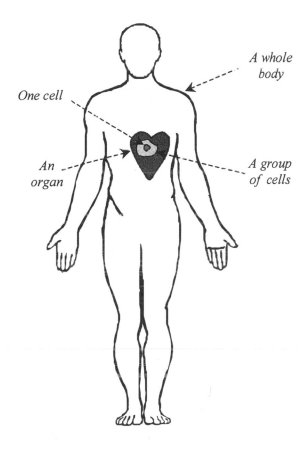

Figure 1.2 *Systems at different scales.*

But the choice of a system that corresponds to a subject of interest is not simply a matter of scale. Systems can overlap with each other, and this is also an issue for boundary setting. Let us think again about a whole person as a system. Within this image, many other systems can be found. For example, medical specialisms each have their own view of what is interesting or important. A neurologist may focus on the nervous system, consisting of the brain, spinal cord and the network of nerves that spreads through the body to just beneath the surface of the skin. While its physical boundary is almost identical to that of the body as a whole, the nervous system contains only various specialized nerve cells. The interests of a haematologist thinking about the circulatory system have a similar boundary, yet this system contains instead blood cells, blood vessels and the heart. A person can also be considered as a variety of non-physical systems. A psychologist might study an individual's cognitive system or emotional system, or may consider a child's intellect as a learning system. If we move along the scale again, a social psychologist may think of a family as many overlapping systems: a child-rearing system, an economic system, a house-maintenance system, etc. These views of the family as a system have purely conceptual boundaries, since family members remain part of their system no matter how distant they are in space.

The boundaries of different systems can overlap or coincide. Indeed, two systems may be closely related, may have identical boundaries, and yet still be distinct. This is potentially a tricky part of the theory. At what point does one system end and another begin? Also, can a thing (e.g. a human organ) simultaneously be part of one system and of another, at one and the same time?

The answer to this lies in the fact that the usefulness of systems is in helping us to understand something about how the world works. They do this by representing selected aspects of the world in an abstract way. In most cases, it is not important whether or not the system corresponds in every detail to the thing it represents. Checkland and Scholes (1990) explain this as follows.

> … it is perfectly legitimate for an investigator to say 'I will treat education provision *as if it were* a system,' but that is very different from saying that it *is* a system. . . Choosing to think about the world as if it were a system can be helpful. But this is a very different stance from arguing that the world *is* a system, a position that pretends to knowledge that no human being can have.[3]

Figure 1.3 illustrates the same point graphically.

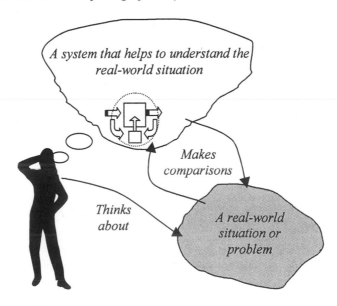

Figure 1.3 *The relationship between system and reality (loosely adapted from Checkland and Scholes, 1990).*

---

[3] Italics in the original.

This does not mean to suggest that no system has a basis in reality. On the contrary, many systems are made of real components. For example, it is quite possible to touch physically all the parts of a central heating system. But it is a matter of choice, based on our interest at a given moment, whether we choose to *think* about it as a system. Any system that we think about necessarily exists in our thoughts, rather than in the world, and such a system, however much it corresponds to the world, is still a subjective view of reality, not the reality itself.

Input, output and interface

Systems have interactions with their environment. For example, human cells take in food and oxygen and excrete waste materials. Some cells produce things of use to the body as a whole. For example some white blood cells produce the antibodies that fight infection. The nervous system receives sensory information in the form of light, sound, touch, etc. and transforms these ultimately into signals that move the body's muscles and generate speech. Each interaction is based on a set of inputs and outputs. Inputs originate outside the system, and are taken in to be used in some way. Outputs are created by the system, and sent into the environment in order to have an effect somewhere else, but always to achieve a purpose of the system. Figure 1.4 shows some inputs and outputs for three systems.

System	Inputs	Outputs
A student	Information Exercises Guidance	New knowledge New ideas Solutions
A family	Money Social standards and norms (e.g. laws) Purchases Daily news	New citizens (i.e. children) Products of family members' work Social influence Votes in elections
A business	Raw materials and labour Investment Information (e.g. customer orders)	Profit and taxes Finished products Information (e.g. the company report)

Figure 1.4 *System inputs and outputs (note that there is not necessarily a one-to-one correspondence between inputs and outputs).*

An important characteristic of purposeful systems, such as a business, is the way they transform inputs into outputs, since this is how they fulfil their objectives. For some purposes, it is enough to know the relationship between inputs and outputs. For example, a customer of the McGregor on-line shopping system described in Box 1.1, is probably not interested in what goes on behind the scenes. Her main focus is on the interaction with the system—in other words, the inputs she must provide (information, money) and the outputs she receives (information, a new fridge/freezer, customer service). This is called a *black box* approach, since it treats the system as an opaque box whose internal workings are completely hidden. All that can be seen is what goes into the system, and what comes out. This is illustrated in Figure 1.5.

Figure 1.5 *A black box view of a system.*

An output from one system can simultaneously be an input to another. The two systems share part of their boundary, across which inputs and outputs pass between them. The shared boundary is an interface. For example, tiny little air sacs in your lungs act as an interface between the circulatory and respiratory systems. Here, freshly inhaled oxygen is output by the lungs and input to the bloodstream for distribution, while spent carbon dioxide is output by the blood and input to the lungs for exhalation. We shall see later, in many places throughout this book, that the identification and understanding of interfaces is of particular importance in the development of information systems.

Sub-systems

Sub-systems are a natural consequence of other systems ideas introduced above. For example, a white blood cell making an antibody can be seen as a system on its own, or as a sub-system that contributes to the larger system. Figure 1.6 illustrates some sub-systems that can be found in the description of the Agate case study in Case Study A1 (this case study is used to provide many of the illustrative examples throughout the book).

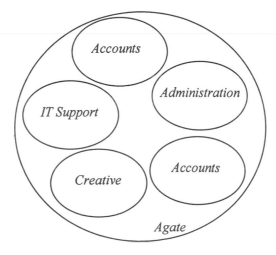

Figure 1.6 *Sub-systems of Agate.*

Sub-systems are at once part of a larger system, and also coherent systems in their own right[4]. Communication between sub-systems is, by definition, through interfaces. This kind of diagram is called a system map. Another way of arranging systems and sub-systems is as a hierarchy. Hierarchies are a very important aspect of systems theory, to which we will return frequently (particularly in Chapter 4).

Control in systems

Many systems have a specialist sub-system whose function is to control the operation of the system as a whole. In fact, general systems theory sprang partly from the science of cybernetics, which studies control in natural and artificial systems. A familiar type of cybernetic control is the family of thermostatic devices that control central heating, hot water and air conditioning systems, and also temperature-controlled appliances such as freezers and ovens. Many are based on a simple feedback loop, as shown in Figure 1.7. System control is usually based on a comparison between two or more input values, whose similarity or difference guides a decision about whether any controlling action is required.

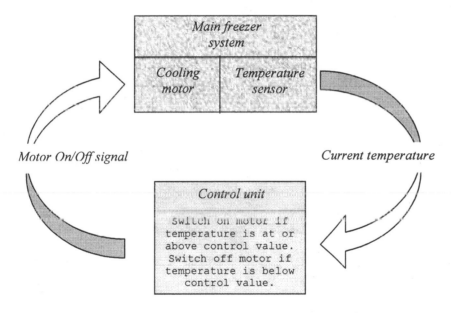

Figure 1.7 *Thermostatic control in a freezer—a simple negative feedback loop.*

Feedback

The term *feedback* indicates that one or more outputs of the system is sampled and literally fed back to the control unit. In Figure 1.7, a sensor samples the temperature in the main freezer compartment (an output measure) and this is fed back to the control unit. Here a

---

[4] Koestler (1967) coined the word 'holon' to express this duality. A holon is something that, seen from one perspective, looks like a whole thing, but, viewed in another way, looks like a part of something else.

logic mechanism (this could be either mechanical or electronic) decides whether action is required. Possible actions are turning the cooling motor on and turning it off. Both are effected by a signal sent to the motor by the control unit, thus 'closing' the 'feedback loop'. In this way, the control unit monitors the current temperature, compares it a reference value (in this case probably set by the owner using a dial or switch), and if necessary adjusts the system's operation to bring the difference between the two values within pre-defined limits.

This is *negative* feedback, because it aims to maintain the system's equilibrium by opposing deviations from some norm. Negative feedback is used widely in physical systems, such as electronic equipment, manufacturing systems, etc. By contrast, *positive* feedback works by reinforcing deviations instead of opposing them and therefore tends to increase movements away from equilibrium. Systems governed by positive feedback are inherently unstable, which is undesirable when the deviation that is reinforced is itself undesirable. For example, the ear-splitting howl sometimes heard from public address systems is the result of a positive feedback loop. It usually occurs when sound output from a loudspeaker is caught by a microphone for re-input to the amplifier. Under certain conditions the signal resonates round and round the loop, growing louder at each turn. The volume then builds quickly until the amplifier reaches its limit.

Both negative and positive feedback occur in social and business systems, but here they are rarely so clear-cut or simple. For example, the Bank of England regularly reviews the performance of the UK economy to determine whether or not it meets the Government target for monetary inflation. When the economy deviates from the target, action is taken to bring it back on track, usually by altering the bank's base interest rate. This would be a simple example of negative feedback, though of course, in reality the situation is much more complex. Another example of negative feedback is when a business adjusts its manufacturing output to maintain a constant inventory. If sales rise, stock falls, and the rate of production is accordingly increased. If sales fall, stock tends to rise, and the rate of production is reduced. Here the stock level acts as a *buffer* between production and sales sub-systems.

Positive feedback is beneficial when a steady state is undesirable. For example, in a soccer game, equilibrium would tend always to result in a goal-less draw. Any competitive edge that one team gains over its opponents is exploited to retain and maximize the advantage. However, this picture is greatly complicated by the fact that both teams continually strive to get the upper hand in a dynamic game, and so it is rare to see the cycle build to an extreme. Instead, the system (i.e. the game) usually tends to continue in a state of dynamic disequilibrium. Negative feedback also plays a role, as teams try to maintain any equilibrium that acts in their favour. A complex mix is seen also in business competition, where firms capitalize on any competitive advantage they can gain over their rivals (positive feedback, aiming to disrupt an unhelpful equilibrium), while also trying to prevent their rivals from gaining an edge (negative feedback, maintaining a beneficial equilibrium).

Positive feedback does not necessarily mean that every deviation from the norm is reinforced. Some may be allowed to die away through lack of encouragement. In audio amplifiers all frequencies are amplified, but only resonant frequencies are disastrously reinforced by the feedback cycle, while others fade naturally without ill-effect. Often a sound engineer need only adjust the amplifier's tone control, reducing slightly the volume in the troublesome frequency band, to cure a feedback problem. A similar type of control can be discerned in businesses that directly link productive output to sales, perhaps because there is a very short-term seasonal market (say, toys at Christmas). In this case, when sales rise, production is increased as far as possible. But when sales fall below a critical level, a product may be dropped altogether.

Feed-forward

Feed-forward information relies on sampling the system's inputs rather than its outputs. For example, the Christmas toy business mentioned in the previous paragraph may find that the use of feedback from sales does not allow them to react quickly enough to changing market conditions. This may leave them with unsold stock, manufactured before they learned that sales had slumped. Ideally, they should adjust their manufacturing to suit the level of orders (an input), rather than sales (which measure an output). This would allow the firm to avoid manufacturing products for which there is no market. While feed-forward control information can help a system to be more responsive to environmental fluctuations, it is not always easy to implement or manage in a business organization. Difficulties still arise if the rate at which conditions change in the environment is faster than the rate at which the business can adapt. The effects of this are apparent to anyone who visits the kind of specialist book shop where remaindered titles are sold cheaply.

In the Agate case study, the agency must employ and train sufficient staff to cope with the anticipated workload. If there is a serious slump in orders for new work, it may not be possible to reduce the number of staff quickly enough to avoid bankruptcy, due to the need to give staff a period of notice before they can be laid off. The company may also not be able to respond quickly enough to a sudden surge of orders, due to the lead time for recruiting and training new staff. Forecasting the level of demand for a service such as Agate's is an important role of information systems in business.

Emergent properties

One of the truly distinctive characteristics of a system is that it is more than just the sum of its parts. The system possesses some feature or ability of its own that is not present in any of its components. These are called emergent properties. For example, a car is only a form of transport if there is enough of it to drive. It then has the property of being a vehicle, but the wheels, windscreen, motor, etc. do not have this property until they are correctly assembled.

This recognition that systems have emergent properties is the main reason why the systems approach is described as *holistic*. This means thinking about each system as a whole, with important aspects that will be overlooked if we think only about its parts in isolation from each other. The holistic world view of the systems approach is manifest in techniques like the rich picture. This aims at capturing in one picture or diagram all that is essential to an initial understanding of a system[5]. An opposing view is taken by reductionists, who begin with the assumption that complex phenomena can be explained by reducing them to their component parts. This approach has a very important place in the physical sciences (e.g. physics and chemistry, although there are dissenters even in those disciplines). We believe it is thoroughly misapplied when used to explain complicated human situations.

---

[5] See Checkland and Scholes (1990) for a full explanation of rich pictures with many examples of their use in practical situations.

1.3.2 Systems of interest to this book

We have used a mixture of natural and artificial systems as examples. General systems theory is applicable to both, but in this book we are principally interested in the artificial kind. Even some of these are more relevant than others. For example, we will probably only be professionally interested in the system of parliamentary democracy if we intend to develop an information system that supports it. Two types of system are of particular importance to us. The first is the *human activity system* (Checkland, 1981). A key feature of this type is that it centres on a purposeful activity, such as we find when we look at a business, a club, a hospital, or any organized activity. There is often wide disagreement among the participants about exactly what that purpose is, and this can be a significant problem for the analysis of information system requirements. Figure 1.8 shows how some of the people who work at Agate see the purpose of one system in which they are engaged.

System	Purpose of system	As seen from the perspective of ...
Agate (a business system).	To become a successful advertising agency on the international stage, thus providing both wealth and prestige for its directors.	A director.
	To provide varied and interesting work with a good salary, and also to be a useful stepping stone towards the next career move.	A copy-writer.
	To provide a pleasant and comfortable life until retirement (5 years away), without the need to make too much effort.	Another director.

Figure 1.8 *A human activity system with multiple purposes.*

The second kind of system in which we have an interest is, naturally, the information system. Information systems are constructed to help people in a human activity system to achieve their goals, whatever these might happen to be. Thus human activity systems are the context of, and provide the meaning for, the activity of information systems development. Anyone who does not understand the meaning and purpose of a human activity system will find that it is impossible to specify, still less to build, an information system that supports it in any useful way.

1.4 Information and Information Systems

Some computing professionals see the development of information systems as essentially a matter of designing and building computer technology (including software) that meets a set of clearly understood needs. We see this as a simplistic view that misses much of importance. It happens that information technology is now the normal technology used to implement an information system, but, as we saw in Section 1.2, information systems were constructed in many other ways prior to the development of digital computers.

Designing and building technology can be the easy part of the job, and the hardest part is often determining the needs that the technology must serve. This involves identifying ways that an information system can support the purposes of a human activity system. An understanding of the information that will be useful to the human actors is an important ingredient in this, as is an understanding of how the information can be used effectively. As a result of these many different concerns, Information Systems has become a multidisciplinary subject that bridges many other fields, in particular computer science and business management, but also psychology, social theory, philosophy and linguistics, among others.

1.4.1 Information

Information is conveyed by messages and has meaning. Meaning always depends on the perspective of the person who receives a message. We are always surrounded by a vast mass of potential information, but only some of this ever comes to our attention, and only some of that is actually meaningful in our present context. Checkland and Holwell (1998) describe the process by which raw facts become useful (i.e. become information) through a sequence of stages.

For an example of how this happens, consider three people watching the evening sky. A plume of smoke is rising in the middle distance. For the first, the smoke is just part of the general view, and she does not consciously notice it. The second sees it, and it evokes a memory of a camping trip long ago. But he is aware that the only connection between past and present smoke is a coincidence of shape and colour, so he moves on to look at something else. The third is thrown into consternation, when she realizes that the smoke is rising from her house, which is on fire. She runs to phone the Fire Service before doing whatever else she can to save her house.

The sight of the smoke is, on the face of it, a single message available to be received by all three, yet its meaning is different in each case. As the first person does not even notice it, Checkland and Holwell call this *data* (from the Latin for 'given'), meaning that this is a fact, but one that has not been selected for attention. The second person notices the smoke but does not relate it to any present context. Checkland and Holwell call this *capta* (from the Latin for 'taken'), meaning a fact that has been selected for attention. The third person not only notices the smoke, but give it a meaning derived from the context in which she sees it (recognizing her house, understanding the implications of the smoke, etc.). This can be called *information* since it has a meaning within the context. The meaning of a fact is thus entirely dependent on its relevance to the observer.

There is a final step where information becomes *knowledge*, by being structured into more complex meanings related by a context. We can see this in the third person's response to the smoke. She integrates information from many sources: the link between smoke and fire, the effect of fire on houses, the existence and purpose of a Fire Service and the location of nearby phones. All this information comes together in a single framework of knowledge that is relevant to the context. In a word, she *knows* what to do.

1.4.2 Information systems

In the past, it has been usual for systems analysis and design textbooks to include a section that broadly classifies the variety of information systems an analyst may encounter in practice. Such a classification is less useful today than it was once, since information systems have tended to become ever more integrated with each other, and thus the boundaries have blurred between categories that, not long ago, were quite distinct. Thus, while it is still helpful to present a brief overview of some of the general types of application in organizations, the following is intended to introduce some aspects of the role of information system, rather than to depict actual systems.

Operational systems

Operational systems automate the routine, day-to-day record-keeping tasks in an organization. The earliest commercial information systems were operational ones, because routine, repetitive tasks that involve little judgement in their execution are the easiest to automate. Accounting systems are one example. These derive from the need in all organizations to keep track of money—the amount coming in, the amount going out, the cash available to be spent and the credit that is currently available. Few modern organizations could survive long without a computerized accounting system. Sensible organizations have a 'disaster recovery plan' that details how they intend to cope with an emergency that destroys data or renders computer systems inoperable.

The flow of information through an accounting system is based on thousands, even millions of similar *transactions*, each of which is an exchange of a quantity of something, usually a money value. For example, when you buy a tube of toothpaste in a supermarket, two separate records are made. One records that a tube of toothpaste was sold, and the other records the money you paid in exchange for it. As this repeats day after day for each item, customer, checkout and branch, an overall picture is built up that allows the company's accountants to compare total income with total costs and determine whether a profit has been made. Real accounting systems are more complicated than this, often with sub-systems to handle wages, taxation, transport, budget planning and major investments.

Organizations use personnel systems to keep track of their employees, partly so that managers can control the money spent on wages, but also for many other reasons. The simplest personnel systems provide information that keeps the internal phone directory up to date, or invites employees to retirement planning seminars when they reach a certain age. More complex applications may calculate each employee's working hours for the next few months, based on the company's current level of orders, how many employees are currently sick, how many have booked holidays and the mix of skills needed for each production team to be effective[6].

Management support systems

Information systems intended to support management usually work at a much higher level of complexity than operational systems. But the blurring of boundaries in recent years has had particular effect here. For example, the personnel planning application discussed at the end of the preceding section could be considered as management support rather than operational. Moreover, much of the information used by management to make decisions is

---

[6] At the time of writing, a major UK car manufacturer has just announced a flexible working agreement with its employees that is similar to this. Such a deal will require a lot of information to be gathered and processed correctly during each planning cycle.

derived directly from information stored at the operational level. In practice, many management support systems are built on top of operational systems, while other systems combine elements of the two, meeting a complex set of needs at different levels of the organization.

Many of the earliest management support systems were developed simply by adding a set of programs (known as a management information system, or MIS) to extract data from existing operational systems, and analyse or combine it to give managers information about the part of the organization for which they were responsible. We can easily see how this happened with an accounting system. Once all routine sales transactions had been stored on a computer, it was a short step to the realization that if this data were analysed appropriately, it could tell managers at a glance which products were not selling well, which checkout operators took too long dealing with a customer, which store had the lowest volume of trade, and so on.

This information is useful to managers because they have a responsibility to maximize the performance of an organizational sub-system. An important part of this is identifying and resolving problems as they occur. Thus, the crucial aspect of a management support system is the feedback or feed-forward that it provides, alerting managers to problems and opportunities, and assisting them in the process of tuning the organization's performance. Relating this to the diagrams in Figures 1.1 and 1.7, the chief difference between operational and management support systems is that they fit into different parts of the overall system. Operational systems are located in the central box of Figure 1.1 (labelled 'what the system does'), or alternatively assist its work by supporting the flow of inputs or outputs. Management support systems are located in the box in the lower part of Figure 1.7 (labelled 'control unit'), or assist its work by supporting the flow of feedback to, or control information from, the control unit.

Real-time control systems

Real-time systems are explicitly concerned with the direct control of a system's operations, often physical in nature. Examples include lift control systems, aircraft guidance systems, manufacturing systems and the robot forklifts in the McGregor system described in Box 1.1. For this reason, they are perhaps best considered as a control sub-system of a physical processing system. Their role is thus very different from both operational and management support systems. Real-time systems usually have human operators (to date, few are completely independent of human supervision, though this may become common in the future), but they are generally insulated from the surrounding human activity system. In fact, many authors would not agree that real-time systems are information systems at all. We do not regard this as an important issue. The techniques used for the analysis, design and implementation of real-time systems are very similar to those used for other computer systems, so in practical terms any distinction is artificial.

1.4.3 Information technology

We have left information technology until last because decisions about information technology should ideally be made last in the cycle of development. Only when the human activity system has been understood, the need for an information system has been identified, and the information system requirements have been defined—only then should the emphasis turn to the information technology that will implement it. Sadly, we cannot pretend that this is how things always happen in the real world. Indeed it is partly for this reason that so many systems in the past have been so unsuccessful. But it is how they *should* happen. Section 1.5 of this chapter examines some of the reasons for this in more detail.

Here we offer only a broad definition of information technology, and this only for the sake of clarity. We understand it to include all the varieties of hardware familiarly known as, based upon, or that include within them, a computer or its peripheral devices. For example, obvious things like desktop PCs, pocket electronic organizers, modems, network cabling, file-servers, printers and computer-controlled machinery in factories and airliners, and also less obvious things like digital mobile phones, the electronic circuits in some cars that calculate fuel consumption, the microchips in some cameras that set the aperture and shutter speed—in other words, everything misleadingly described in marketing literature as 'intelligent'. Due to the unprecedented rate of technical progress, the range of devices that can be described as information technology increases almost daily, and the boundaries between them blur. For example, consider the devices that combine a mobile phone, a modem and email software, or the add-on cards that let a PC display broadcast TV or the new digital TV sets that behave in many ways like network computers.

All the examples of information technology that have just been mentioned are really just tools that, like any tool, can be used for many different tasks—and not only those for which they were intended. There is a saying that, if your only tool is a hammer, the whole world looks like a nail. The corollary is also true: if you can see only nails, you will be inclined to use any tool that comes to hand as a hammer, regardless of whether it is actually a wrench, a book or a can of beans. So how a tool will *actually* be used matters much more than how it is *meant* to be used. A modern word-processing package provides a skilled user with the facilities to automate many complex tasks by using macro programs, mailing lists and embedded objects like spreadsheets, sound clips and hyperlinks to the Web. Yet many users have no need of all this and would be just as happy with an electronic typewriter. The question is, then, if an electronic typewriter is all that is required, why install a powerful PC running all the latest software?

1.5 Strategies for Success

Business strategy is an essential part of the context for this book, since all information systems analysis and design projects begin with the identification of a business issue that can be aided or a problem that can be solved by the use of an information system. The underlying assumption is that information systems are only worthwhile if they meet the needs of the organization in which they are installed. In this section, we consider some ways that business needs can be identified at a very high level, suggesting possible application areas for information systems and information technology.

1.5.1 Identifying a business strategy

The development of a business strategy essentially begins with the question: 'Where would we like our organization to be in (say) ten years' time?' Answers to this question identify a set of goals for the organization. Logically, the next question is then: 'How do we get from where we are now to where we want to be?' In effect, how do we meet those business goals? Answers to this question will ideally suggest some practical steps that can be taken towards achieving the strategic goals.   A number of different techniques have been developed to help with this process. One example is the well-known SWOT approach[7], but many others have been used by organizations to develop their business strategies, including the Value Chain Model, which we introduce in the next section.

The detailed contents of any particular strategy (i.e. the actual goals and steps it contains) depend very much on the characteristics of the organization, its environment, the skills of its workforce, and many other factors. In the Agate case study, we are told that the strategy is 'to continue to grow slowly and to develop an international market' (these are the goals). The directors also have a view about how to achieve these objectives: they want to get more business from multinationals, and they hope to succeed in this through the quality of their work and by developing 'campaigns that have a global theme but are localized for different markets around the world' (these are the steps). We can safely assume that these elements have been included in the strategy because the directors are confident of certain other things. For example, they may believe that the technical quality of Agate's work and the creativity of their staff are both strengths of the company, and will meet the demands placed on them.   They probably also believe that their current client base and contact list is extensive enough for them to win the kind of work they are seeking.

1.5.2 The contribution of information systems

Information systems can contribute to the achievement of business goals in so many different ways that it can seem daunting to decide which are the systems that really matter. Many techniques have been developed to help arrive at useful answers, some adapted from their original purpose as tools for developing a business strategy. One example is Porter's Value Chain Model (1985). While not the only technique for information system planning, and not necessarily the best for all situations, it is a useful way of structuring this discussion because of the systemic view of an organization that it presents.

The model is illustrated in Figure 1.9, and depicts an end-to-end 'chain' or flow of materials through the organization. The flow is transformed at each stage, beginning with raw inputs that become a product through operations applied to them, then are shipped and sold to a customer who finally receives an after-sales service. In practical applications, a group of managers and other staff allocate activities and departments to appropriate compartments of the model, and identify the value each adds to the overall product or service.

Porter used the chain metaphor to reinforce the point that all primary activities are essential, but any weak link negates the value of work done at every other stage.   For example, if a business is good at selling its products, but the products themselves are of poor quality, it is unlikely to be successful. Alternatively, a business that makes excellent products but has very poor arrangements for obtaining its raw materials will also not be successful.   In a successful organization, each primary activity (the lower part of the

---

[7] This stands for Strengths, Weaknesses, Opportunities, Threats. Usually carried out as a group brain-storming exercise, it involves identifying and categorizing everything about the organization's current circumstances that falls into these categories. The resulting strategy is based on finding ways of exploiting the strengths and opportunities, while counteracting the weaknesses and threats.

Functions in the Organization

Administration and Infrastructure				
Human Resource Management				
Product and Technology Development				
Information Systems Development				
Inbound Logistics	Operations	Outbound Logistics	Sales and Marketing	Services

Support activities do not directly add value

Primary activities create or add value

Direction of flow through the system. Ideally, value is added at each stage.

Figure 1.9 *The Value Chain Model (adapted from Porter, 1985).*

diagram) adds value to the products (i.e. it benefits the company more than it costs). Activities in the upper part of the diagram provide services, but do not directly add value to products. They are not essential, and are therefore only worth doing if they make a contribution to the efficiency or effectiveness of primary activities. Their role must therefore be tuned to support primary activities.

This tool can be useful in information systems planning since it focuses attention on activities that are critical to a business, either because they are currently a problem, or because they represent a major source of profit or competitive edge. Information system development projects can then be targeted at assisting those operations that can directly contribute to the success of the organization as a whole.

1.5.3 Information systems and information technology strategies

The best managed organizations separate their strategic thinking into the three layers illustrated in Figure 1.10. The key idea is that the development of new information systems should only be considered in the context of a well-thought-out business strategy, while purchases of information technology hardware should only be considered in the context of specific information systems that are planned for development. Thus, the business strategy drives the information system strategy, which in turn drives the information technology strategy. Information flows in the diagram are not only in one direction. In formulating a business strategy, managers need to be advised on those areas of the business where information system can contribute to fulfilling business goals. Thus the planning cycle is iterative. A similar two-way communication takes place between the information system and information technology strategy planning functions. The role of the information technology strategy is to enable the successful implementation of the systems defined in the information system strategy, while also informing the information system strategists about what is feasible.

Informs and enables   *Drives and sets goals*

Business Strategy

Where IS can help   *What must be done*

Information Systems Strategy

Hardware capabilities   *System requirements*

Information Technology Strategy

Figure 1.10 *The relationship between business, IS and IT strategies.*

1.6 Summary

In this chapter we have introduced the main concepts that underpin information systems. These include some important ideas from general systems theory, such as the ideas of control, communication and emergent properties. We have also discussed the relationship between information and meaning, the central point here being that information only makes any sense within a specific context. For systems analysts and designers this context is usually the human activity system of an organization. This leads to a necessary sequence of relationships between the goals of an organization, the steps it must take in order to fulfil them, the information that its staff must have to carry out their activities, the information systems that provide the information, and finally the information technology that supports the information systems. While we recognize that there are many real organizations where this does not happen in the way we describe, we can only wonder how much more could be achieved, and express the hope that readers of this book will start work from firmer foundations.

The idea of information systems is far from new, yet information technology has changed the field almost beyond recognition in the last few years. This process will undoubtedly continue in the future, but however much information systems may change, valuable lessons may still be learned from historical information systems, both those developed since the introduction of the computer and those that pre-date it.

Review Questions

1.1   What is the difference between an *information system* and *information technology* ?

1.2   Identify some things that a computerized *information system* can do, which are difficult or impossible for a non-computerized equivalent.

1.3   Why does it not matter whether a system is real, or exists only in someone's mind?
1.4   Why are boundary and environment important for understanding a system?
1.5   What is the difference between feedback and feed-forward?
1.6   Why has a human activity system more than one purpose?
1.7   What is the purpose of a management support system?
1.8   What is meant by disaster recovery? Why is it important for a business organization?
1.9   What are the relationships between business goals, information systems strategy and information technology strategy?
1.10   Define information. How does it differ from data?
1.11   Define knowledge.
1.12   Give an example of some knowledge that you possess. What is its purpose?

Case Study Work, Exercises and Projects

1.A   Think of three or four information systems that are not computerized (either historical or contemporary). Identify (or imagine) a computerized equivalent. For each pair, write a brief description of the boundary and main inputs and outputs. What are the main differences between the computerized and non-computerized versions?

1.B   Re-read the description of the on-line shopping system in Box 1.1, and assume that everything described (computer software, hardware, human activities, etc.) is a single system. Identify its main sub-systems and control mechanisms. What feedback and feed-forward information do you think they would use? Don't be constrained by the description in Box 1.1—use your imagination too. And remember that some control may not be computerized.

1.C   Read the first part of the FoodCo case study in Chapter B1, up to and including Section B1.2.1. What do you think are FoodCo's business goals for the next ten years? Make any assumptions that you feel are justified.

1.D   Roughly copy out the diagram from Figure 1.9 (make it large, with plenty of space in each compartment). Try to map FoodCo to the value chain model by identifying people, activities, sections, etc. that fit in each compartment. What flows through your value chain model?

1.E   Identify the value that you think is added by each primary activity to FoodCo's products. Which do you think are the weak links?

Further Reading

Checkland and Holwell (1998) is an accessible and up-to-date account of the related subjects of systems, information and information systems.

Zuboff (1988) is a personal view of the way that information technology and information system are changing the nature of work and management based on research in many American companies during the 1980s. One of Zuboff's conclusions is that managers who do not understand how information technology changes an organization, may undermine all its potential benefits.

Webster (1995) is a scholarly debunking of the more exaggerated claims about how the 'information revolution' is changing social relationships.

Fidler and Rogerson (1996) describe the role of information systems in providing support to managers, and include an introduction to Strategic Information Systems Planning. There are many practical case studies throughout the book, but the approach taken is not an object-oriented one.

Koestler (1967) is a classic text that applies systems concepts to many aspects of life. This book ranges widely over human history and society, and presents some early speculations on the systemic role of evolution in modern social behaviour.

2

Problems in Information Systems Development

Objectives
In this chapter you will learn:
- who the main players are in an information system project;
- the problems in information systems development;
- the underlying causes of these problems;
- how the concept of a stakeholder helps identify ethical issues in information systems development;
- the costs of problems and ethical issues.

2.1 Introduction

It is only by understanding what can go wrong during a system development project that we can hope to avoid failure. Some authors use the metaphor of a journey to describe the process of information systems development (Connor, 1985), highlighting the fact that the main concern in any project is to avoid wrong turns on the way to the destination. At the start of a project, many different routes are available. Some lead to the planned destination, while others lead to a destination that may surprise everyone by turning out to be satisfactory, but is nevertheless unintended. Both are acceptable, although the latter can be nerve-wracking along the way. There are also routes that arrive at the wrong destination, while others are dead ends that lead nowhere. We must recognize and avoid these. This is the explicit purpose of carrying out analysis and design before developing a new system. Analysis is a way of understanding what needs doing before we begin trying to do it, and design is a way of checking that the planned action or solution really meets the needs of the situation, before it is put into practice. An understanding of potential problems is an essential precursor to systems analysis and design.

Projects fail for many different reasons, and in varying degrees. As we shall see, some failures are more visibly catastrophic than others. In this chapter, we look at the question of project failure from the perspective of each of the main players (Section 2.1). We discuss the causes that can lead to failure. These are broadly categorized into those that relate to the quality of what is delivered, and those that relate to the productivity of the project (Section 2.2). While not usually a direct cause of project failure, we also consider some ethical issues associated with systems development (Section 2.3). Some of these may act as a delay or an obstacle for the developer, while others only appear as problems when viewed from a wider context. Finally, we mention the costs of the problems, whether or not they result in outright system failure (Section 2.4).

2.2 What are the Problems?

Many information systems are very successful, and failure is the exception rather than the rule. But the consequences of failure can be severe. One report estimated that failed information systems projects in the USA accounted for $81b annually—almost a third of total information systems development expenditure (Standish Group, 1995). A similar picture probably prevails in the UK, where total IS/IT expenditure for 1995 was estimated at £33.6b (Willcocks and Lester, 1996). The bill for one aborted project (the London Stock Exchange Taurus system) is estimated at £480m. All potential causes of failure are at least to some extent under the control of the developer. A professional must take the possibility of failure seriously, and work hard to avoid it, even if this sometimes is limited to an awareness of the risks, followed by a damage-reduction exercise.

One difficulty with the question 'What can go wrong?' is that the answer depends partly on who gives it. But this is not surprising, as information systems development is a complex activity that always involves people. This is not to say that all people are difficult—this only applies to a very few! But, in any organization, people have varying perspectives that influence their view of a situation, and what, if anything, is to be done about it. Although some attitude differences derive from psychology and personal history, which are beyond the scope of this book, others relate to an individual's position within the organization and their link to an information system project. It is useful to discriminate between three categories of people with important relationships to a project. The first is the group of employees who will become end-users of the information system when it is completed. The second is the group of managers, here called 'clients', who have control (or at least influence) over the initiation, direction or progress of a project. The third is the group of professionals responsible for the development of the information system. For the moment, we ignore differences within each group.

2.2.1 An end-user's perspective

End-users come in many varieties, and can have varying relationships to an information system. The examples that follow concentrate on the experiences and frustrations of those who either consume the products of an information system (i.e. they use its outputs to help achieve a task, such as despatching an ambulance to an emergency), or are responsible for entering raw data into an information system.

'What system? I haven't seen a new system'

One problem that can be experienced by an end-user is vividly expressed by a term that gained widespread usage in the 1980s. *Vapourware* describes a software product that is much talked about, but never released to its intended users. In other words, instead of arriving, it evaporates. While many businesses are reluctant to talk about information system project failures in public, vapourware may be surprisingly common.

Some surveys have found that an astounding proportion of information systems development projects fail to deliver any product at the end. In the US it has been suggested that 'as many as 25% of all projects in large MIS departments are never finished' (Yourdon, 1989). In the UK, total costs on a Regional Information System for the Wessex Regional Health Authority amounted to £63m, yet no system was delivered (Collins, 1998b). This example indirectly hurt patients, as the aim of the system was to help manage hospital resources more effectively throughout the region, and thereby deliver a better, more responsive service. When a project is not completed, the expected benefits to users and other beneficiaries are not achieved.

'It might work, but it's dreadful to use!'

This relates to systems that are unpleasant or difficult to use. Systems may fail to meet the criterion of usability in a number of ways, including: poor interface design, inappropriate or illogical sequence of data entry, incomprehensible error messages, unhelpful 'help', poor response times and unreliability in operation. Figure 2.1 gives some examples (what these mean, and how to avoid them, is explained in more detail in Chapter 15).

System characteristic	Example
Poor interface design	A web page with yellow text on a white background.
Inappropriate data entry	A system where the backspace key sometimes deletes whole words.
Incomprehensible error messages	A system message that says 'error #13452'.
Unhelpful 'help'	A system message that says 'wrong date format—try again'.
Poor response times	Nurses in an Intensive Care Unit complained that their new computerized patient chart system took longer to store and retrieve data, compared with the manual system. This took them away from patients for longer[1].
Unreliability in operation	A national motor vehicle insurance company lost most of its digital records of customers' policies due to a system error. Staff were unable to send renewal notices, but were compelled instead to write to customers asking them to phone in with their policy details[2].

Figure 2.1 *Aspects of poor system usability with examples.*

'It's very pretty—but does it do anything useful?'

A system may be beautifully designed and easy to use, but still not do the 'right' things. This may be a question of the tasks that should be carried out by the system. For example, a library catalogue enquiry system would be of limited use if it could only retrieve shelving information about a book when provided with the title and the author's name, in full and spelt correctly. Readers often do not know the title of the book for which they are searching. Even if the author's name is known, it may be spelt incorrectly. Another way that a system may fail to meet its users' needs is through poor performance (this overlaps with the question of usability, discussed in the previous sub-section). An extreme example was the London Ambulance Service emergency despatch system, abandoned shortly after delivery in 1992, after incurring a total development cost estimated at £43m.

---

[1] Source: Goss et al. (1995).

[2] One of the authors is a customer of this company.

Far from speeding up the dispatch of London's ambulances and paramedic teams, response times became slower than ever.

After a series of high-profile failures, the system was canned amid claims—never officially substantiated—that it was responsible for several needless deaths.

(Barker, 1998)

A system may also be of doubtful value to its users because it requires them to work in a way that seems nonsensical. One report describes a warehouse management system, designed partly to increase managers' control over the routine activities of warehouse workers. The workers found that the new system removed much of their discretion in their work, and this prevented them being able to maximize the use of scarce storage space.

... because they could see how improvements in these areas would save money for the company they found ways of working around the system.

... They were reproved by management for their bad attitude, and yet, it was their commitment to the company as a major employer in their local community which led to their frustration with what they regarded as unnecessarily wasteful rules and procedures.

(Symons, 1990)

2.2.2 A client's perspective

By *client* we mean that person, or group of people, responsible for paying for the development of a new information system. A client usually has influence over whether or not approval is given to a project before it starts. Some clients (but not all) also have the power to stop a project once it is under way. A client may also be a user. If so, we can assume that they share the user's perspective on the kind of things that can be a problem. They may make only indirect use of the system's outputs, which insulates them from the immediate experience of a badly designed interface for example. While the concerns of a client may overlap with those of an end-user, they also include distinct matters relating to payment, ownership and value-for-money.

'If I'd known the real price, I'd never have agreed'

It is almost routine in many organizations for information system projects to exceed their budget. A survey of approximately 200 large US organizations found that 'the typical project was one year late and 100% over budget' (Capers Jones, 1986). Any project that over runs its budget can reach a point where the total costs outweigh the benefits that will be provided on completion. This point is not always recognized when it is reached; this may result in the expensive completion of a system that might have been better cancelled. In other cases the project is cancelled, either because it is recognized that costs are escalating out of control, or because it becomes apparent the benefits will not be as great as originally promised. The rationale is summed up in the saying: 'Don't throw good money after bad'. One of the most striking examples in recent years was the London Stock Exchange electronic trading system known as Taurus, cancelled after three years development.

> The City of London was shocked by the sudden cancellation of Project Taurus. The London Stock Exchange had spent [...] £80 million in the venture. The securities industry had invested over £400 million developing their own system in readiness for Taurus. On Wednesday 11 March 1993 the Stock Exchange publicly admitted it had all been for nothing.
>
> (Drummond, 1996)

'It's no use delivering it now—we needed it last April!'

A project that is completed late may no longer be of any use. At present (late 1998), many organizations are working hard to fix all the manifestations of the so-called 'millennium bug'. The deadline for many of these projects is critical, and absolutely immovable. Some Year 2000 projects that are not completed by 31 December 1999 will be judged a waste of time and effort, since their purpose is to prevent the systems failures expected to commence at midnight on that date.

Many other kinds of project are time-critical. This can be due to new legislation that affects the organization's environment. An example of this was the de-regulation of the UK electricity supply market in April 1998. This required electricity companies to make extensive modifications to their computer systems so that they would be able to handle customers' new freedom to switch between suppliers. A few years earlier, all Local Authorities in the UK faced a similar challenge twice in three years, when Central Government changed the basis for local tax calculations. Each change required hundreds of councils to specify, develop (or purchase) and successfully install new computer systems that allowed them to produce accurate invoices and record income collected from local tax-payers. Failure to implement the new systems in time risked a massive cashflow problem at the beginning of the new tax year.

Commercial pressures can also have an effect. This often translates into whether a business succeeds in being the first to market a new product or service. The success of the Internet bookstore Amazon.com™ derives, at least in part, from the considerable competitive advantage of being the first of its kind. If another company wished to jump on the bandwagon now, it might not need to take the same risks with new technology. But to attract customers away from Amazon, it would need to differentiate itself in some way, perhaps offering new services, or perhaps by being even better at what Amazon already does well. Clearly, given Amazon's lead, either strategy would be an uphill struggle.

'OK, so it works—but the installation was such a mess my staff will never trust it'

Once a new system gets a bad press it can be very difficult to overcome the resistance of the people who are expected to use it, particularly if there is an existing alternative available to them. The following scenario is based on a real situation, recently observed at first hand by one of the authors.

> A small company introduced a local area network (LAN) to connect the PCs in its office. Staff were encouraged to store files on the LAN disk-drive, where other staff could also access them (previously, all data was stored on local hard drives, accessible only from one PC). Most saw the mutual benefit of sharing information, and complied. Management claimed that the daily back-up of the LAN drive onto tapes was a further benefit, since there was no longer a need to

keep personal back-ups on floppy disks. Then a mechanical fault occurred. This erased all data on the LAN drive, and when the engineer tried to restore it from the tape, it emerged that the tape drive had not operated correctly for several weeks. All tapes recorded over the previous six weeks were useless. Staff were told that all data stored in that time was permanently lost. Re-entering it all took many person-days. The faulty disk and tape drives were replaced, and tapes are now checked after every back-up, but many staff reverted to keeping all important data on their local hard drives. Most keep personal back-ups on floppy disks too. Perhaps nothing will persuade them to trust the LAN again.

'I didn't want it in the first place'

Organizations are complex and political by nature. The politics with which we are concerned here are to do with conflicting ideals and ambitions, and the play of power within the organization. At the end of Section 2.1.1 we mentioned one potential area of disagreement between management and workers. There can also be contention between individual managers, and between groups of managers. One result can be that a manager is sometimes an unwilling client in relation to a project. The following scenario is based on another real-life situation recently observed by one of the authors.

> The head office of a multinational company decided to standardize on a single sales order processing system in all its subsidiaries throughout the world. But the Hong Kong office already had information systems that linked electronically with customers in Singapore, Taiwan and other places in South East Asia. It became apparent that the existing links would not work with the new system. For the Hong Kong management, this meant extra costs to make the new system work, and disruption to established relationships which, in their view, already worked smoothly and did not need to be changed. They therefore had little desire to see the project succeed in their region, but felt they had no other choice. Had they been less scrupulous, they might have tried to find ways of sabotaging its progress, either in the hope that it would be abandoned altogether, or at least that they might be exempted from the global rule.

'Everything's changed now—we need a completely different system'

It is almost inevitable for any information system project that, by the time the system is completed, the requirements are no longer the same as they were thought to be at the beginning. Requirements can change for many reasons.
- Project timescales can be very long (the Taurus project ran for three years), and business needs may change in the meantime.
- Users naturally tend to ask for more, as they learn more about what is available.
- External events can have a dramatic impact—for example a currency crisis that led to the collapse of a significant market might render a new information system pointless.

This does not apply only to new systems currently under development. Systems that have been in operation for some time may also be affected. This is part of the natural, ongoing process of maintenance, modification, upgrading and eventual replacement that all information systems undergo, and a topic to which we return in Chapters 3 and 19. From a client's perspective, the motivation is usually to make an information system fit better with the business, and therefore provide better support for business activities.

2.2.3 A developer's perspective

The perspective of the developer is quite different both from that of an end-user and that of a client. This is because the developer adopts the role of 'supplier' to the 'customer' (i.e. client or end-user). For this reason, when problems occur the developer may feel forced into a defensive position of justifying the approach taken during the project. Since at this stage we are discussing only problems, many of the problems identified by a developer tend to centre on blame and its avoidance.

'We built what they _said_ they wanted'

Changes to the requirements for a system, based on sound business reasons, always seem perfectly reasonable from a client's point of view. However, for a developer, given the responsibility for building a system to meet those requirements, they can be a real headache. If we were able to distil the essence of how many developers feel about this, it would read something like the following.

> No matter how skilled you are, you can't achieve anything until the users, clients, etc., tell you what they want, and at the start they don't even agree with each other. Eventually, with skill and perseverance, you produce a specification with which everyone is reasonably happy. You work for months to produce a system that meets the specification, and you install it. In no time at all, users complain that it doesn't do what they need it to do. You check the software against the specification, and you find that it does exactly what it was supposed to do. The problem is that the users have changed their minds. They just don't realize that it's not possible to change your mind late in a project. By then, everything you have done depends on everything else, and to change anything you would almost have to start all over again. Or it turns out that they didn't understand the specification when they accepted it. Or there is some ambiguity about what it meant, and you've interpreted it differently from them. Whatever the reason, it's always your fault, even though all you ever tried to do was to give them what they wanted.

In reality, of course, analysts, programmers etc. often do understand why users and clients change their minds during a project, or after delivery. But this doesn't always make it less frustrating when it happens.

'There wasn't enough time to do it any better'

In every project, there are pressures from outside that limit the ability of the development team to achieve excellence. In the first place, projects almost invariably have a finite budget, which translates directly into a finite amount of time to do the work. There may also be an externally-imposed deadline (in a Year 2000 project for example). Another external pressure results from the impatience of users and clients to see tangible results.

This, too, is often understandable, since they are not so much concerned with the information system itself, as with the benefits it can bring them—an easier way to do a tedious job, a quicker way to get vital information, and so on. But it can be very counter-productive, if it becomes a pressure within the project team to cut short the analysis and get on with building something (anything!) quickly to keep the users happy. The result of haste in these circumstances is usually a poor product that meets few of the needs of its users. Developers know this, but they don't always have the power to resist the pressure when it is applied.

'Don't blame me—I've never done object-oriented analysis before!'

In a successful information system development team, the members must possess a harmonious blend of skills that are appropriate to the needs of the project. These may include the use of techniques (such as object-oriented analysis), knowledge of methodologies (such as Objectory™), skill in programming languages (such as Java™) or detailed knowledge of hardware performance (such as networking devices). There must be a complementary set of skills within the team for a project to succeed. For example, the analysts must all use the same or related techniques otherwise the results could not be coherent (e.g. the products of object-oriented analysis are entirely incompatible with those of structured analysis). Similarly, all programmers must be able to interpret the designs or else they will not know what the class operations are meant to do. Problems occur when the available staff do not possess adequate expertise in the particular skills required for a project. Many highly skilled and experienced systems analysts may know SSADM very well (a structured methodology used widely in the UK), but have little experience of object-oriented analysis. Many excellent programmers know COBOL inside out, but have little experience in Java or Smalltalk. As a consequence, some projects with highly skilled staff are still carried out rather poorly, because the staff are inexperienced with the particular techniques they must use.

'How can I fix it?—I don't know how it's supposed to work'

This complaint is often heard from programmers who have been asked to modify an existing programme, and who have then discovered that there is no explanation of what it does or how it works. To modify or repair any artefact, whether a computer system or a bicycle, it is usually necessary to understand how it was intended to work, and thus what the consequences are of changing this or that feature. Anyone who has ever tried to repair anything electronic or mechanical, such as a motor vehicle, washing machine, or VCR, will know that much of the time is spent trying to understand what the various parts do, and how they interact with each other. This is true even when a maintenance manual is to hand. The situation is no different for computer software. While software may be more intangible in form than a VCR, it is no less mechanistic in its operation.

'We said it was impossible, but no-one listened'

Just like client managers, systems developers can sometimes be overwhelmed by organizational politics. At times this means that a project is forced on an unwilling team, who do not believe that it is technically feasible to achieve the project's goals. Alternatively, the team may not believe the project can be completed within the time made available. But if opposing views prevail, the team may find itself committed to trying to achieve what it said could not be done. In these circumstances, it can be very hard for team members to become enthusiastic about the project.

'The system's fine—the users are the problem'

A few information systems professionals, usually those who understand least about business and organizations, are sometimes prone to blame 'the user' for everything. Those who hold this attitude believe that problems that occur in the use of software inevitably result from the fact that most users are far too stupid or ignorant to make proper use of the system. On the other hand, they believe that the design and execution of the software is not open to serious question. Many of these technocrats are undoubtedly very talented, but this view is patently absurd since it assumes that the answer to a problem is known before the situation has even been investigated. In a word, it is a prejudice. We will say no more about it, other than to comment that anyone who hopes to learn the truth about a situation must also be prepared to examine critically their own preconceptions.

2.3 Why Things Go Wrong

Flynn (1998) proposed an analytical framework to categorize project failures, and this is illustrated in Figure 2.2.

Type of failure	Reason for failure	Comment
Quality problems	The wrong problem is addressed.	System conflicts with business strategy.
	Wider influences are neglected.	Organization culture may be ignored.
	Analysis is carried out incorrectly.	Team is poorly skilled, or inadequately resourced.
	Project undertaken for wrong reason.	Technology pull or political push.
Productivity problems	Users change their minds.	
	External events change the environment.	New legislation.
	Implementation is not feasible.	May not be known until the project has started.
	Poor project control.	Inexperienced project manager.

Figure 2.2 *Causes of IS project failure (adapted from Flynn, 1998).*

Complete failure is the most extreme manifestation of a problem, but Flynn's framework can also be applied to less catastrophic problems. In Flynn's view, projects generally fail on grounds of either unacceptable quality, or poor productivity. In either case, the proposed system may never be delivered, it may be rejected by its users or it may be accepted yet still fail to meet its requirements.

These categories are what are sometimes called 'ideal types'. This means that they are intended to help explain what is found in reality, but that does not imply that any real example, when examined in all its detail, will precisely match any one category. Real projects are complex, and their problems can seldom be reduced to one single cause. Many of the examples in the following sections show some of the characteristics of more than one category of cause.

2.3.1 Quality problems

One of the most widespread definitions of the quality of a product is in terms of its 'fitness for purpose' (see any standard text on software quality assurance, e.g. Ince, 1994 or Sanders and Curran, 1994). In order to apply this to the quality of a computer system, clearly it is necessary to know (i) for what purpose the system is intended, and (ii) how to measure its fitness. As we shall see, both parts of this can be problematic at times.

The wrong problem

Here the emphasis is on the purpose for which a new system is intended. The argument goes like this: if an information system does not contribute anything worthwhile to the aims of the organization, then it is, at best, a waste of resources and a distraction from those things which really matter. At worst, it may do real harm if the objectives of an information system are in direct conflict with the organization's business strategy. The difficulty is in knowing the right viewpoint to take when defining the aims of a project. If the aims of the organization as a whole are unclear, or are not communicated to those responsible for planning information system projects, there is always a risk that a project will fall into this error. It may then be regarded as a complete success by its developers and users, yet appear a failure when seen in a wider frame of reference.

A primary cause of system failures in general is that some projects are started with no clear idea about exactly what are the nature and goals of the client organization. This means that failure, or at least lack of success, is almost inevitable. If an organization itself is not understood, then it is extremely difficult to identify and develop information systems that support it in fulfilling its aims.

Neglect of the context

This emphasizes the fitness of an information system to fulfil its purpose. This can take the form of a system that is too difficult to use, since the designers have taken insufficient account of the environment in which its users work, or the way that they like to work. Some examples given earlier in this chapter can be interpreted in this way, depending on assumptions about the situation. For example, in one case cited earlier (Section 2.2.1) workers and managers held different views about the purpose of a warehouse management system. Managers believed they needed to control the activities of workers more closely. Yet the system designed to do this had also the side effect of obstructing the workers from carrying out their work in an efficient way, to the detriment of the whole company.

Incorrect requirements analysis

This, too, focuses on a system's fitness to fulfil its purpose, rather than the purpose itself. Even if the aims are clear at the outset, many pitfalls lie along the route, particularly if the development team do not have the right skills, the right resources or enough time to do a good job. However, even when none of these present a difficulty, the project can still fail if the techniques being applied by the team are inappropriate to the project.

The results of this category of failure are generally among the most visible systems problems, when seen from a user's point of view. This is because they directly affect either the external design of the system (e.g. the content or layout of its screens), or the selection of tasks that the system performs (e.g. an essential function may not be included).

Project carried out for the wrong reason

Here, the emphasis is once more on the intended purpose of the system. To give a recent example, many organizations today seem to be rushing to create an 'Internet presence'. While this is clearly a great success for some, others appear to have derived little or no benefit from their first move into cyberspace. It has been pointed out that this is sometimes because an organization has not thought carefully enough about some key questions (McBride, 1997).

- What does the business aim to achieve by a presence on the Internet?
- How must the business re-organize to exploit the opportunities offered by the Internet?
- How can the business ensure that its presence on the Internet is effective?

The sensible conclusion must be that there is a great deal more to successful trading on the Internet than just writing a few web pages in HTML, and placing them on a web server where surfers can find them. Some organizations, in moving onto the World Wide Web, are simply following a trend. They do so, not because they understand what it can do for their business, but precisely because they do not understand, and therefore fear the consequences of being left out of something good.

Two possible underlying reasons explain why this can happen. One of these is political push within the organization. For example, a powerful group of managers may wish the business to look modern, even when no clear benefit has been identified. The other is the pull of new technology. An organization is very vulnerable to this if the most senior managers have little understanding of information technology, and therefore no rational basis for evaluating the exaggerated claims that vendors are prone to make about their newest products. In practice these two reasons often combine into a single force that can be irresistible.

2.3.2 Productivity problems

Productivity problems relate to the rate of progress of a project, and the resources (including time and money) that it consumes along the way. If quality is the first concern of users and clients, then productivity is their other vital concern. The questions that are likely to be asked about productivity are as follows.

- Will the product be delivered?
- Will it be delivered in time to be useful?
- Will it be affordable?

Requirements drift

Requirements almost always drift if they are allowed to do so. In the simplest case, this just means that users' requests change over time. In principle, it would be unreasonable to prevent this from happening, but, in extreme cases, change requests can bedevil a project and even prevent its completion. The longer a project proceeds, the more complex both its products and its documentation become. To compound this, each part of the final system probably depends on many others, and the inter-dependencies grow more numerous and complex. It becomes progressively more difficult over time to make changes to a system under development, because any change to one part requires changes to many other parts so that they will continue to work together. A limit is reached when a project is stalled by an escalating workload of making all the changes required as a consequence of *other* changes that were requested by the users. At this point, management have only two choices. They can cancel the project and write off the money spent so far (this happened at the London Stock Exchange in 1993). Alternatively, an effort can be made to bring the project back on track. This is almost always both difficult and expensive, and requires highly skilled management.

External events

This is one cause of failure that is normally beyond the control of both project team and higher management. Depending on the environment in which the organization operates, decisive external events can even be impossible to anticipate. Nevertheless, it is prudent on any project at least to assess the vulnerability of the project to external events, since some are much more at risk than others. For example, a project to build a distributed information system that is to operate on new, state-of-the-art computers communicating over public telephone circuits may be sensitive to external factors such as the reliability of the telephone network and call pricing. By contrast, a project to build an information system that will operate on existing, tried and tested hardware within one building can safely ignore these factors.

Poor project management

The manager of a project is ultimately responsible for its successful completion, and it could therefore be argued that any project failure is also a failure of the project management. To some extent this is true, but there are also some cases where the only identifiable cause of failure overall is a failure of management. This is almost always due to either poor planning at the start, or a lack of care in monitoring progress against the plan. As a result, the manager allows the project to falter, or permits its costs to grow in an uncontrolled way.

Implementation not feasible

Some projects are over-ambitious in their technical aims, and this may not become evident until an attempt is made to implement the system. This is particularly the case when a system is intended to work together with other systems, whether or not these are already in use. The problems of testing and debugging a new system can grow steadily more complex as attention is focused on larger and larger sub-systems. Sometimes the task of interfacing several large, complex software systems, written in different programming languages, installed at different sites and running on different makes of computer hardware, can turn out to be impossible to achieve.

An example of the kind of complex installation routinely tackled in many modern projects is a £25m system called Caseman, which has computerized many administrative tasks in the 250 County Courts of England and Wales. This system is based on an Oracle® database combined with the WordPerfect® word-processing package, running on the UNIX® operating system on DEC® computers. Although each of these components, taken independently, is thoroughly tried and tested, nevertheless implementation was delayed because the company responsible for the project encountered problems in making the different parts work together (Schneider, 1997). A further difficulty occurs when each supplier insists that their product is performing exactly as specified, and any problem must lie elsewhere. It can then be very hard to locate the exact origins of the problem.

Technical problems with the implementation do not always become evident until after the system is implemented. One example of this was with the LAS ambulance despatch system. A practical difficulty that staff encountered in its use was with the deliberately high-tech design that used a digital on-screen map:

> … which showed the location of the ambulances. The staff hated it and said it was dangerous […] The problem with a map system is accuracy.
>
> (Barker, 1998)

This was a contributory factor to the system's overall failure to speedily despatch ambulances to emergencies, resulting from the application of a new technology to a critical task which was not sufficiently understood.

2.4 The Ethical Dimension

Chambers Twentieth Century Dictionary (1977) defines ethics as:

> *n.* the science of morals, that branch of philosophy which is concerned with human character and conduct: a system of morals, rules of behaviour.

With this definition in mind, it is hard indeed to think of a computer-based information system which does not have a significant ethical dimension to its design, construction or use. As with any system of rules, it is necessary to consider the consequences if they are broken. It is not overstating the case to say that a breach of the ethical rules that apply to any aspect of an information system project is necessarily a problem for someone. Sometimes these problems are noticeable at the time and must be handled within the project, otherwise they can lead to a failure of the project overall. For example, a system intended to automate some activities of a business may result in job losses among the staff. At other times there is hidden damage to a project, an organization, the information systems profession as a whole or even in extreme cases to society at large. For example, there is a debate about whether individual users of email systems have a right to privacy from surveillance. Some evidence in the prosecution case that led to the impeachment of President Clinton was gathered from personal emails. Investigating officers were apparently able to retrieve copies of certain sensitive messages, even though their author believed that all copies had been safely destroyed (Wall Street Journal, 1998).

One of the difficulties in assessing the ethical issues in a project is that the person who may have a problem is not necessarily the developer of the system, its user, its client—or indeed anyone at all who is obviously connected with the project. In fact, one of the first problems that must be faced in this area is the identification of all the people who may be affected by the system. These are often called *stakeholders*, since, in their different ways, each has a stake in the outcome of the project.

To illustrate the diversity of stakeholders who may be associated with a project, consider the introduction of a network of bank ATM machines in the branches of a supermarket chain. Figure 2.3 shows a preliminary tracing out of the possible effects of this system, together with the groups of people affected.

Stakeholder affected	Possible consequence of system	Nature of effect on stakeholder
Bank clerks.	Automation of bank activities currently carried out manually.	Reduced need for staff—redeployment or redundancies.
Bank customers.	More convenient access to bank services.	Improved service.
Supermarket customers.	More people using supermarket car park.	Reduction in service.
Bank shareholders.	More people attracted to use bank, so greater commercial success.	Increased dividends.
Supermarket shareholders.	More people attracted to use supermarket, so greater commercial success.	Increased dividends.
Local citizens.	More journeys to supermarket to use ATM.	Increased pollution.

Figure 2.3 *Possible stakeholders in a bank ATM network.*

You may not feel that every group identified in this analysis is equally affected by the project, and in some cases you may feel that the effect is quite minor. However, until such an analysis has been carried out, it is not possible to speak with any authority about the nature or the extent of impact that a new information system will have.

2.4.1 Ethical issues within a project

The issue of professionalism is at the forefront of any discussion of ethics within information systems development. By their very nature, information systems projects have deep consequences for the lives and work of many people who will either use the software, or will be affected in some way by its use. There are several reasons for this. Firstly, information systems projects are often major investments for the client organization, and money spent on these is necessarily money that now cannot be spent on other worthwhile projects. Secondly, they are often concerned with the way that important business activities are carried out, and they can therefore have a direct effect on the overall success or failure of the organization. Thirdly, the introduction or modification of information systems often causes radical changes in the way that employees carry out their work, and how they relate to their managers, colleagues and customers. Given these responsibilities, it is important that project team members behave in a professional manner. To some extent this is just a matter of being aware of the rules for behaving in a professional way, and then following them.

Behaving ethically is usually not just a matter of applying a straightforward rule in a mechanical way. We often find ourselves confronting ethical dilemmas, where no one course of action seems entirely right, yet we must find some basis for making a decision. This is compounded when our actions affect different stakeholders in different ways.

Some ethical effects arise from aspects of a project that are not under the direct control of an individual developer. For example, Sachs (1995) describes a 'Trouble Ticketing System' that was intended to improve the work efficiency of testers (telephone repair staff) by tracking the progress of repairs and allocating the next step in a job to any tester that was available at the time. In practice, the system disrupted informal communication between employees, thus damaging their efficiency instead of improving it. According to Sachs, the fault lay in a failure to develop an adequate understanding of the testers' working practices, in particular the way that their informal communications helped in troubleshooting difficult problems. But individual developers are rarely free to choose which aspects of a situation should be analysed.

Legislation adds a further ethical dimension for the members of an information system project team. In the UK the relevant legislation includes the Data Protection Act 1998, the Computer Misuse Act 1990 and the Health and Safety (Display Screen) Regulations 1992.

2.4.2 Wider ethical issues

Among the wider ethical issues associated with information systems development, one of the longest-running debates is over the effect that IT and IS have had on levels of unemployment throughout the world. Some authors claim that, within the foreseeable future, one effect of the spread of information technology will be an unprecedented, sustained rise in global unemployment that will leave countless millions idle and impoverished (Rifkin, 1995). Others argue that, while IT has certainly destroyed some jobs (tens of thousands of jobs lost in the UK banking sector over the last decade), it has also created many new jobs to replace them, often bringing new opportunities to neglected backwaters in the world economy.

For many years, an ethical debate has raged between the proponents of freedom (some would say anarchy) on the Internet, and the big business interests who already own its infrastructure and may soon dominate most of its content. In its brief history, the Internet has been at the centre of many debates revolving around freedom of access, including the widespread concerns that have been expressed about the ready availability of pornography, political materials and other contentious content. Yet for many others, this chaotic situation has brought unprecedented opportunities to meet (in a virtual sense) and communicate with similarly minded people all over the world.

One issue that is likely to receive more attention in the next few years is the way that some companies use new applications of information technology to gather unprecedented quantities of detailed personal information. For example, a South African bank issues its customers with mobile phones that display their current account balance each morning. The phones also enable the bank to monitor all phone numbers dialled, and this adds to the bank's profile of data about individual customers. Customers are categorized, and those that the bank feels are least profitable are 'encouraged' to leave the bank, e.g. by higher service charges. The bank is considering plans to add a geographic information system capability, which would also allow customers' daily movements to be tracked, and a link with Internet providers, which would allow the bank to ascertain which web sites are visited by their customers (Collins, 1998a).

Questions of access to computers and computerized information have wide implications. There have been arguments over the years about whether the use or the availability of computers have tended to favour men over women, the middle class over the working class, and those in the affluent north over those from the poor south of the globe. Some even suggest that computers and the Internet have already begun to create a new division in the world's population, cleaving people everywhere into the information-rich and the information-poor.

So many things are changing so fast at present that it is not possible to be sure how these questions will seem in a few years time. What is clear is that future generations will look back on our time as one of great change. Such an era is inevitably one that raises profound ethical questions about the effects on our fellow citizens of the way that we design and apply technology.

2.5 Costs of Failure

Many projects discussed in this chapter, particularly the more famous ones (Taurus and the LAS despatch system), have very large price tags attached. Some companies are widely thought to be reluctant to admit their failures, since this can reflect badly on the business as a whole, perhaps damaging customers' or investors' confidence. For this reason, it is thought that the known, high profile failures represent only the visible tip of a much larger iceberg. But these are not the only costs associated with project problems. Many costs are associated with projects that have not failed outright, but which, for a variety of reasons, do not fully meet their requirements.

A system that is poorly designed or functionally limited has many consequences for its users. If we take interface design as an example, an ill-considered screen layout (for example one that compels users to switch back and forth between two screens as they enter data) can have effects that range from mild irritation, through increased error rates, absence due to sickness, to faster staff turnover. Each has an associated cost—but it may be hard to measure accurately, and in some cases may not be recognized at all. If an employee leaves to take another job one reason may be his or her frustration with a poorly designed computer system.

System reliability can be important in determining overall costs. One of the authors was recently contacted by his motor insurance company, and asked to resubmit his personal details, because their computer system had crashed and most customer data had been lost. This is not a reassuring picture for any insurance company to give its customers, and it is a safe conclusion that many will have switched to another insurer as a result. Some will also have told friends and colleagues, who in turn will be less likely to use this company in the future. It is very unlikely that anyone could accurately determine the full cost to this business in terms of lost customers.

Some of the effects that can occur are summarized in Figure 2.4.

Design aspect	Example	Immediate effects	Other consequences
User Interface.	Illogical screen layout. Difficult to read screens. Unhelpful help messages.	Wasted time. Increased frustration. Increased error-rate.	Loss of confidence in system. Increased sickness. Increased absenteeism. Greater staff turnover.
Program Execution.	System response is slow.	As above.	Increased operating costs.
Data storage.	Lost data. Inaccurate outputs.	Extra work re-entering data. Extra work checking outputs.	Reduced income. Loss of customer confidence. Lost sales.

Figure 2.4 *Some hidden costs of poor design.*

The list of effects in Figure 2.4 is by no means exhaustive. But, in showing how poor design and operational characteristics of an information system can affect the organization as a whole, it reinforces the importance of getting things right before the problems have a chance to appear. This is true even if we ignore the ethical consequences of some information systems applications under development or already in use today. The full social cost of some ethical issues outlined in Section 2.3 is probably incalculable.

2.6 Summary

In this chapter we have looked at the issue of project failure from many different perspectives, including those of the people who use information systems, the people who purchase them and the people who build them. We have also outlined some of the deeper causes of project and system failure, and considered the costs and wider ethical issues.

The failures in information systems development can teach valuable lessons, and, moreover, ignoring a difficulty does not make it go away. Rather, it increases the likelihood of repeating past mistakes. So it is important to understand as much as possible about what can go wrong with an information system development project, the better to be able to avoid it. But after so much concentrated attention on the many dead ends that our unluckier (or more careless) predecessors have encountered, it seems appropriate to end the chapter on a final upbeat note. Two of the failures discussed earlier have since become success stories.

The London Stock Exchange has successfully installed an on-line share trading system. The new system is much simpler than Taurus—and not all users are entirely happy with its functionality—but it does meet its basic requirements. It was introduced in April 1997, on time and within budget (Philips, 1997).

The London Ambulance Service also has a new ambulance despatch system, introduced five years after its disastrous predecessor was scrapped in 1992.

> Now the service's emergency call room plays host to visitors from around the world who come to see the successful implementation of an emergency response system in one of the biggest emergency call centres in the world. Indeed, the

new system won the prestigious BCS Elite Group Information Systems Management Award in May [1997].

(Barker, 1998)

We can also remind ourselves of the many positive contributions that modern information systems make to our lives, enabling so many things that were not possible for earlier generations. It is difficult to imagine an aspect of life in modern society that is not facilitated in some way by a computerized information system. Our goal, then, should be to ensure that tomorrow's systems build on the successes of the past, while avoiding the problems and failures, as far as it is in our power so to do.

Review Questions

2.1 Why do users, clients and developers disagree on the nature and causes of the problems in information systems development?

2.2 What are the main underlying causes of problems in information systems development?

2.3 Define quality.

2.4 What are the main differences between quality problems and productivity problems?

2.5 Why do the requirements drift once a project is underway?

2.6 What can be the results of ignoring the organizational context of an information system?

2.7 Define stakeholder.

2.8 What ethical issues might be involved in setting up an on-line shopping system that has links to an organization's management information systems?

Exercises, Case Study Work and Projects

2.A Do some research in computing trade journals, and find examples of recent projects that failed or ran into difficulties. Draw up a table with four separate columns. Give these titles like: 'Nature of problem', 'Who sees it as a problem', 'Probable cause' and 'Flynn category'. Enter the projects into your table, using your own intuition to complete column 3. Then complete the 4th column using Flynn's categories (summarized in Figure 2.2). How do your causes compare with Flynn's categories?

2.B The British Computing Society (BCS) publishes a Code of Conduct for its members, who include thousands of computing and information systems professionals in the UK. In the US, similar codes are published by the Association for Computing Machinery (ACM) and the Institute of Electrical and Electronic Engineers (IEEE). In many other countries around the world, including India, Zimbabwe and Singapore, there is a national professional society with an equivalent code of professional ethics. Write down some ethical problems associated with the development of an information system to support staff who despatch ambulances to medical emergencies, and use this to identify a list of issues you would expect to be covered in a professional code of conduct for information systems developers. Then obtain a copy of the BCS Code (or, if you are a reader in another country, your nearest local equivalent). Compare it to your list of issues. What are the main differences?

2.C Write down all the stakeholders who you think are associated with an emergency ambulance despatch system. How are they affected?

2.D Review the ethical issues you identified in Exercise 2.B, and identify one or more issues that appear as a problem from the perspective of one stakeholder, but do not appear as a problem from the perspective of another stakeholder.

2.E Find out what legislation applies to information systems development activity in your country, and what implications it has for developers. Would the South African bank customer profiling system described in Section 2.4.2 be fully legal under these laws? Which particular aspects of the system make it ethically questionable, and from whose perspective?

Further Reading

Trade magazines such as *Computing* and *Computer Weekly* regularly publish articles that report on problems in current projects as they occur.

Sauer (1993) discusses reasons for project failures in terms of the project's organizational environment, and includes several practical case studies that illustrate the problems in practice.

Ince (1994) is one of many books that deal with software quality assurance (SQA), a discipline that endeavours to ensure that software systems fully meet their defined requirements. The latest edition of Pressman's standard software engineering text (1997) also has a good chapter on SQA.

De Montfort University's Centre for Computing and Social Responsibility runs a web site that focuses on ethical issues in information systems development. This has many links to other sites, and also carries the full text of a number of academic papers in this field. It can be found at www.ccsr.cms.dmu.ac.uk.

3

Avoiding the Problems

Objectives
In this chapter you will learn:
- the stages in the waterfall life cycle;
- about prototyping and incremental life cycles;
- the importance of project management;
- how users may be involved in a project;
- the role of CASE tools in systems development.

3.1 Introduction

Problems that commonly affect the development of computerized information systems have been described in Chapter 2. These problems can significantly reduce the success of a software development project, and it is clearly important to adopt strategies and procedures that will minimize their occurrence. In this chapter we consider the steps that can be taken to make a computerized information system development project as successful as possible.

The problems identified in Chapter 2 fall into two major categories, those that are concerned with the management of the project and those that relate to poor quality in the delivered system. These are the two major areas on which we should focus if we wish to produce information systems within budget, on time and providing the required functionality.

There is much debate about what approach to information systems development is most likely to produce a quality system. One major source of difficulty is the inherent complexity of software development. In an object-oriented approach complexity is managed in such a way that it helps to address this problem.

From a more general perspective, building computerized information systems can be viewed as a form of problem solving. Figure 3.1 shows a general problem solving model adapted from Hicks (1991). The stages *Data Gathering* and *Problem Redefinition* are concerned with understanding what the problem is about; the *Finding Ideas* stage attempts to identify ideas that help us to understand more about the nature of the problem and possible solutions. *Finding Solutions* is concerned with providing a solution to the problem and *Implementation* puts the solution into practice. This approach to problem solving divides a task into sub-tasks, each with a particular focus and a particular objective.

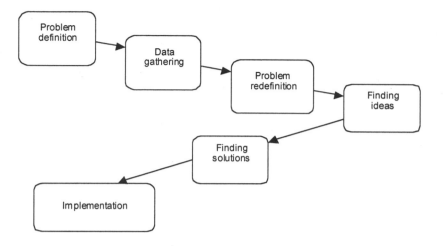

Figure 3.1 *General problem solving model (adapted from Hicks, 1991).*

The information systems development process may be subdivided simply into three main tasks, identifying what is required, planning how to deliver what is required and delivering what is required. There are many other ways of subdividing an information systems project but they all include an analysis activity that identifies what the system should do, a design activity that determines how best to do it and some construction that builds the system according to the design. The phases that contain these activities are given various names but the core activities remain the same. For example, Fowler (1997a) suggests a development process that has the phases inception, elaboration, construction and transition.

An important consequence of subdividing the development process is that techniques and skills specific to the different phases can be identified. Teams of developers with these specialized skills can be allocated to the particular phases or activities, maximizing the chance that the activities are completed as well as possible. Subdividing the development process results in smaller tasks that can be managed more easily to achieve the appropriate quality standards and to stay within the allocated resource budget.

A further source of help to overcome the problems the software developer faces is computer support. Software development is an activity that, like many others, may benefit from computerized information systems. It is interesting that computerized support for software development did not became realistic in terms of cost and technology until the mid-1980s. It is now accepted that computerized support is an important component of a development project (Allen, 1991).

We have already alluded to the benefits to be gained from managing the software development process effectively and have explicitly identified poor project management as a source of many of the problems. Building a software system is very different from building almost any other human artefact. Software is intangible, it cannot be weighed, its strength cannot be measured, its durability cannot be assessed, its resistance to physical stress cannot be gauged. Of course we try to find (and with some success) measures of a software system that enable us to make judgements about its size, its complexity, the resource required to build it, and so on. But these measures are much less well understood than their counterparts in the design and construction of such tangible artefacts as buildings and cars.

3.2 Project Life Cycles

Subdividing the process of software development produces what is known as a life cycle model. Just as an animal goes through a series of developmental stages from its conception to its demise so, it is argued, does a computerized information system. Various project life cycles can be applied to computerized information systems development. We will discuss some of the most commonly used life cycle models. Arguably two activities precede the information systems development project, strategic information systems planning and business modelling. The successful completion of these activities should ensure that the information system that is developed is appropriate to the organization. It can be argued that these are part of the information systems development life cycle. However, their focus is not on computerization, per se, but rather the identification of organizational requirements. Whether they are viewed as part of the life cycle or not, their importance is almost universally accepted for commercially oriented computer systems development. If a system is not commercially oriented (e.g. an embedded control system) strategic information systems planning and business modelling are not precursors to systems development. There is a distinction to be made between systems development, where a system may incorporate human, software and hardware elements, and software development which focuses primarily on software construction, although it involves the human users and the hardware upon which it executes. It is perhaps a matter of perspective. A software development project is based on the assumption that a software system will satisfy the user requirements whereas a systems development project is not.

Strategic Information Systems Planning. As we saw in Chapter 1, information systems work within the context of an organization and must satisfy its current requirements as well as providing a basis from which future needs can be addressed. In order to do this, strategic plans are developed for the organization as a whole and within their context a strategic view of information systems needs can be formed. For example, in the Agate case study a strategic decision may be made to target multinational companies for international advertising campaigns. This has consequences for campaign management and its supporting information systems.

Business modelling. In order to determine how an information system can support a particular business activity it is important to understand how the activity is performed and how it contributes to the objectives of the organization. Campaign management is an important business function for Agate and it should be modelled in order to determine how it is carried out, thus providing some of the parameters for subsequent information systems development.

3.2.1 Traditional life cycle

A diagrammatic view of the traditional life cycle (TLC) for information systems development is shown in Figure 3.2. This model is also known as the waterfall life cycle model (the difficulty of returning to an earlier phase once it is completed has been compared to the difficulty of swimming up a waterfall). The individual phases of a TLC are described in the following sections.

System Engineering. An information system involves human, software and hardware elements. The first stage of an information systems project is to identify the major requirements for the whole system and then to identify those parts of the system that are best implemented in software, those parts that are best implemented in hardware and those components that should be allocated to human participants. This phase produces a high-level architectural specification that defines how these major parts of the system will interact with each other.

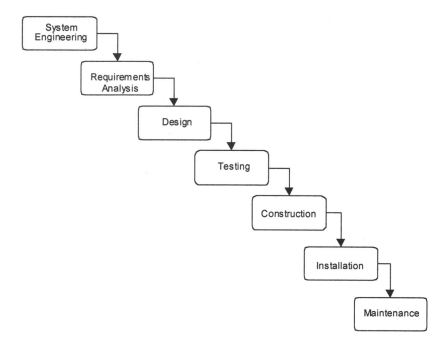

Figure 3.2 *Traditional waterfall life cycle model.*

Requirements Analysis. All the system's requirements need to be defined clearly and requirements gathering becomes more intensive. If the project is concerned largely with the development of elements that will be implemented in software then these are the main focus of requirements analysis. The objective is then to identify what the users would require from the software elements of the system. Requirements gathering for large or complex systems benefits from fact-finding techniques that help to identify the requirements, and further techniques to document them appropriately and to analyse them[1]. User requirements can also be used to develop the criteria for user acceptance tests.

Design. Once it is known what is required from the system, the design process determines how best to construct a system that delivers these requirements. Design is first concerned with the specification of a software architecture that defines the major software components and their relationships. Design involves reaching a balance between requirements that conflict with each other (e.g. maintainability and performance) within implementation environment constraints. Human activities may also be designed at this stage.

Construction. The completed design is now translated into program code. Depending on how the design phase has been completed, part of the construction may be automated. Construction may utilize different programming languages and database management systems for different parts of the system.

---

[1] We describe the techniques provided by UML for these purposes in Chapters 5–7.

Testing. The system is tested to ensure that it satisfies the user requirements accurately and completely. Typically, several levels of testing are performed. Individual components are tested independently, then are tested together as a sub-system and then the sub-systems are tested together as a whole system. Often users perform some form of acceptance testing before the system is finally accepted as complete.

Maintenance. It is most likely that the system will be subject to change during its operating life. The delivered system may operate erroneously and corrections may have to be made to the software (corrective maintenance). Certain aspects of the system's behaviour may not have been fully implemented (because of cost or time constraints for instance), but are then completed during the maintenance phase (perfective maintenance). The operating environment may also change in various ways causing requirements changes that have to be accommodated (adaptive maintenance).

There are many variations of the waterfall model (e.g. Pressman, 1994 and Sommerville, 1992), differing chiefly in the number and names of phases, and the activities allocated to them. A major justification for life cycle models like this is that their phases have explicitly defined products or *deliverables*. Sommerville (1992) suggests a series of deliverables produced by different phases of development, shown in Figure 3.3.

Phase	Output deliverables
System Engineering	High level architectural specification
Requirements Analysis	Requirements specification
	Functional specification
	Acceptance test specification
Design	Software architecture specification
	System test specification
	Design specification
	Sub-system test specification
	Unit test specification
Construction	Program code
Testing	Unit test report
	Sub-system test report
	System test report
	Acceptance test report
	Completed system
Installation	Installed system
Maintenance	Change requests
	Change request report

Figure 3.3 *Life cycle deliverables (adapted from Sommerville, 1992).*

These products can be used to monitor productivity and the quality of the activity performed. Several phases have more than one deliverable. If we need to show a finer level of detail to assist in the monitoring and control of the project, phases can be split so that each sub-phase has only one deliverable. Alternatively, a phase may be viewed as comprising a series of activities each of which has a single deliverable and can be managed individually. Project management issues are discussed in Chapter 21. An organization that embarks upon systems development should specify phases in the systems development process and their associated deliverables in a way that is applicable to the organizational context and the nature of the systems being developed.

The traditional life cycle has been used for many years but is the subject of several criticisms.

- Real projects rarely follow such a simple sequential life cycle. Project phases overlap and activities may have to be repeated.
- Iterations are almost inevitable, because inadequacies in the requirements analysis may become evident during design, construction or testing.
- A great deal of time may elapse between the initial systems engineering and the final installation. Requirements will almost inevitably have changed in the meantime and users find little use in a system that satisfies yesterday's requirements but may hamper current operations.
- The TLC tends to be unresponsive to changes in client requirements or technology during the project. Once architectural decisions have been made (e.g. during systems engineering) they are difficult to change. A technological innovation that may make it feasible to automate different parts of the whole system may become available after the project has been running for some time. It may not be possible to incorporate the new technology without re-doing much of the analysis and design work already completed.

The diagram in Figure 3.4 shows possible paths for iteration within the TLC but these iterations can be very costly. For example, if a major deficiency in the requirements analysis is discovered during construction iterating back to requirements analysis may result in significant redesign making some of the construction so far completed inappropriate. If problems such as this do occur, project constraints may result in the missed requirements being addressed without a suitable redesign but instead using an ad hoc coding solution. This approach tends to produce unmaintainable systems. However, despite its drawbacks the TLC does provide a very structured approach to systems development that has the following advantages.

- The tasks in a particular stage may be assigned to specialized teams. For example, some teams may specialize in analysis, others in design and yet others in testing.
- The progress of the project can be evaluated at the end of each phase and an assessment made as to whether the project should proceed.
- The controlled approach can be effective for managing the risks on large projects with potentially high levels of risk.

3.2.2 Prototyping

Many approaches to systems development incorporate some iteration. For instance, analysis may involve a series of tasks that are repeated iteratively until the analysis models are deemed complete. The waterfall model discussed so far aims to deliver a final working system as the finished product. However such approaches have a significant difficulty, in that the user only actually experiences how the system operates once it is delivered. Sometimes users find it difficult to imagine how their requirements will be translated into a working system and it is certainly the case that different implementations can be produced from the same set of requirements. The prototyping approach overcomes many of the potential misunderstandings and ambiguities that may exist in the requirements.

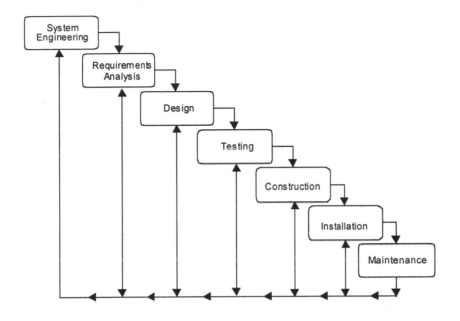

Figure 3.4 *Waterfall life cycle with iteration.*

In software development a prototype is a system or a partially complete system that is built quickly to explore some aspect of the system requirements and that is not intended as the final working system. A prototype system is differentiated from the final production system by some initial incompleteness and perhaps by a less resilient construction. If the prototype is to be discarded once it has fulfilled its objectives the effort required to build a resilient prototype would be wasted. A prototype will typically lack full functionality. It may have limited data processing capacity, it may exhibit poor performance characteristics or may have been developed with limited quality assurance. Prototype development commonly uses rapid development tools though such tools are also used for the development of production systems.

Prototypes may be constructed with various objectives in mind. A prototype may be used to investigate user requirements. For example, a prototype may be focused on the human–computer interface to determine what data should be presented to the user and what data should be captured from the user. A prototype might also be used to investigate the most suitable form of interface. A prototype may be constructed to determine whether a particular implementation platform can support certain processing requirements. A prototype might be concerned with determining the efficacy of a particular language, a database management system or a communications infrastructure.[2] A life cycle for prototyping is shown in Figure 3.5.

---

[2] For example, an Object Request Broker (these are described in Chapter 19).

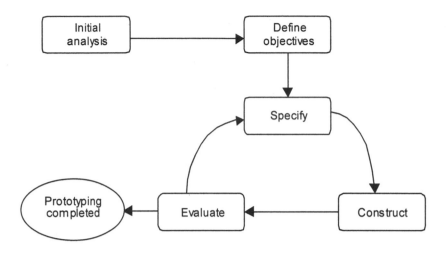

Figure 3.5 *A prototyping life cycle*

The main stages required to prepare a prototype are as follows:
- perform an initial analysis;
- define prototype objectives;
- specify prototype;
- construct prototype;
- evaluate prototype and recommend changes.

These are described in more detail below.

Perform an initial analysis. All software development activity utilizes valuable resources. Embarking upon a prototyping exercise without some initial analysis is likely to result in an ill-focused and unstructured activity producing poorly designed software.

Define prototype objectives. Prototyping should have clearly stated objectives. A prototyping exercise may involve many iterations, each iteration resulting in some improvement to the prototype. This may make it difficult for the participants in a prototyping exercise to determine if there is sufficient value to continue the prototyping. However, with clearly defined objectives it should be possible to decide if they have been achieved.

Specify prototype. Although the prototype is not intended for extended operation it is important that it embodies the requisite behaviour. It is almost certainly the case that the prototype will be subject to modification and this will be easier if the software is built according to sound design principles.

Construct prototype. Since it is important that prototype development is rapid, the use of a rapid development environment is appropriate. For example, if an interactive system is being prototyped, environments such as Delphi™ or Visual Basic® can be most effective.

Evaluate prototype and recommend changes. The purpose of the prototype is to test or explore some aspect of the proposed system. The prototype should be evaluated with respect to the objectives identified at the beginning of the exercise. If the objectives have not been met then the evaluation should specify modifications to the prototype so that it may achieve its objectives. The last three stages are repeated until the objectives of the prototyping exercise are achieved.

Prototyping has the following advantages:

- early demonstrations of system functionality help identify any misunderstandings between developer and client;
- client requirements that have been missed are identified;
- difficulties in the interface can be identified;
- the feasibility and usefulness of the system can be tested, even though, by its very nature, the prototype is incomplete.

Prototyping also has several problems and their impact on a particular project should be estimated before engaging in prototyping:

- the client may perceive the prototype as part of the final system, may not understand the effort that will be required to produce a working production system and may expect delivery soon;
- the prototype may divert attention from functional to solely interface issues;
- prototyping requires significant user involvement;
- managing the prototyping life cycle requires careful decision making.

3.2.3 Incremental development

Several suggestions on how to overcome the problems with the TLC have been made. Gilb (1988) suggests that successful large systems start out as successful small systems that grow incrementally. An incremental approach performs some initial analysis to scope the problem and identify major requirements. Those requirements that will deliver most benefit to the client are selected to be the focus of a first increment of development and delivery. The installation of each increment provides feedback to the development team and informs the development of subsequent increments.

Boehm's (1988) spiral model can be viewed as supporting incremental delivery. However, Gilb (1988) argues that the spiral model does not fully support his view of incremental development as there are aspects of systems development that it does not emphasize or include. These are as follows:

- the production of high-value to low-cost increments;
- the delivery of usable increments of 1% to 5% of total project budget;
- a limit to the duration of each cycle (e.g. one month);
- a measure of productivity in terms of delivered functionality or quality improvements;
- an open-ended architecture that is a basis for further evolutionary development.

Figure 3.6 shows how Boehm's spiral model can be adapted to suit incremental delivery. Note that prototyping may be used either during the risk analysis or during the software development part of the development cycle.

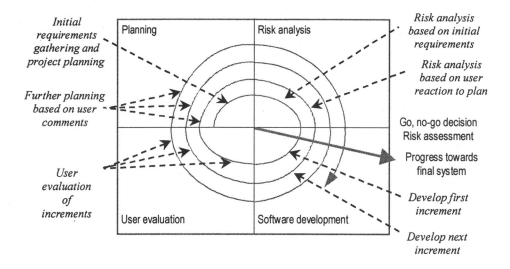

Figure 3.6 *Spiral model for incremental delivery (adapted from Boehm, 1988).*

3.2.4 The OMG object-oriented life cycle

An object-oriented life cycle has been proposed by the OMG and a variation is shown in Figure 3.7. This model focuses on the activities involved rather than the sequence within which they occur. The diagram illustrates that many activities may occur at the same time and that there is a need to co-ordinate these activities if we are to achieve reuse of models or of software.

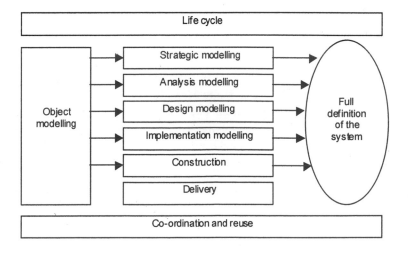

Figure 3.7 *Object-oriented life cycle model (adapted from Yourdon, 1994)*

Texel and Williams (1997) suggest a detailed object-oriented life cycle model that includes 17 stages. A life cycle with such a large number of delineated stages is probably more suited to large projects. If used for small projects the management overhead is likely to be excessive. Booch (1994) suggests an iterative incremental approach to object-oriented software development.

Review Questions

3.1   What are the advantages of the traditional waterfall life cycle?
3.2   What are the disadvantages of the traditional waterfall life cycle?
3.3   How are some of the disadvantages listed in Question 3.2 overcome?

3.3 Managing Information Systems Development

An information system development project has many attendant risks. Late delivery may result in lost income and missed opportunities for the client and punitive legal action against the software developers. The continuing ability of an enterprise to compete effectively may be dependent upon a new information system. Failure to deliver a new system may even cause a company to fail. There are also risks associated with adopting a new information system. For example, suppose an organization chooses to automate the payment of all invoices. An automatic invoice payment system may be designed to balance the commercial and moral duties to pay suppliers promptly with a requirement to maximize the interest that is earned on the organization's assets. Should the system fail to achieve this balance in practice, the company may suffer serious consequences. On the one hand, late payment of an invoice might lead to damaging legal action. Of course, one way to avoid this particular risk is not to adopt the new information system but the alternatives will also have attendant risks and costs. For instance, a manual system for processing invoices may have higher costs and greater risk of error. On the other hand, if invoices are paid too quickly the company's interest earnings will be reduced. Use of any new information system must be controlled and monitored to ensure that it continues to operate within the parameters defined by the company.

Organizations are increasingly dependent on the operation of computerized information systems to manage the business, to the point where a new system that interferes with the successful operation of existing systems may prevent the company from being able to operate at all. As a project proceeds, more detail becomes available as to what the proposed system should do and how it will operate. It is important at each stage to identify risks to the organization and determine their significance so that judgements can be made on the continuing viability of the project. If a new system is thought likely to place the organization at risk of a catastrophic failure then either the system should be redesigned to avoid this (perhaps at significant extra cost) or the project should be cancelled. Note that in Figure 3.6 each iteration includes a risk assessment that determines whether the next cycle should proceed.

3.4 User Involvement

A key factor in maximizing the chance of success is ensuring that there is continued and effective user involvement throughout the project. The traditional waterfall life cycle is less amenable to user involvement throughout the whole project and hence is less likely to stay in line with the changing user requirements. A prototyping approach is normally dependent upon continuing user involvement and by its very nature encourages user involvement. However, care has to be taken to ensure that the users have sufficient time to perform their roles effectively. In prototyping, the evaluation of the prototype requires significant time from the users.

Users can be involved in projects at various levels and can play many different roles. Some approaches to software development (e.g. DSDM[3]) directly involve users in the development team, so that they then have a significant opportunity to influence the way the project proceeds, perhaps by identifying difficulties and suggesting more acceptable alternatives. It is important that users who have such a significant influence on the direction of the project should understand the organizational requirements and the needs of fellow users. Direct involvement of users is more likely to be successful if they are considered as full members of the project team and if they are genuinely empowered to represent the organization and make decisions within clearly defined operating parameters. There is always a danger that users who become members of a project team cease over time to represent the user perspective effectively, as they begin to view themselves more as a team member and less as a user representative. One way of overcoming this tendency is to rotate membership of the development team among a group of users. However, this can result in a loss of continuity. A more satisfactory approach is to scope each activity so that a user team member can see it through to completion in a reasonably short time, say within three months.

At the other end of the spectrum are users whose only involvement in the project is as participants in fact gathering. Such a user may provide information about current working practice but they have little or no influence on the design of the new system. In this situation users are likely to be concerned about the project and may fear the effect it will have on their jobs. As a result, they may be less co-operative and will be less willing to take ownership of the new system when it is installed.

Even when users are not invited to join the project team, effective participation can still be encouraged through a consultative approach. Procedures are set up so that users are able to review certain aspects of the systems development, provide feedback, give their views of the system and then be able to see the response of the project team to their feedback. In some circumstances a large part of the task of requirements capture may be delegated to users. They are likely to feel a strong affinity to the delivered system, provided that it satisfies the requirements they specified.

Whatever form of involvement users have with the project it is important that their role is carefully explained and that training is given as required. Some large organizations have gone to the trouble of training users so that they understand the terminology and models used by the systems developers. Furthermore, users must be given the time to participate. It is no good expecting users to review requirements documents effectively if the only time they have available is during their lunch break.

---

[3] DSDM is introduced in Chapter 21.

Users who participate in a systems development project can be selected in various ways. They can be designated by management as being the most appropriate representatives or they can be selected by their peers. In either circumstance they must be genuine representatives of the user perspective.

3.5 Methodological Approaches

One of the major influences on the quality of the systems developed is the software development approach adopted. If the approach used is not appropriate for a particular type of application then it may limit the quality of the system being produced. In part at least, the object-oriented approach provides a mechanism for mapping from real-world problems to abstractions from which software can be developed effectively. Furthermore object-orientation provides conceptual structures that help to deal with modelling complex information systems. As information systems requirements are becoming increasingly complex the use of an object-oriented approach is more necessary. It is a sensible strategy to transform the development of a large, complex system into the development of a set of less complicated sub-systems. Object-orientation offers conceptual structures that support this sub-division. Object-orientation also aims to provide a mechanism to support the reuse of program code, design and analysis models.

A methodology consists of an approach to software development (e.g. object-orientation), a set of techniques and notations (e.g. the Unified Modelling Language—UML) that support the approach, a life cycle model (e.g. spiral incremental) to structure the development process and a unifying set of procedures and philosophy. In this text we do not espouse a particular methodology but apply object-oriented techniques in a co-ordinated fashion using UML. Methodologies are discussed in more detail in Chapter 20.

3.6 CASE

Computer Aided Software Engineering (CASE) tools have been widely available since the mid-1980s and now provide support for many of the tasks the software developer must perform. CASE tools have been categorized in various ways according to the phase in the life cycle in which they should be used. Upper-CASE tools provide support for the analysis and design while lower-CASE tools are concerned with the construction and maintenance of software. These two categories of tools directly support the overall development process. Other computer tools address the tasks involved in the management of information system development projects.

Early CASE tools could be categorized clearly as upper-CASE or lower-CASE but this distinction is much less true these days when CASE tools offer facilities that support many life cycle phases. Typically CASE tool vendors offer a series of products that are integrated to offer extensive life cycle coverage. Such CASE tools are known as I-CASE (the I stands for Integrated). CASE tools tend to be designed for use with a particular approach to software development, particular notations and implementation environments. In the 1980s and early 1990s CASE tools were being developed to support an apparently ever increasing range of different development approaches. The growth in popularity of object-oriented approaches has exacerbated the situation with the emergence of many new methodologies with different notations. Although it is acknowledged that no single approach to software development is appropriate for all projects it has become accepted that a common notation set can be used for requirements and design modelling. The move to a common notation has resulted in the development of the Unified Modelling Language which, it must be emphasized, is not a methodology. Modern I-CASE tool-sets provide an increasingly wide range of facilities and cover most life cycle activities. These are discussed turn below.

3.6.1 Diagram support

Most approaches to information systems development utilize diagrams to represent the relationships between different elements of the requirements and to document the design structure before the system is constructed. Analysis and design diagrams are models of the system and of its requirements. Diagrammatic analysis and design techniques were used in information systems development for many years before computerized support became available. However the difficulties in preparing diagrams by hand and then ensuring that they remain up to date whenever the system is modified meant that in practice they quickly became inconsistent. The automated diagram support offered by CASE tools appears to overcome this problem. However, although the process of creating and maintaining diagrammatic models of the system are much eased by CASE tools, they must still be enforced by quality assurance procedures.

Computer support for preparing and maintaining diagrams requires various features which include:

- checks for syntactic correctness;
- data dictionary support;
- checks for consistency and completeness;
- navigation to linked diagrams;
- layering;
- traceability;
- report generation;
- system simulation;
- performance analysis.

These features are described in turn below.

Syntactic correctness. The CASE tool checks that the correct symbols are being used on the diagrams and that they are being linked in permissible ways. This ensures that the correct vocabulary is being used but does not ensure that it is meaningful or relevant to client requirements.

Data dictionary. In the same way that a language dictionary provides clear definitions of the words in a language, a data dictionary contains definitions of all the elements in an information system. Originally data dictionaries were only designed to hold the definitions of the data attributes of a system. However, increasingly it became clear that all elements in a system development project are open to ambiguous interpretation unless clearly defined. Some CASE vendors use the terms repository or encyclopaedia for data dictionaries that can hold the descriptions of all the elements involved in the project and their relationships.

Consistency and completeness. Most CASE tools support various diagrammatic models that capture different aspects of the system. As all relate to the same system it is important that any one element that appears on several diagrams (perhaps viewed from different perspectives) should be consistent with itself. Most approaches to analysis and design stipulate that certain diagrams must be completed and that the elements referred to in those diagrams must all be documented in the data dictionary. To manually check the consistency and completeness of a system of any significant size is a task that is very onerous, time-consuming and error-prone. A good CASE tool may check the consistency and completeness of a large model in seconds and provide the developer with a comprehensive report on any inconsistencies found or omissions identified.

Navigation to linked diagrams. A complex system is likely to require many diagrammatic models to describe its requirements and its design. For a CASE tool to be usable, easy navigation between linked diagrams is essential. For example, double-clicking on a component at one level of abstraction may automatically open up a diagram that describes it at a more detailed level. It is also helpful to be able to move directly from one view that contains a particular element to another view that contains the same element.

Layering. An information system of any significant size is by nature complex and it is unlikely that all relationships between its components can be shown on a single diagram. Just as maps are drawn at different scales with different levels of detail, system models are produced at various levels of abstraction. A high-level diagram may represent the relationships between large components such as sub-systems. A diagram drawn at a lower level of abstraction may describe the elements within a particular component in detail. In order to cope with complexity, we divide the system into manageable chunks and link them in layers. A good CASE tool provides a capability to layer the models of the system at different levels of abstraction. The consistency and completeness checking discussed earlier should also check that the representations of one element at different levels of abstraction are consistent.

Traceability. Some of the elements created during the development of an information system are derived from other elements, and the connections between them should be maintained. It must be possible to trace through from the data dictionary entries that describe a particular requirement to the program code that provides the functionality that satisfies the requirement. If a requirement changes, the maintenance activity is easier if all the code that implements that requirement can be readily identified. It should be possible to trace all requirements from the analysis documentation, through the design documentation to the implemented code. This feature is known as requirements traceability.

Report generation. Complex systems involve modelling many elements. Comprehensive reporting capabilities improve the usability of a CASE tool by ensuring that the developer can easily obtain information about the models for a system in suitable formats. In fact a CASE tool would be of little use if the information it held about a project were not readily available, no matter how effective it was in other respects.

System simulation. When a CASE tool has been populated with models of an application it should be possible to simulate some aspects of system behaviour. For example, how does the system respond to a particular event? Some CASE tools provide capabilities that enable a software developer to examine the consequences of a design decision without the need to actually build the software.

Performance analysis. The performance of a system is an important ingredient in its success. For example, a system that supports staff who deal directly with customer enquiries should be able to respond quickly to a query about the availability of a certain product. If customers are kept waiting for too long, this will probably result in lost sales. The analysis of performance is particularly difficult for an application that runs on multiple processors and uses a complex communications infrastructure. Some CASE tools even provide the capability to perform a 'What if' analysis to examine the implications of alternative implementation architectures.

3.6.2 Software construction

CASE tools can offer a range of features to support software construction and maintenance. These include code generation and maintenance tools.

Code generators. The capability to generate code directly from a design model is a major benefit to the developer for several reasons. Firstly, a working software system is likely to be produced more quickly. Secondly, one source of error is largely removed when the code is produced automatically and consistent with the design. Thirdly, when requirements change, a consequent change to the design documentation can be followed by automatic code generation. If the application logic is defined completely and precisely in the design model, full code generation is possible. If a design model contains process descriptions then it is likely that a code framework can be generated to which further code can be added. In order to reduce the level of detail required for the design model, code generators may make certain assumptions concerning the implementation. Code generators are available for many different languages and development environments and are likely to include the capability to generate database schemata for the major proprietary database management systems.

Maintenance tools. Software maintenance is a major issue, particularly for those who are responsible for systems that may be affected by the Year 2000 problem. Various tools are available to help with systems maintenance. For some programming languages, reverse engineering tools may also be available that can generate design documentation directly from program code (although if the program code is poorly structured the resulting design documentation may be of little use). Tools are also available that can analyse program code and identify those parts that are most likely to be subject to change.

3.6.3 Benefits and difficulties of using CASE

CASE tools can bring many benefits to the development activity. They help to standardize the notation and diagramming standards used within a project, and this aids communication among the team members. They can perform automatic checks on many aspects of the quality of the models produced by analysts and designers. The report generation capabilities of a CASE tool reduce the time and effort that needs to be spent by analysts and designers in retrieving data about the system upon which they are working. Where a CASE tool can carry out automatic code generation, this further reduces the time and effort that is required to produce a final system. Finally, the electronic storage of models is essential to the reuse of models, or components of them, on other projects that address similar analysis or design problems.

Like any other technology, CASE tools also have their disadvantages. These include limitations in the flexibility of the documentation that they can provide. The development approach may also be limited by the need to work in a particular way in order to fit in with the capabilities of the CASE tool. The ability of a CASE tool to check all models for their consistency, completeness and syntactic correctness can in itself give rise to a danger. Developers may make the erroneous assumption that, because their models are correct in those specific senses, they are therefore also necessarily relevant to user requirements. There are also certain costs attached to the installation of a CASE tool. Aside from the cost of the software and manuals, there is also likely to be a significant cost in additional training for developers who will be expected to use the CASE tools.

On balance CASE tools can provide useful and effective support for the software development activity, but it requires appropriate management for this to be achieved without any damaging side-effects.

3.7 Summary

We have considered how to avoid the problems that typically arise during information systems development. Several strategies have been discussed. Life cycle models are used to provide structure and management leverage for the development process itself. User involvement is crucial to ensure relevance and fitness for purpose of the delivered system. Furthermore, many of the difficulties that occur during installation are limited if ownership of the proposed system has been fostered by effective participation. The modelling requirements of the activities involved in the software development are well supported by object-orientation. Finally we considered the importance of CASE tool support for the software developer.

Review Questions

3.4 What is prototyping?

3.5 How does prototyping differ from incremental development?

3.6 What are the different ways of involving users in the systems development activity? What are potential problems with each of these?

3.7 How do 'syntactic correctness', 'consistency' and 'completeness' differ from each other?

3.8 What does requirements traceability mean?

3.9 Why is it not enough for a diagram to be syntactically correct, consistent and complete?

3.10 What is the purpose of a data dictionary?

Case Study Work, Exercises and Projects

3.A What life cycle model would you recommend for the development of the production control system for FoodCo? Justify your decision.

3.B For a CASE tool with which you are familiar, explore and critically assess the consistency and completeness checking facilities available.

3.C For a CASE tool with which you are familiar, explore and critically assess its system generation capabilities.

3.D In your library find references for three life cycle models not discussed in this chapter. Briefly review each of these life cycle models.

Further Reading

Hicks (1991) provides a comprehensive introduction to problem solving skills that are valuable to the software developer. Sommerville (1992) and Pressman (1994) provide good discussions of life cycle issues. Gilb (1988) contains much good advice concerning software development and is well worth reading. Texel and Williams (1997) offer a detailed object-oriented methodology specifying deliverables and tasks clearly.

Many CASE tools are in widespread use but, so far, only a few support the UML standard. Among these are SELECT Enterprise® (we used Versions 5 and 6 of this tool to prepare some of the diagrams and screenshots in this book) and Rational Rose™ from Rational Corporation, the company that developed UML. More information can be obtained from their respective web sites: www.selecst.com and www.rational.com.

4

What Is Object-Orientation?

Objectives
In this chapter you will learn:
- the fundamental concepts of object-orientation;
- the justifications for an object-oriented approach;
- how object-orientation is used in practice.

4.1 Introduction

It is a major premise of this book that the object-oriented approach to systems development helps to avoid many of the problems and pitfalls described in earlier chapters. In this chapter, we lay the foundations for understanding object-orientation by presenting an explanation of the fundamental concepts. This is important for all readers, but particularly so for those who are already familiar with other approaches to systems analysis and design, for example, structured analysis. There are important differences between the object-oriented approach and earlier approaches. It is necessary to have a sound grasp of the basic concepts in order to be able to apply the techniques of object-orientation in an effective way.

At this stage the most important concepts to grasp are object, class, instance, generalization and specialization, message passing and polymorphism. We explain each of these concepts in turn, showing both what the parts of an object-oriented system are, and how they use message passing to effectively isolate different parts of a system from each other, thus controlling the complexity of the interfaces between sub-systems (Section 4.2). The discussion is necessarily conceptual in places, but practical examples and analogies are used to illustrate the theoretical points wherever this is appropriate. We then place object-orientation in its historical context, and look at some of the reasons why it has been able to make a contribution to successful information systems development (Section 4.3). Finally, we conclude by summarizing the state of evolution that object-oriented languages have reached today (Section 4.4).

4.2 Basic Concepts

The most important concept addressed in this section is the object itself, and it is to this that we first pay attention. The other concepts explained in this section are strongly dependent on each other, and all contribute to an adequate understanding of the way that objects interact, and thus to their significance for information systems. The following explanations concentrate on giving an understanding of the territory that later chapters will explore in greater depth.

4.2.1 Objects

In an early book on object-oriented analysis and design, Coad and Yourdon (1990) define *object* as follows:

> Object.   An *abstraction* of something in a problem domain, reflecting the capabilities of the system to keep information about it, interact with it, or both.

This may not immediately appear to help very much, as parts of the definition are themselves a little obscure and raise further questions. In particular, what exactly do we mean by 'abstraction', and which 'system' are we talking about? A useful definition of abstraction in this context might be: 'A form of representation that includes only what is important or interesting from a particular viewpoint.' To give a common example, a map is an abstract representation. No map shows every detail of the territory it covers (impossible, in any case, unless it were as large as the territory, and made from similar materials!). The intended purpose of the map guides the choice of which details to show, and which to suppress. Road maps concentrate on showing roads and places, and often omit landscape features unless they help with navigation. Geological maps show rocks and other sub-surface strata, but usually ignore towns and roads. Different projections and scales are also used to emphasize parts of the territory or features that have greater significance. Each map is an abstraction, partly because of the relevant features it reveals (or emphasizes), and also because of the irrelevant features it hides (or de-emphasizes). Objects are abstractions in much the same way. An object represents only those features of a thing that are deemed relevant to the current purpose, and hides those features that are not relevant.

This brings us to our second question. To which system are Coad and Yourdon referring? The context of the quotation makes it reasonably clear that their answer would be something like this: 'The proposed object-oriented software system, whose development is under consideration'. This seems appropriate—after all, this book is about the development of object-oriented software—but we should note that other systems are also involved. The most important of these is the human activity system that we must understand before beginning the construction of any software.

We should also note that we are using one tool to serve two purposes. As we shall see particularly in Chapter 6, objects are used to model an understanding of the application domain (essentially part of a human activity system), yet we shall also see in later chapters that objects are also understood as parts of the resulting software system. These are distinct purposes and there will be some occasions when we may need to take care to be clear about which meaning is intended.

Rumbaugh et al. (1991) take a slightly different slant on defining an object. These authors also explicitly recognize the dual purpose noted in the last paragraph.

> We define an *object* as a concept, abstraction, or thing with crisp boundaries and meaning for the problem at hand. Objects serve two purposes: They promote understanding of the real world and provide a practical basis for computer implementation.

This definition contains a repetition (although this does no harm)—by 'concept' we actually mean an abstraction, something that is logical rather than physical. A concept may be intangible, even imaginary (in the sense that not every participant agrees on its meaning, or sometimes even its existence). In the Agate case study, one concept is the 'campaign'. While they are clearly important, campaigns are intangible and difficult to define with precision. They really exist only as a relationship between a client (say FoodCo), the Account Manager, some other staff, some advertisements and various tasks and components that go into creating advertisements.

Concepts include many kinds of relationship between people, organizations and things. In information systems development, it is often necessary to recognize relationships such as contracts, sales or agreements. While intangible, some of these relationships can be quite long-lasting, and can have a complex influence on how people and other things in the application domain are able to act.

Let us take a simple transaction as an example. Imagine buying a tube of toothpaste in your local supermarket. On one level, this is just a sale, an exchange of money for goods. On a deeper level you might be entering into a complicated relationship with the shop and the manufacturer. This probably depends on other factors; e.g. the warranty may vary depending on the country that you are in at the time of the purchase, and perhaps the sale will earn points for you on a loyalty card. Perhaps the packaging includes a money-off coupon for your next purchase, or a contest entry form which must be accompanied by a valid proof of purchase. Now suppose you find something wrong with the toothpaste—you might be able to claim a refund or replacement. Maybe you can even sue the shop for damages. The point here is that we cannot understand the business without understanding these possible consequences of the sale in some appropriate way. In this case, 'sale' will almost certainly become an object in the system.

At a fairly abstract level, when choosing objects—in fact, at the level that corresponds to the mapmaker—we need to ask: 'What sort of map is this; what details should it show, and what should it suppress?' All objects have certain similarities to all other objects, summarized by Booch in the statement that an object 'has state, behaviour and identity' (Booch, 1994). Here, 'state' represents the particular condition that an object is in at a given moment, 'behaviour' stands for the things that the object can do (and that are relevant to the model) and 'identity' means simply that every object is unique.

For example, in Figure 4.1 the following are considered as objects: a person, a shirt, a sale, a bottle of ketchup. Some possible identities, behaviours and states for these objects

Object	Identity	Behaviour	States
A person.	'Hussain Pervez'.	Shout, grow tired.	Angry, busy, unemployed.
A shirt.	'My favourite button-down white denim shirt'.	Shrink, stain, rip.	Pressed, dirty, worn.
A sale.	'Sale no 0015, 15/12/98'.	Earn loyalty points.	Invoiced, cancelled.
A bottle of ketchup.	'*This* bottle of ketchup'.	Spill in transit.	Unsold, opened, empty.

Figure 4.1 *Characteristics of some objects.*

are shown, although note that these are for illustration only, and do not indicate any particular system perspective.

In some texts (e.g. Wirfs-Brock et al., 1990), objects are deliberately characterized as if each were a person, with a role in the system that is based on its answers to three questions.

- Who am I?
- What can I do?
- What do I know?

This approach defines an object in terms of its responsibilities and its knowledge, themes that we will encounter again in later chapters.

4.2.2 Class and instance

Let us begin with the definition of an object given in the UML Notation Guide (Rational, 1997). While not the clearest definition given in the UML documentation (the glossary to the UML Semantics volume has a better definition, which we give later in Box 4.1), it is worth quoting here as it illustrates the strength of the connection between the three concepts of object, instance and class.

> An object represents a particular instance of a class.'

Objects that are sufficiently similar to each other are said to belong to the same *class*. *Instance* is another word for a single object, but it also carries connotations of the class to which that object belongs: every object is an instance of some class. So, like an object, an instance represents a single person, thing or concept in the application domain. A class is an abstract descriptor for a set of instances with certain logical similarities to each other.

Figure 4.2 shows some classes that might be identified from the Agate case study (Chapter 6 describes a practical approach to identifying classes).

Figure 4.2 *Some possible object classes in the Agate case study.*

A class and its instances are related in the following manner. For staff at Agate, the idea of 'a campaign' is an abstraction that could represent any one of several specific campaigns. In a class diagram, the class Campaign represents the relevant features that all campaigns have in common. There is one instance of the class to represent each real-world campaign. Some examples of campaigns are: a new commercial for FoodCo's *World Tradition* range, shown on British TV throughout October 1998; a series of magazine adverts for various Yellow Partridge products placed throughout Europe in January 1999; a national series of

TV, cinema and magazine adverts for the launch of the Soong Motor Co's *Granda* estate in Summer 1999.

Each instance of a class is unique, just as every living person is unique, however closely they resemble someone else. This is true even when two instances have identical characteristics. For example, there could be two members of staff at Agate with the same name—say 'Ashok Patel', as this is a fairly common name. It is even possible (though unlikely) that both joined the company on the same date and work in the same section. Yet each remains a separate individual, and would therefore be represented by a separate instance of the class StaffMember.

4.2.3 Class membership

The idea that instances belong to a class logically implies that there must be a test that determines to which class an instance belongs. Since membership is based on similarity, such a test will also be capable of determining whether any two instances belong to the same class. There are two distinct types of logical similarity which must be tested. Firstly, all objects in a class share a common set of descriptive characteristics.

For example, the staff at Agate record a client's company name, address, telephone number, fax number, email address and so on. Each item is included in the list because it is useful in some way to the users of the system[1], and the full list gives a complete description of a client. The value of each item (e.g. the actual company name) will vary from one client to another. But, while the values may differ, this information *structure* is the same for every client.

To take another example, users of the Agate system also need to know about the member of staff assigned as staff contact to a client. A staff contact might be described by a staff name, staff number and staff start date. Again, the full list gives a complete description of a staff member. Here, too, the value of each item (e.g. the staff name) will normally vary from one person to another, but again the structure is the same for all Staff Contacts.

Now compare the two descriptions, summarized in Figure 4.3.

Class	Characteristics	Class	Characteristics
Staff Member	Staff name Staff number Staff start date	Client	Company name Company address Company telephone number Company fax number Company email address

Figure 4.3 *Information structures for two classes.*

Both staff members and clients have a name, but otherwise there is little in common. The information structure used to describe a client would not be capable of describing a member of staff, and vice versa. It is reasonable to consider all clients as members of one class, and

---

[1] What appears in the list depends on the needs of the application. Another team may be developing a system to monitor whether client companies comply with environmental laws. Their list may share some characteristics with ours, but may also add others of no interest to our model.

all staff as members of one class, but it would not be justified to consider them as members of the *same* class. When two objects cannot be described by a single set of features, they cannot belong to the same class.

A second logical similarity must also be tested. All objects in a class share a common set of valid behaviours. For example, clients may initiate a campaign, may be assigned a staff contact, may pay for a campaign, terminate a current campaign, and so on. Perhaps no one client will actually do all this in reality, but that does not matter. The point is that any client could do any of these things, and the information system must reflect this.

If we consider staff members, we find a different set of valid behaviours. Staff can be assigned to work on a campaign, be assigned to a campaign as staff contact, change grade and maybe other things we do not yet know about. It may be more likely for staff members than for clients that they will actually go through the same sequence of behaviours, but this, too, does not matter. The point again is that a member of staff could do these things.

All clients, then, have a similar set of possible behaviours, and so do all staff members. But clients can do some things that a member of staff cannot do, and vice versa. On this count, too, we would consider clients as a class, and staff as a class, but we cannot make a case for considering clients and members of staff as instances of the same class. To summarize the examples, Client is a valid class, and StaffMember is a valid class. We can also note that, while the discussion here has been rather formal, an informal version of this dual test also works. In fact, we have already met this in Section 4.2.1: all members of a class give the same answers to the questions 'What do I know?' and 'What can I do?'

4.2.4 Generalization

In this section, we look at the concept of generalization and its application within object-orientation. In the UML Semantics Guide (Rational, 1997), generalization is formally defined as a:

> ... taxonomic relationship between a more general element and a more specific element that is fully consistent with the first element and that adds additional information.

This is perhaps a little much to grasp in one take, so in the sections that follow we will dismantle the definition into its main components, and examine each part on its own. We will highlight the general principles by looking first at an example of species classification (note that this is intended only to illustrate the concept of generalization, and is not a technical presentation of modern biological taxonomy). Later, in Chapter 7, we present an extended practical example from the Agate case study.

Taxonomic relationship

'Taxonomy' is defined by The Concise Oxford Dictionary (1995) as: '*n.* (1) the science of classification, esp. of living and extinct organisms. (2) a scheme of classification.' In other words, it means either an applied set of classifications, or the principles by which that set is constructed. The word was originally used for the hierarchic system of classification of plants and animals, hence the example shown in Figure 4.4.

Box 4.1 Class and object

Different books use the same terminology in different ways, and this can be confusing for the novice. For example, some authors use 'object' and 'class' interchangeably to mean a group of similar objects. But 'object' can also mean a single individual object, with 'class' reserved for the definition of a group of similar objects.

A further confusion arises if it is not clear whether 'object' refers to application domain things and concepts, or to their representation in an object-oriented model. The map is not the territory, even when the same words are used in both.

A third source of confusion can occur when the project focus shifts from analysis to design. During this transition, what was a model of the application domain is transformed into a model of software components. However, as we have not yet reached that point in the life cycle, we defer further discussion until later.

The UML Semantics Guide (Rational, 1997) defines the terms as follows.

A **class** is *'A description of a set of objects that share the same attributes, operations, methods, relationships and semantics.'*

An **object** is *'An entity with a well-defined boundary and identity that encapsulates both state and behavior. State is represented by attributes and relationships, behavior is represented by operations, methods, and state machines. An object is an instance of a class.'*

We follow these definitions. From now on 'object' is used consistently and interchangeably to mean a single instance of a class in a model. 'Class' is used consistently to mean the descriptor of (specification of) a set of similar object instances in a model. When we need to refer explicitly to people, things or concepts in the application domain, we will say so explicitly, as in this sentence.

A further distinction should be made between a *class* and a *type*. The UML Semantics Guide defines a **type** as *'A stereotype of class that is used to specify a domain of instances (objects) together with the operations applicable to the objects. A type may not contain any methods.'*

What does this rather obscure definition mean? Cook and Daniels (1994) express the distinction with more clarity, noting that they prefer to use the term type rather than class to denote a collection of similar real-world objects. A class has connotations of a software implementation. Hence the UML constraint that a type may not contain any methods, for a method is an implementation of an operation. An operation is a logical specification of behaviour, while a method is a program that implements the specification. In the same way, a type is a specification for a set of similar objects, while a class is an implementation of a type.

We prefer to take a pragmatic approach and use 'class' for the most part as interchangeable with 'type'. Wherever the context does not make it sufficiently clear which meaning is intended, further explanation will be given.

The taxonomic relationship between the two elements in this hierarchy labelled 'domestic cat' and 'mammal' can be simply rephrased as 'the domestic cat belongs to the class of mammals', or even more simply 'a domestic cat *is a kind of* mammal'. Many other relationships are also identified in the diagram. For example, a domestic cat is also a cat in the more general sense that also includes a tiger, and it is a living thing. We can summarize this by saying that in each case the common relationship is that one element 'is a kind of' the other element.

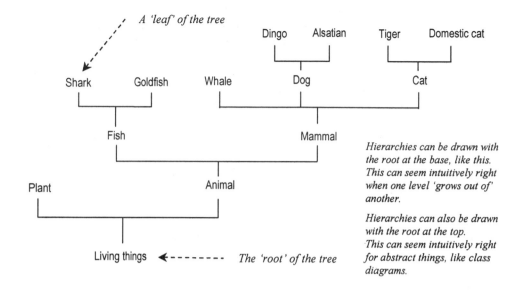

Figure 4.4 *A simple taxonomy of species.*

A more general element

The mammal is a more general element than the domestic cat. Thus any description of a mammal must apply to many different animals: domestic cats, tigers, dogs, whales, etc. A description of a general mammal really means just a description of common characteristics shared by all mammals. This may be rather brief, perhaps consisting of only one characteristic, e.g. that all mammals suckle their young. In the tree diagram, the more general an element is, the nearer it is to the root of the tree.

A more specific element

As an element of the hierarchy, 'domestic cat' is more specific than 'mammal'. In other words, the former conveys more information. Knowing that an animal is a cat, we can guess something about its diet, general body shape, size (within certain limits), number of legs (barring accidents, etc.), and so on. If we know only that it is a mammal, we can guess little, if anything at all, of its description. The more specialized elements of a hierarchy are those which are further from the root of the tree and closer to the leaves. The most specialized elements of all are those that actually form the leaves of the tree. In Figure 4.4, the leaves are individual species[2].

---

[2] If 'species' is considered as a class, then this in turn is an abstraction of its members. We could as easily extend the hierarchy one step further to show individual cats as 'leaves', rather than their species. The choice of where to terminate the branches of any hierarchy depends entirely on context.

Fully consistent with the first element

Whatever is true for a mammal is also true for a domestic cat. If the defining characteristic of a mammal is that it suckles its young, then a domestic cat also suckles its young. This is true for every mammal, whether a tiger, dog or whale. This is an important feature of any hierarchic taxonomy. As an illustration, we can think about what would happen if a zoologist discovers that a defining mammalian characteristic does not apply to an animal previously thought to be a mammal. For example, suppose a research project found conclusive evidence that common field mice lay eggs instead of bearing live young. This alone would be enough grounds either to re-classify the field mouse, or to re-define what distinguishes a mammal from other animals. Perhaps a brand new classification would be invented specifically to accommodate egg-laying mice.

Adds additional information

A full description of the domestic cat would contain a great deal more information than that for a general member of the mammal class. For example, we might define a domestic cat by saying that it suckles its young, has a certain skeletal structure, a particular arrangement of internal organs, carnivorous teeth and habit, the ability to purr, and so on. Apart from suckling its young, none of these characteristics apply to *all* other mammals. A full zoological description of any species contains at least one characteristic (or a unique combination of characteristics) that differentiates it from all other species. Otherwise it would not make sense to consider it a species in the first place[3].

Using generalization

The main application for *generalization* in object-orientation is to describe relationships of similarity between classes. Object classes can be arranged into hierarchies much the same as the species example. This has two main benefits.

The first results from the use of object classes to represent different aspects of a real-world situation that we wish to understand. Using generalization, we can build logical structures that make explicit the degree of similarity or difference between classes. This is an important aspect of the *semantics* of a model—in other words, it helps to convey its meaning. For example, knowing what hourly-paid and monthly-salaried employees in a business have in common with each other may be just as important as knowing how they differ. The former may help to understand that some types of information must be recorded in identical ways for both types of employee. Figure 4.5 illustrates this with an example that might be suitable for a payroll system.

In this model, each employee is represented by their date of appointment, date of birth, department, employee number, line manager and name, together with some details that depend on whether they are paid weekly or monthly (these are the only significant differences modelled). A hierarchically structured model allows the close similarity to be shown clearly.

---

[3] Note that a species is not just distinct from other members of its zoological class—membership of the class already makes it distinct from all species in other classes.

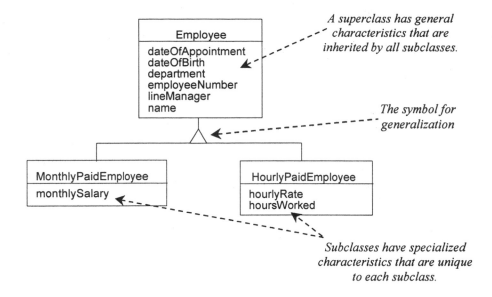

A superclass has general characteristics that are inherited by all subclasses.

The symbol for generalization

Subclasses have specialized characteristics that are unique to each subclass.

Figure 4.5 *Hierarchy of employee types.*

A second benefit comes from the relative ease with which a hierarchy can be extended to fit a changing picture. If this company were to decide that a new, weekly-paid type of employee is required, it is a simple matter to add a new subclass to the hierarchy to cater for this, as shown in Figure 4.6.

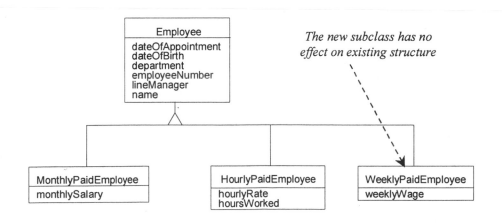

The new subclass has no effect on existing structure

Figure 4.6 *Hierarchies are easy to extend.*

Other features of generalization

Three features of generalization that are not mentioned explicitly in the definition given earlier are sufficiently important to deserve some discussion at this point. These are: the mechanism of *inheritance*, the *transitive* operation of inheritance and the *disjoint* nature of generalization hierarchies.

Inheritance. This is a mechanism for implementing generalization and specialization in an object-oriented programming language. When two classes are related by the mechanism of inheritance[4], the more general class is called a *superclass* in relation to the other, and the more specialized is called its *subclass*. At a first approximation, the rules of object-oriented inheritance work more or less as shown below, although this is very simplified and will need to be refined shortly.

(1)   A subclass inherits all the characteristics of its superclass[5].

(2)   The definition of a subclass always includes at least one detail not derived from its superclass.

Inheritance is very closely associated with generalization. Generalization describes the logical relationship between elements that share some characteristics, while inheritance describes an object-oriented mechanism that allows the sharing to occur.

Transitive operation. This means that the relationship between two elements at adjacent levels of a hierarchy 'carries over' to all more specialized levels. Thus, in Figure 4.4, the definition of an animal applies in turn to all mammals, and thus by a series of logical steps to a domestic cat. So we can rewrite the rules of inheritance given above, as follows

(1)   A subclass always inherits all the characteristics of all its ancestors.

(2)   The definition of a subclass always includes at least one detail not derived from its ancestors.

Disjoint nature. In a hierarchic system, the branches of the tree diverge as they get further away from the root and closer to the leaves. They are not permitted to converge. This means, for example, that a cat cannot be both a mammal and a reptile. In other words, each element in a hierarchy can only be a member of one classification at any given level of the hierarchy. It may, of course, be a member of other classifications at other levels of the hierarchy. In fact, as we have just seen, this is just what transitivity means.

The disjoint aspect of generalization means that we sometimes need to be careful about the characteristics chosen to express a generalization. For example, we could not use 'Has four feet' as the only defining characteristic of a mammal, even supposing that it were true of all mammals—because many lizards also have four feet, and this would make it possible to classify a lizard as a mammal. A class must be defined in terms of a unique set of characteristics that differentiate it from all other classes at its level of the hierarchy.

It is worth stressing here that generalization structures are abstractions that we choose to apply, since they express our understanding of some aspects of an application domain. This

---

[4] The name comes from a similarity to biological inheritance, emphasized by the use in UML of the term 'ancestor' to denote a more general class in a generalization hierarchy. But object-orientated inheritance is very different from inheritance in living things, and care should be taken not to confuse the two.

[5] We shall see in Chapter 7 that inherited characteristics are sometimes over-ridden in a subclass.

means that we can also choose to apply more than one generalization structure to the same domain, if it expresses a relevant aspect of the situation. Thus for example, a human might be simultaneously classified as a creature (*Homo Sapiens*), as a citizen (a voter in a city electoral division) and as an employee (an Account Manager in Agate's Creative Department). If each of these were represented as a hierarchy in an object-oriented model, the position of a human might be an example of multiple inheritance, which means that a subclass is at once a member of more than one hierarchy, and inherits characteristics from every ancestor in each hierarchy.

We should also note that real-world structures are not compelled to follow the logical rules applied in object-oriented modelling. Sometimes they are not disjoint or transitive, and therefore not strictly hierarchic. This does not detract from the usefulness of hierarchic structures in object-oriented development.

4.2.5 Message passing

In an object-oriented system, objects communicate with each other by sending messages. This might seem a little obvious if we think of the system as a kind of simulation, with objects that represent things in a real-world system collaborating to carry out a collective task. After all, this is how people communicate with each other: we send messages too. In the case of human communication, the means by which we transmit our messages are diverse, and sometimes it is not obvious who is the sender or receiver for a particular message, but all human communication can be seen as made up of messages. For example, everything we say to our friends and family, the emails we read when we log onto the network, advertising posters on the bus, games shows and cartoons on TV, even the clothes we wear, our tone of voice and our posture—these are all messages of one sort or another. If this is so, then just what is special about saying that objects only communicate via messages?

What makes it special is the fact that software was not constructed in this way until quite recently. Earlier approaches to systems development tended to separate data in a system from the processes that act on the data. This was done for sound analytical reasons, and is still appropriate for some applications, but it can give rise to difficulties. Chief among these is the need for a process to understand the organization of the data that it uses; for such a system, processes are said to be dependent on the structure of the data.

In itself, this is not necessarily a problem—in fact, it is to a certain extent inescapable. But if taken too far, and for insufficient reason, dependency of process upon data structure can cause many problems. For example, if the data structure were changed for any reason (a not uncommon occurrence in business), those processes that use the data may also need to be changed. This is an example of coupling between sub-systems, and we must minimize coupling if we are to construct systems that are reliable, that can be upgraded or modified, and that can be easily repaired if they break down (coupling is discussed in Chapters 11 and 12). Object-oriented systems avoid these problems by locating each process, as far as is practicable, with the data it uses. This is another way of describing an object: it is really little more than a bundle of data together with some processes that act on the data. These processes are called *operations*, and each has a specific *signature*. An operation signature is a definition of its interface. In order to *invoke* an operation, its signature must be given (signatures are sometimes also called *message protocols*).

In practice, it is not usually possible for all processes to be located with all the data that they must access, and data and processes are distributed among many different objects. Message passing is a way of insulating each object from needing to know any of the internal details of other objects. Essentially, an object knows only its own data, and its own operations. But in order for collaboration to be possible, the 'knowledge' of some objects includes knowing how to request services from other objects, which may include the retrieval of data. In this case, an object knows which other object to ask, and how to formulate the question. But it is unnecessary for an object to know anything about the way that another object will deliver the service. Such knowledge would introduce an undesirable and unnecessary degree of coupling into the relationship between the two objects.

When an object receives a message it can tell instantly whether the message is relevant to it. If the message includes a valid signature to one of its operations, the object can respond. If it does not, the object cannot respond. So we can think of operations as residing within objects, only able to be invoked by a message that gives a valid operation signature. Meanwhile, the object's data lies even deeper inside, since the data can only be accessed by an operation of that object. Clearly, then, if the signatures to an object's operations are not changed, it makes no difference what changes are made to the way that the operations run or the way that the data is stored. Neither of these changes would be visible from the outside. For this reason, the complete set of signatures for an object are sometimes known as its *interface*. This is called *encapsulation*, and is illustrated in Figure 4.7.

Other objects send messages requesting services.

An object's operations can only be called by a message with a valid operation signature.

An object's data can only be accessed by its own operations.

The representation of an object's data is hidden inside.

Figure 4.7 *Encapsulation: the layers of protection that surround an object.*

Consider a simple system to print pay cheques for employees in a business. One way of designing this would be to have a class Employee, with an instance that represents each person on the payroll. Each Employee object is responsible for knowing about the salary earned by the real employee it represents. Next, there might be a PaySlip object that is responsible for printing each employee's payslip each month. In order to print a payslip, the PaySlip object must know how much that employee has earned. The object-oriented approach to this is for the `PaySlip` object to send a message to each `Employee` object, asking how much salary to pay. The `PaySlip` object need not know how the `Employee` object works out the salary, nor what data it stores. All it needs to knows is that if it asks an `Employee` object for a salary figure, an appropriate response is given. Message passing allows objects to encapsulate (i.e. hide) their internal details from other parts of the system, thus minimizing the knock-on effects of any changes to the design or implementation.

4.2.6 Polymorphism

When one person sends a message to another, it is often convenient to ignore many of the differences that exist between the various people that might receive the message. For example, in some situations a mother may use the same phrasing regardless of the age or gender of her child. 'Go to bed now!' carries much the same import whether the child is a five-year-old boy or a thirteen-year-old girl. But the precise tasks to be carried out by each child in going to bed may be very different. The five-year-old may set off towards bed by himself, but perhaps then requires help with washing his face, brushing his teeth, putting on his pyjamas and he may also expect to be read a bedtime story. The thirteen-year-old may not require any further help, once convinced that it really is bedtime.

This is rather like *polymorphism*, which is an important element in the way that object-oriented approaches encourage the decoupling of sub-systems. Polymorphism literally means 'an ability to appear as many forms', and it refers to the possibility of identical messages being sent to objects of different classes, each of which responds to the message in a different, yet still appropriate way. This means the originating object does not need to know which class is going to receive the message on any particular occasion. The key to this is that a receiving object is responsible for knowing how to respond to messages.

Figure 4.8 uses a collaboration diagram to illustrate polymorphism in a business scenario (the notation for collaboration diagrams is described fully in Chapter 8). The diagram assumes that there are different ways of calculating an employee's pay, depending on whether they are paid a monthly salary, a weekly rate, or an hourly rate based on the number of hours worked. An object-oriented system to calculate pay for these employees might include a separate class for each type of employee, each able to perform an appropriate pay calculation. Suppose one of the outputs from this system is a print-out showing the total pay-bill for the company: to assemble the total, a message is sent to each employee object, asking it to calculate its pay. The requesting object (here called `PayBillPrint`) need not even know the class of each receiving object, still less anything about how the calculation is carried out for each employee. It need only know the proper format for the message.

Polymorphism has proved to be a tremendously powerful concept for the information system developer. It permits a clearer separation between different sub-systems than any earlier approach to systems development. This means that a system can easily be modified or extended to include extra features, since it is only the interfaces between classes that need to be known. The way that each part of the system is implemented (its internal structure and behaviour) is hidden from all parts of the system that do not absolutely need this knowledge.

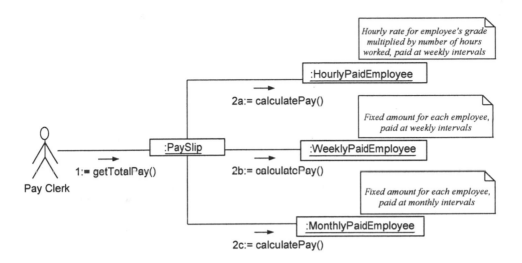

Figure 4.8 *Polymorphism allows a message to achieve the same result even when the mechanism for achieving it differs between different objects.*

Program designers and programmers have struggled for many years to achieve such a high level of modularity in software. Object-orientation gives the greatest promise yet of practical success.

4.3 The Origins of Object-Orientation

Object-orientation is only one stage, and perhaps not the last, in a long development path. We review below some strands in the history of computing that have led to object-oriented analysis and design.

Increasing abstraction

From the very earliest days of digital computers, there has been a steady increase in the level of abstraction at which programmers have been able to operate. The increase in abstraction applies both to the activity of programming itself, and to the tasks that computer programs are expected to perform. In a relatively short time, the purposes to which computers are applied have become enormously more complex and demanding, thus greatly increasing the complexity of the systems themselves. Some of the waymarks in this progress are summarized in Figure 4.9.

Event-driven programming

Early work on computer simulation led directly to the object-oriented paradigm of independent, collaborating objects that communicate via messages. A typical simulation task is to model the loading of vehicles onto a large ferry ship, in order to determine the quickest safe way to do this. This simulation would be run many times under differing

Period	Programming languages	Interface technology	Applications
1940s	Machine code.	Hardwired input.	Military and scientific.
1950s	Assembly language. Early third-generation languages (3GLs).	Punched tape and cards.	Simple commercial applications.
1960s	More mature 3GLs, e.g. Fortran, and COBOL.	Punched cards, teleprinters.	More complex commercial and scientific applications.
1970s	Non-procedural languages: Simula, ADA.	VDU terminals.	Early mainframe networks.
1980s	Object-oriented languages, e.g. Smalltalk, Eiffel, C++.	PCs. Early GUIs.	Desktop applications. Emergence of client-server computing.
1990s	Java, Python.	GUIs, Web applets.	Distributed systems.

Figure 4.9 *The increasing abstraction of programming (greatly simplified).*

assumptions, for example, the sequence of loading, the arrangement of vehicles on the storage decks, the speed at which vehicles are driven onto the ship, the time separation between one vehicle and the next, and so on. The real-world situation that is being modelled is very complex, and consists of a number of independent agents, each responding to events in ways that are easy to model at the level of individual agents, but very difficult to predict in the large, with many agents interacting with each other all at once. This kind of task is very difficult to program effectively in a procedural 3GL, since designs for these languages are based on the underlying assumption that the program structure controls the flow of execution. Thus for a program to tackle the simulation task described above in a 3GL, it must have separate routines that test for, and respond to, a vast number of alternative conditions.

One solution to this problem, that evolved into early simulation languages such as Simula-67, was to structure the program in a similar way to the problem situation itself: as a set of independent software agents, each of which represents an agent in the real-world system that is to be simulated. This remains one of the key ideas in object-oriented software development: that the structure of the software should mirror the structure of the problem it is intended to solve. In this way, the tension between the model of the application domain and the model of the software (mentioned in Section 4.2.1) is resolved, turning a potential weakness in fact into a strength.

The spread of GUIs

The rapid spread of Graphical User Interfaces (GUIs) in the 1980s and 1990s posed particular difficulties for contemporary development methods. GUIs brought some of the problems encountered earlier in simulation programming into the world of mainstream business applications. The reason for this is that users of a GUI are presented on their computer screen with a highly visual interface, that offers many alternative actions all at once, each one mouse-click away. Many other options can be reached in two or three more clicks via pull-down menus, list boxes and other dialogue techniques. Interface developers naturally responded by exploiting the opportunities offered by this new technology, with the result that it is now almost impossible for a system designer to anticipate every possible route that a user might take through a system's interface. This means that the majority of desktop applications are now very difficult to design or control in a procedural way. The object-oriented paradigm offers a natural way to design software, each of whose components offers clear services that can be used by other parts of the system quite independently of the sequence of tasks or the flow of control.

Modular software

In an object-oriented system, classes have two kinds of definition. From an external perspective, a class is defined in terms of its interface, which means that other objects (and their programmers) need only know the services that are offered by objects of that class and the signature used to request each service. From an internal perspective, a class is defined in terms of what it knows and what it can do—but only objects of that class need to know anything about this internal definition. It follows that an object-oriented system can be constructed so that the implementation of each part is largely independent of the implementation of other parts, which is what modularity means. This contributes to solving some of the most intractable problems in information systems development.

- It is easier to maintain a system built in a modular way, as changes to a sub-system are much less likely to 'ripple' through the rest of the system.
- For the same reason, it is easier to upgrade a modular system. As long as replacement modules adhere to the interface specifications of their predecessors, other parts of the system are not affected
- It is easier to build a system that is reliable in use. This is because sub-systems can be tested more thoroughly in isolation, leaving fewer problems to be addressed later when the whole system is assembled.
- A modular system can be implemented in small, manageable increments. Provided each module is designed to provide a useful and coherent package of functionality, they can be introduced one at a time.

Life cycle problems

Most structured methodologies of the 1980s and early 1990s were based on the waterfall life cycle model originally designed for large engineering projects, such as constructing new jet aircraft and building road bridges. The difficulties with this life cycle model have already been discussed in Chapter 3. Object-orientation addresses them by encouraging a cyclic development approach. In this, there is little difficulty in revisiting earlier stages and revising earlier products in an iterative process that can repeat, if necessary, until everyone is satisfied with the quality of the software—subject, of course, to time and budget constraints.

This aspect is strongly linked to the highly modular character of an object-oriented system, described in the previous section, and also to the 'seamless' development of models throughout an object-oriented life cycle. The latter point is discussed in the next section.

Model transitions

In structured approaches to information systems development, the models of process that are developed during the analysis phase (e.g. data flow diagrams) have only an indirect relationship to the process models developed during the design phase (e.g. structure charts or, in SSADM, update process models). This has meant that designs for new systems, however good in their own right, have been hard to trace back to the original requirements. Yet, what makes a design successful is that it meets the requirements in a way that is functional, efficient, economic, and so on (see Chapter 11). This is particularly frustrating when one considers that a major strength of the structured methods was their concentration of effort on achieving a good understanding of users' requirements.

Object-oriented analysis and design avoid these transition problems by using a core set of models throughout analysis and design, adding more detail at each stage, and avoiding the awkward discontinuities that arise when one model must be discarded to be replaced by another with a different, incompatible structure. In UML, the fundamental analysis models are the use case and the class diagram (described in Chapters 5 and 6, respectively), and these continue as the backbone of the design, with other design models derived directly or indirectly from them.

Reusable software

Information systems are very expensive, yet in the past their developers have tended to reinvent new solutions to old problems, many times over. This wastes time and money, and has led to the demand for reusable software components, which can eliminate the need to keep reinventing the wheel. Object-oriented development methods offer great potential, not yet fully realized, for developing software components that are reusable in other systems, for which they were not originally designed. This is a result of the highly modular nature of object-oriented software, and also of the way that object-oriented models are organized. Inheritance is of particular importance in this context, and we say more about this in Chapter 7.

Object-oriented languages today

A number of object-oriented programming languages are available today, with some significant differences between their capabilities and the extent to which they adhere to the object-oriented paradigm. Figure 4.10 lists some of the main characteristics of the most widely used of these languages. This is not a programming textbook, so we offer only a very brief description of each feature, as follows.

Strong typing refers to the degree of discipline that a language enforces on the programmer when declaring variables. In a strongly typed language (most modern languages are strongly typed), every data value and object that is used must belong to an appropriate type for its context. Static typed languages enforce this with type-checking at compile time. Dynamic typed languages check types at run-time, but some languages offer a hybrid approach that allows the flexibility of loading classes at run-time. Garbage collection is important for memory management in systems that create and delete many objects during their execution. If objects are not removed from memory when they are deleted, the system may run out of memory in which to execute. Multiple inheritance refers to the capacity of an object to acquire features from more than one hierarchy. This is important as it minimizes the amount of code duplication and hence reduces inconsistencies that can cause maintenance problems. In static typed languages, multiple inheritance can allow a new class to stand in for any of its ancestors, and this reduces the amount of special-case programming required elsewhere in the system. Languages in which all constructs are implemented as classes or objects are said to be 'pure' object-oriented languages. Some languages permit data values that are not objects, but this introduces extra complexities for the programmer. Other languages allow un-encapsulated types, but this gives the sloppy programmer opportunities to bypass the safe encapsulation of classes. Both of these circumstances can cause a system to be difficult to maintain and extend. Dynamic loading refers to the ability of a language to load new classes at run-time. This can be used to allow software to reconfigure itself, for example, to cope with hardware or environment changes. It can also be used in a client–server environment to allow the client software to evolve by loading it with different classes from the server. Dynamic loading can help to propagate improvements and bug fixes, by concentrating maintenance efforts on the server side. Standardized class libraries give the programmer access to classes that are known to run successfully on a variety of hardware platforms, and under a variety of operating systems. When these are not available, it can be difficult to modify an application so that it will run on another platform, or in conjunction with applications that have used a different library. Finally, correctness constructs include pre-conditions and post-conditions on methods, forming a contract between any client–supplier pair. Contracts are important to the development of robust software and are discussed in Chapter 9.

Feature	Smalltalk	C++	Eiffel	Java
Strong typing	✓	optional	✓	✓
Static/dynamic typing (S/D)	D	S	S	S+D
Garbage collection	✓	✗	✓	✓
Multiple inheritance	✗	✓	✓	✗
Pure objects	✓	✗	✓	✗
Dynamic loading	✓	✗	✗	✓
Standardized class libraries	✓	✗	✓	✓
Correctness constructs	✗	✗	✓	✗

Figure 4.10 *Characteristics of some widely used object-oriented languages.*

Limitations of object-orientation

Some applications are not ideally suited to object-oriented development, and in this section we make a few comments about these. There are two main examples. The first kind includes systems that are strongly database-oriented. By this we mean both that they have a record-based structure of data that is appropriate to a relational database management system (RDBMS), and also that their main processing requirements centre on the storage and retrieval of the data (e.g. a management information system used mainly for querying data in the database). Such applications cannot easily be adapted to an object-oriented implementation without losing the many benefits of using a RDBMS for data storage. Commercial RDBMS are a very mature form of technology that organize their data according to sound mathematical principles. This ensures a good balance of efficiency of retrieval, resilience to change and flexibility in use. However, RDBMS are limited in their capabilities for storing and retrieving certain kinds of complex data structure, such as those that represent multimedia data. Graham et al. (1998) mention the spatial (map-based) data that forms the basis of a geographic information system (GIS) as a particular example of data structures to which RDBMS are not well suited, but which are ideally suited to object-oriented development. We return to the relative advantages and disadvantages of RDBMS as compared to object-oriented database systems in Chapter 17.

Applications that are strongly algorithmic in their operation are less suited to an object-oriented development approach. For some scientific applications that involve large and complex calculations (for example, satellite orbit calculations) it may be neither feasible nor desirable to split the calculation down into smaller parts. Such a system, if developed in an object-oriented manner, might contain very few objects, but each would be extremely complex. This would not be sound object-oriented design, and so either a procedural or a functional approach (these are alternative styles of programming) is recommended instead.

4.4 Summary

In this chapter we have introduced the most important concepts in object-orientation, in particular, object and class, generalization and specialization, message passing and polymorphism. Understanding these gives an essential foundation for later chapters that deal with the practical application of object-oriented analysis and design techniques. There is a great deal of synergy in the way that the different fundamental concepts contribute to the success of object-orientation. For example, message passing and polymorphism both play a significant role in achieving sound modularity in a system. But there is no clean break with the past; instead the characteristics of object-orientation are best seen as the result of a gradual process of evolution that can be traced back to the earliest days of electronic digital computers. This evolutionary process is by no means finished yet, but rushes onwards into the future. As applications and computing environments grow ever more complex, there is a continuing need for reliable, maintainable, modifiable information systems.

Review Questions

4.1   Define object, class and instance.

4.2   What do you think is meant by 'semantics'?

4.3   How does the object-oriented concept of message passing help to encapsulate the implementation of an object, including its data?

4.4   What is polymorphism?

4.5   What is the difference between generalization and specialization?

4.6   What rules describe the relationship between a subclass and its superclass?

4.7   What does it mean to say that an object-oriented system is highly modular?

4.8   Why is it particularly hard for a designer to anticipate a user's sequence of tasks when using a GUI application?

4.9   What is an operation signature?

Case Study Work, Exercises and Projects

4.A Section 4.2.1 mentions the human activity system and the proposed software system as particularly important systems to consider, but these are not the only systems that an analyst will encounter or work with. Make a list of any other systems you can think of that might be involved in the process of software development. What interfaces exist between them?

4.B Re-read the description of inheritance in object-orientation given in Section 4.2.4. How does object-oriented inheritance differ from inheritance between a parent and a child: (i) in biology, and (ii) in law?

4.C Arrange the following into a hierarchy that depends on their relative generalization or specialization: person, thing, green, shape, Maria, cub, polar bear, square, law, child, colour, animal. Add more classifications as necessary so that it is clear what *is* generalized or specialized at each level.

4.D Read the first section of the case study material for FoodCo (Section B1.1), and identify classes that represent FoodCo's whole business environment.

4.E List all FoodCo's products that are identified in the case study material in Case Study B1 and arrange these into a hierarchy. Imagine some more products that make your hierarchy more interesting, and add these to your diagram.

Further Reading

Most standard texts on object-oriented analysis and design contain a section that introduces the fundamental concepts of object-orientation, but to date few use UML notation. With this proviso, Jacobson et al. (1992) and Rumbaugh et al. (1991) provide good general introductions. However, it would make sense to look at Booch's, Jacobson's and Rumbaugh's new books based on UML Version 1.3. At the time of writing (March 1999) these have just appeared (Booch et al., 1999; Jacobson et al., 1999; Rumbaugh et al., 1999).

Agate Ltd Case Study—Introduction

Agate Ltd

A1.1. Introduction to Agate

Agate is an advertising agency in Birmingham, UK. Agate was formed as a partnership in 1982 by three advertising executives, Amarjeet Grewal, Gordon Anderson and Tim Eng (the name is a combination of their initials). Amarjeet and Gordon had previously worked for one of the UK's largest and most successful advertising companies in London, but felt frustrated at the lack of control they had over the direction of the campaigns they worked on. As a result, they moved to the West Midlands region of the UK and set up on their own business in 1981. Shortly afterwards they were joined by Tim Eng, with whom they had worked on a project in Hong Kong, and Agate was formed.

In 1987, the three partners formed a UK limited company and between them own all the shares in it. Gordon Anderson is Managing Director, Amarjeet Grewal is Finance Director and Tim Eng is Creative Director. They now employ about fifty staff at their office in the centre of Birmingham (see Figure A1.1) and a further hundred or so at seven offices around the world. Each of these other offices is set up locally as a company with the shares owned jointly by Agate and the local directors.

Initially the company concentrated on work for the UK motor industry, much of which is located in the West Midlands region. However, as it has expanded and internationalized, the type of work it takes on has changed, and it now has clients across a wide range of manufacturing and service industries.

The company strategy is to continue to grow slowly and to develop an international market. The directors would like to obtain business from more large multinational companies. They feel that they can offer a high standard of service in designing advertising campaigns that have a global theme but are localized for different markets around the world.

The company's Information Systems strategy has a focus on developing systems that can support this international business. Currently, as well as considering the development of a new business information system, the directors are deciding whether to invest in hardware and software to support digital video editing. They have until now sub-contracted video editing work, but recently have invested in ISDN connections for video-conferencing with other offices, and this means that they have the capability for fast file-transfer of digital video between offices.

Senior Management

3 Directors

Creative

4 Account managers

6 Graphic designers

2 Photographers

4 Copy writers

2 Editors

1 Audio technician

1 Resource librarian

IT Support

1 Computer manager

2 Network support technicians

Administration

1 Office manager

3 Personal assistants (to directors)

2 Receptionist/telephonists

4 Secretaries (to account managers)

2 Clerk/typists

1 Filing clerk

Accounts

1 Accountant (and company secretary)

1 Credit controller

2 Accounts clerks

2 Purchasing assistants

Figure A1.1 *Staff at Agate Ltd UK office.*

A1.2. Existing Computer Systems

Agate uses computers extensively already. Like most companies in the world of design and creativity, Agate uses Apple Macintosh® computers for its graphic designers and other design-oriented staff. The secretaries and personal assistants also use Apple Macs®. However, the company also uses Intel®-based PCs to run accounts software in Windows®.

Last year, Agate had a basic business system for the UK office developed in Delphi for Windows 95. However, after the system was developed, the directors of Agate decided that it should have a system developed in Java, the object-oriented language originated by Sun Microsystems Inc., which is designed to run over the Internet. One of the reasons for the choice of Java was that it is portable across different hardware platforms, and the company wants software that could run both on the PCs and on the Macs. (Another reason is that Agate foresees the possibility of using Java to deliver information to clients via the World Wide Web.) Unfortunately, the person who developed the Delphi software for the company (and was going to rewrite it in Java) was headhunted by an American software house, because of her skills in Java. She has moved to the USA, and is now earning twice what she could earn as a self-employed system developer in the UK! Fortunately, this developer, Mandy Botnick, was methodical in her work and has left Agate with some object-oriented system documentation for the system she designed and developed.

This existing system is limited in its scope: it only covers core business information requirements within Agate. It was intended that it would be extended to cover most of Agate's activities and to deal with the international way in which the business operates.

A1.3. Business Activities in the Current System

Agate deals with other companies that it calls clients. A record is kept of each client company, and each client company has one person who is the main contact person within that company. His or her name and contact details are kept in the client record. Similarly, Agate nominates a member of staff—a director, an Account Manager or a member of the creative team—to be the contact for each client.

Clients have advertising campaigns, and a record is kept of every campaign. One member of Agate's staff, again either a director or an account manager, manages each campaign. Other staff may work on a campaign, and Agate operates a project-based management structure, which means that staff may be working on more than one project at a time. For each project they work on, they are answerable to the manager of that project, who may or may not be their own line manager.

When a campaign starts, the manager responsible estimates the likely cost of the campaign, and agrees it with the client. A finish date may be set for a campaign at any time, and may be changed. When the campaign is completed, an actual completion date and the actual cost are recorded. When the client pays, the date paid is recorded. Each campaign includes one or more adverts. Adverts can be one of several types:

- newspaper advert—including written copy, graphics and photographs;
- magazine advert—including written copy, graphics and photographs;
- TV advert—using video, library film, actors, voice-overs, music etc.;
- radio advert—using audio, actors, voice-overs, music etc.;
- poster advert—using graphics, photographs, actors;
- leaflet—including written copy, graphics and photographs.

Purchasing assistants are responsible for buying space in newspapers and magazines, space on advertising hoardings, and TV or radio air-time. The actual cost of a campaign is calculated from a range of information. This includes:

- cost of staff time for graphics, copy-writing etc.;
- cost of studio time and actors;
- cost of copyright material—photographs, music, library film;
- cost of space in newspapers, air-time and advertising hoardings;
- Agate's margin on services and products bought in.

This information is held in a paper-based filing system, but the total estimated cost and the final actual cost of a campaign are held on the new computer system.

The new system also holds the salary grades and pay rates for the staff, so that the cost of staff time on projects can be calculated from the timesheets that they fill out. It has been partially implemented and is not used in the existing system.

A1.4. Summary System Specification

The full specification of the system is not provided here, but an overview of the system requirements is provided by the analysis documentation. This includes use cases[1], which are shown in Figure A1.2, and a class diagram[2], which is shown in Figure A1.4. The use case descriptions are provided in Figure A1.3.

---

[1] Introduced in Chapter 5.
[2] Introduced in Chapter 6.

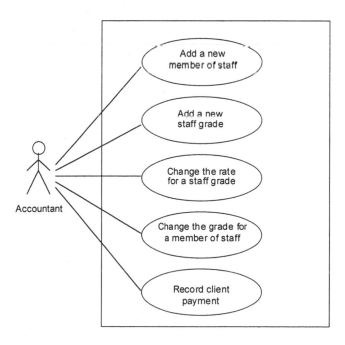

Figure A1.2 *Use case diagrams.*

Add a new member of staff.	When a new member of staff joins Agate, his or her details are recorded. He or she is assigned a staff number, and the start date is entered. Start date defaults to today's date. The starting grade is entered.
Add a new staff grade.	Occasionally a new grade for a member of staff must be added. The name of the grade is entered. At the same time, the rate for that grade and the rate start date are entered; the date defaults to today's date.
Change the rate for a staff grade.	Annually the rates for grades are changed. The new rate for each grade is entered, and the rate start date set (no default). The old grade is retrieved and the rate finish date for that grade rate set to the day before the start of the new rate.
Change the grade for a member of staff.	When a member of staff is promoted, the new grade and the date on which they start on that grade are entered. The old staff grade is retrieved and the finish date set to the day before the start of the new one.
Record client payment.	When a client pays for a campaign, the payment amount is checked against the actual cost and the date paid is entered.
Add a new client.	When Agate obtains a new client, the full details of the client are entered. Typically this will be because of a new campaign, and therefore the new campaign will be added straight away.
Add a new campaign.	When Agate gets the business for a new campaign, details of the campaign are entered, including the intended finish date and the estimated cost. The manager for that campaign is the person who enters it.
Assign staff to work on a campaign.	When a new campaign starts, members of staff are assigned to work on it.
Add a new advert to a campaign.	A campaign can consist of many adverts. Details of each advert are entered into the system with a target completion date.
Record completion of a campaign.	When a campaign is completed, the actual completion date and cost are entered. A record of completion form is printed out for the Accountant as the basis for invoicing the client.
Assign a staff contact.	Clients have a member of staff assigned to them as their particular contact person.
Change a client contact.	Records when the client's contact person with Agate is changed.
Record completion of an advert.	When an advert is completed, the actual date is entered.

Figure A1.3 *Use case descriptions.*

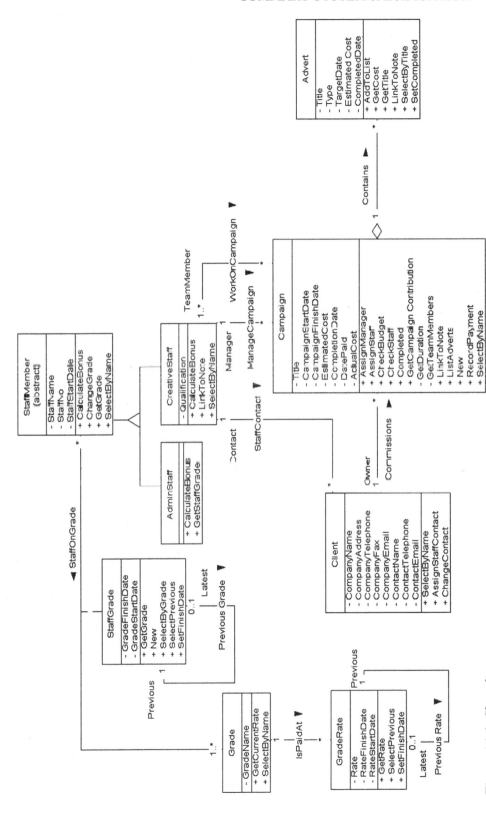

Figure A1.4 *Class diagram.*

B1

FoodCo Ltd Case Study—Introduction

FoodCo Ltd

B1.1 Introduction to FoodCo

FoodCo produces a range of perishable foods for supermarkets, and is based in the flat agricultural lands of the East Anglian region of the UK. John Evans, the present Chairman, started the company when he was demobilized from the Royal Air Force at the end of World War II. He borrowed money to buy 200 acres of arable farmland, but his ambition was to be more than a farmer. As soon as Home Farm was running he opened a factory in a converted barn.

The first product was a pickle made to a traditional family recipe. It sold well, and success financed expansion. Two years later, John acquired derelict land next to the farm and the company moved into a larger, purpose-built factory. The product range soon extended to pre-packed vegetables and salads, and later a wide range of sauces, pickles and sandwich toppings, in fact almost anything made of vegetables that can be sold in jars. FoodCo's traditional customers are major UK supermarket chains. Some lines (e.g. washed salads) sell to all customers, while others (most of the cooked products) are produced for a specific supermarket chain. Most are packaged under the supermarket's 'own brand' label.

The pickle started a company tradition that, as far as possible, ingredients are grown on the company's own farm. This now covers 1,500 acres, and includes a market garden growing tomatoes, peppers, courgettes, chillis and other exotic vegetables under glass, and an extensive herb garden. Ingredients which do not grow well in the UK climate are supplied from carefully selected farms abroad, in Mediterranean Europe, East Africa, the United States and the Far East.

The company's annual turnover and employee numbers are summarized in Figure B1.1.

FoodCo Limited	
Number of employees (actual)	278
Number of employees (full-time equivalent)	213
Annual turnover (current projection for 1999/2000)	£5.3m

Figure B1.1 *FoodCo's current staff complement and annual turnover.*

There are now three factories on the site. Beechfield is the oldest, and this is where raw vegetables are prepared. This work is relatively unskilled. The newer Coppice and

Watermead factories concentrate on the more complex cooking processes involved in making sauces, pickles and the like. These need more skilled and experienced staff. A bottling plant is also located in Watermead, and there are two warehouses in separate buildings. One is refrigerated and stores fresh vegetable and salad packs, while the other stores dry and bottled products. Figure B1.2 shows a recent organization structure chart.

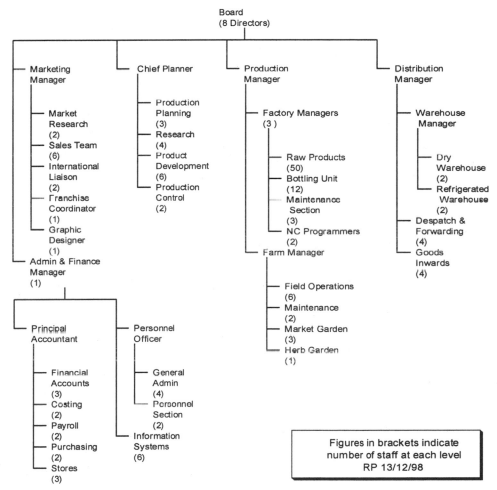

Figure B1.2 *FoodCo's current organization structure, showing the distribution of staff.*

The company is still privately owned and managed. In 1993 John made way for his elder son Harold to become Managing Director. At the time, it was generally agreed by the Board that the company must improve profitability but there was no consensus on how to achieve this. In May 1996, John persuaded the Board that FoodCo must bypass its traditional supermarket customers and market directly to consumers. Partly as a result of John's speech to the Board (Box B1.1 gives an edited version), the *World Tradition* range was launched in 1997. This now sells successfully at the quality end of the UK market. Helped by the growing reputation of British food and cooking, the range has also begun to penetrate continental European and North American markets.

Box B1.1 Scene: the Board Room, May 1996. John Evans speaks to the Directors about Foodco's problems and a possible solution.

'Fellow directors, we all know the company faces great difficulties. This year's profits will be the lowest ever. If we don't do something drastic, we'll be bankrupt in another year. But if we are to turn the situation round, we must know why things got so bad.

'I believe the reason goes right back to our beginnings. Those of you who were with me in 1947, when I started this company, will remember we had a degree of control that seems incredible now. Everything ran the way we wanted: farm, production, sales, distribution. We made consistently high quality goods, and by 1962 the new supermarkets were clamouring to buy. That was all a long time ago, but I think our early success is a direct cause of our present predicament. Let me explain.

'Remember 1968? When we borrowed heavily to finance expansion to meet that demand? Those loan repayments squeezed our profits hard. And then in 1974? When the TrustMart chain emerged as our dominant customer, and began driving down prices? We simply hadn't the financial muscle to fight them. We were still paying off a huge loan! That's why for ten years or more, TrustMart has dictated prices that are so low they have crippled us. And we've been unable to do a thing, because we've simply been scared they'll buy elsewhere. So last year TrustMart bought 65% of our total production— altogether over £3m in sales—and we'll be lucky to clear £100,000 profit on it!

'That's also why TrustMart call all the shots on new products. We don't have the cash to develop products for other customers. Now, I know we've grown in spite of that. It's not all been bad, but let's not kid ourselves it's good. We haven't really run the game since 1980. We all know it! We've been towed along behind TrustMart—and the supermarket sector— like a child dragged along by its father. We've only survived this long because TrustMart had no other source of supply. Now that's changing. We have serious new rivals for the supermarket supply business, and TrustMart have driven our prices still lower, to the point where we may make no profit at all this year.

'We can beat off this attack, but only if we develop new products and sell in a wider market. There is no argument about that, but there is a problem. Our *real* customers are not the supermarkets, but *their* shoppers. And they don't know we exist, because our name is hidden behind all the TrustMart own brand labels on all our packs and jars.

'The answer is to reach the consumers directly. Our market can only expand if they know our name, if they ask for our products. So here's what we do. We're going to launch our own brand name, and promote it so well that everyone knows about us. Customers will begin to insist on our brand, and TrustMart will have to pay *our* price, for a change.

'It won't be cheap. We'll need serious market research. We'll need more staff in the Product Development team, and we'll need time. We'll need a new corporate image. We'll need TV advertising. But it will be worth it. There's a vast market out there, and I'm not just thinking of the UK.

'So can we finance it? Certainly! It means heavy borrowing again, but we can repay the loans out of profits on increased sales. Sure, it's a risk, but we'll sink if we don't take it. There are many details to work out, but this plan can succeed. It *will* succeed! When we started, we were the best in the business. I believe we can be the best again.

'Thank you. Are there any questions?'

B1.2 FoodCo Today

B1.2.1 Current thinking

John Evans still believes that the company's major difficulty was over-reliance on one customer, and that this will be solved over time as the *World Tradition* market expands. His son, Harold, feels that management procedures are now the main problem, particularly management information. He sees the systems as hopelessly inadequate and thinks the company has simply outgrown them. For him, this is an extremely serious issue since it will inevitably worsen as the company grows. But father and son are each as stubborn as the other, so they have never settled their differences on this vital point.

Late in 1995 a new Finance Director, Clare Smythe, was appointed. Less than one year later she achieved a compromise that averted open war in the family. Firstly, she championed the *World Tradition* brand that successfully met John's concerns. This is a range of international condiments, prepared to traditional recipes from many cultures, and using only the finest ingredients. Growing numbers of people in the affluent world want to be able to prepare authentic dishes from world cuisine, ranging from aloo brinjal (Indian potato and eggplant curry) to Yucatan-style cod (Mexican fish cooked in orange, lime and coriander). The new range allowed the company to reposition itself in a new international market, where growth has been highly profitable. It also helps FoodCo to free itself from dependence on TrustMart, still by far their largest customer.

Secondly, Clare helped Harold to persuade the Board that the introduction of a new product range compelled the company to manage its information more effectively and efficiently. The Board agreed to undertake a major review and updating of all information systems, and a national firm of consultants was commissioned to recommend a strategy.

B1.2.2 Information systems

The current systems are a mixed set of applications, some dating back to the late 1960s, that run on diverse hardware platforms. An ageing mini-computer runs an inflexible suite of accounting programmes, a sales order processing system and a stock control system. The stock control system also generates product barcodes for the jar and bottle labels. The mini-computer is accessible from VT100 character-only terminals dotted throughout the factories and offices. Payroll is run off-site on a local computer bureau's mainframe.

Some managers and other staff have networked PCs with standard office software. A handful of proprietary packages include the Computer Aided Design programme used to design production line layouts. The PC network is not linked to the mini-computer or to the outside world for Internet, email, or Electronic Data Interchange.

Some production is automated, including some washing and chopping operations in Beechfield Factory. The automated machines are of the numerical-control type, and are now obsolescent. Although they still do a reasonable job, by modern standards they are awkward to re-program, and maintaining them requires particular specialist skills.

The consultants' report

After some months of investigation the consultants submitted their report early in 1999. This identified serious failings in a number of areas, and recommended a phased approach to change. The top priority was to develop new product costing and production planning systems that would interface to a new in-house payroll package. The improved product costing and production information would give tighter control of production costs. Price negotiations with customers could be conducted on a more realistic basis, and better management information would help managers respond to the volatile international market for *World Tradition* products.

The second main recommendation in the report was for a substantial investment in upgraded hardware. Many more PCs were to be installed, networked to each other and to the mini-computer. These needed to be in place prior to phase 2 of the information systems plan, which called for a rapid spread of automation through the production and distribution departments. Finally, all new software development was to fit in with a medium-term plan to make the most of new technology opportunities. For this reason, a move to an object-oriented development method was seen as a critical aspect. This would help later with building an integrated set of systems, ultimately to include factory automation, management information, electronic links with suppliers and customers and an exploration of on-line Internet marketing. The Board accepted all the recommendations, and a detailed investigation into the requirements for the first systems was begun by FoodCo's in-house IS team, who had recently been trained in object-oriented development methods. Two staff were seconded from the consultants to act as mentors to the first two or three projects.

Product costing: current operations

This section describes the way that product costing activities are currently carried out at FoodCo. It concentrates particularly on Beechfield factory, as this was an area identified by the consultants' report as a priority for action. Further information is given within chapters, where necessary, as part of case study exercises or review questions. Some information that is not strictly necessary to the completion of the exercises in this book has been included, in order to give a broader view of the overall operations.

Line operations. The nature of production control varies between the various factory and farm departments, depending on the operations undertaken and the nature of the product. At Beechfield, the main products are packs of washed salads and prepared raw vegetables, and some uncooked products such as coleslaw and Waldorf salad. There are three production lines. Each can be adapted to produce different products as the need arises, but only one at a time. Operatives' pay and the overall production costs for these lines are based on the entire batch produced during a single run, which often, although not always, equates to a single eight-hour shift. The line is switched on at the beginning of the run and temporarily halted for coffee breaks and lunch, or when a problem occurs. When a line is switched to a different product, this is treated as a separate run. If operatives are required to wait while the line is changed over to another product, or while a problem with the line is sorted out, they are paid a standing rate to compensate them for lost earnings.

Payroll and costing. For workers on the older lines at Beechfield, earnings are calculated using an algorithm that has as its input variables: the piecework rate for each item, the quantity of that item produced, the number of productive hours spent on the line by each employee and the employee's grade. For each run, the line supervisor completes a *daily production record sheet* (see Figure B1.3). These are sent to the costing section for analysis before being passed on to the payroll section.

```
  ●                                    ●
        Daily Production Record Sheet No:

Line: ............................   Supervisor: ................................   Date: ...............................
                                                                        Circle Day:  M  T  W  T  F  S  S
Product: ..........................   Job No: .............................

Run Start: .......................   Run Finish: .......................

Problem Report        When & where problem occurred:..................................
Brief Description:




Fault Log No: ........................   Downtime (H/m): ..................   (Only if applicable)
                                         Production Control use only
Signed: ..............................   Total Qty Produced: ..........................
             (Supervisor)                Checked by: ...............................
                                                                        PR5/1.3
```

Figure B1.3 *Daily production record sheet for Beechfield Factory.*

The supervisors also complete a *weekly timesheet* (see Figure B1.4) for each employee. These are passed direct to payroll section. Each Tuesday, the entire week's production record sheets and timesheets are batched up and sent to the computer bureau. Data from the production sheets and timesheets is input to a piecework system at the bureau to produce a weekly earnings figure for the payroll calculation. After the payroll has been run all paperwork is returned to FoodCo's costing section for analysis. In practice, however, only a limited amount of analysis is carried out.

Some parts of the overall product costing function are outside the scope of this initial project, and will be either be included in a later increment, or possibly in phase 2 of the plan. These are Coppice and Watermead factories, where the problems with product costing are not as significant as at Beechfield, and Home Farm, where the operations are very different in nature.

Problems in product costing

The mini-computer accounting system includes a product costing module. This meets only some of the information needs of the Finance Director and very few of the requests of any other managers, least of all the factory and farm managers who have direct control of most operations. Since the existing product costing system cannot answer most of the queries that are put to it, the costing clerks attempt to provide additional reports by using a spreadsheet application. But the sheer volume of data available for input each week is impossible to process accurately for all products. Due to ongoing staff shortages in the office, it has only usually been possible to produce actual costs for one production line each week. Making the best of a bad job, each line is costed accurately every fifth week and estimates produced for the other four weeks in between. As a result, the 'actual' costs quoted in management reports are often really no more than estimates derived from samples of the data available.

Both Harold Evans and Clare Smythe are convinced that their inability to get accurate costs is a major contributory factor to the company's decline in profitability. In effect, it means that senior management cannot say with confidence which operations are profitable and which are not. Until they have better information, they cannot even tell where their real problems are.

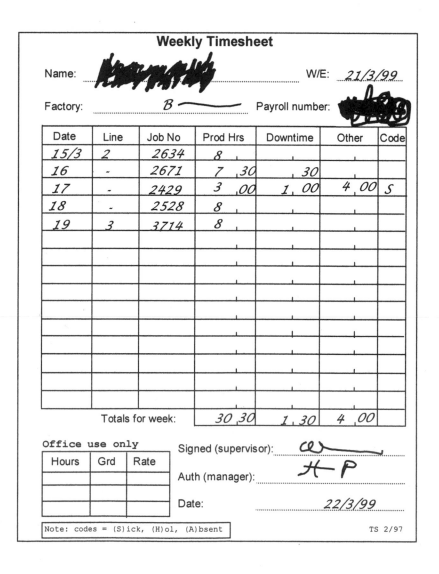

Figure B1.4 *The timesheet used in Beechfield factory.*

B1.3 The Proposal

The first system proposed for development is one to automate product costing. In Box B1.2, some staff from FoodCo's IS team are heard during a meeting early in the project. Louise Lee is the project manager, Ken Ong and Rosanne Martel are both analyst/developers on the team, and Chris Pelagrini is a consultant in object-oriented development (he works for the consultancy firm that produced the IS report).

Box B1.2 Scene: FoodCo's IS Team Meeting Room.

Louise Lee: I'll begin by welcoming Chris Pelagrini. Chris is a consultant on object-oriented development, and he will be working closely with us on this project.

Chris Pelagrini: Thank you, Louise. Yes, that's right, I've been seconded to you for the duration. Provided, that is, we complete in six months (laughs).

LL: Don't worry, we can hit that deadline. OK, let's get started. Today I mainly want to set the scope for the project. Rosanne, you had a meeting with the Beechfield factory manager, Hari Patel. What did you find out?

Rosanne Martel: Yes, I met Hari on Thursday. He's the principal client for this system, and he'll be a user too. He confirmed the reasons why we picked this as our first object-oriented project. It's strategically important, but not so critical that the whole company will fold if we mess up. It's tightly scoped, and only really affects operations in Beechfield and the costing office. But it does need access to a lot of data currently on the mini computer and it's a feeder system for payroll and production planning. If we develop a system that integrates with these, we'll have a sound basis for re-engineering the entire IS provision.

LL: Good. This confirms the consultants' report too. Did you get any idea of the main functionality of the system? We'll need this to estimate timescales with any confidence.

RM: Ken, you've done some work on this. How far did you get?

Ken Ong: Well, it's too early to be precise, but I've sketched out some use cases and a rough class diagram. Users include Hari, his line supervisors, the sales team, production control and the costing office. The main system inputs are staff timesheets and production record sheets, and data we can import from payroll records and the existing costing system. The main system outputs will be production run costs. One obvious problem is that we don't hold any payroll data electronically, so we'll need access to the bureau's files at some point. I would say that as a whole it is not highly complex. My first class diagram has about a dozen classes. There are a few interactions that involve most classes—for example, producing the final cost for a production line run, but most are simpler.

LL: So this is a fairly small system with relatively few users, but lots of interfaces to other systems. Can you show us some of this on the whiteboard?

KO: Yes, of course. Just give me a few minutes (goes to whiteboard and starts to draw).

LL: (While Ken draws) What do you think so far, Chris? Perhaps you could say a little about how you see your role.

CP: My task is to help you apply the object-oriented techniques that you have all learned on the training courses to this project. You all know there is a big difference between knowing the techniques, and understanding how they fit together. I'm here to help when you're unsure about anything. Rosanne's summary suggests this project is an ideal start and I'm confident we will make it a complete success.

LL: That's great, Chris, coming from the expert. OK Ken, now let's take a look at your diagrams.

5

Requirements Capture

Objectives
In this chapter you will learn:
- the distinction between the current and required systems;
- when and how to apply the five major fact finding techniques;
- the need to document requirements;
- how to draw *use case diagrams* to document requirements;
- how to write *use case descriptions*.

5.1 Introduction

Part of the job of the systems analyst is to find out from users what they require in a new information system. Indeed, identifying what a new system should be able to do is one of the first steps in its development, whether you are developing some simple programs for your own use or embarking on the development of a large-scale system for a commercial client. The user requirements can be classified in different ways (Section 5.2), and analysts use a range of techniques to identify and document the requirements (Sections 5.3 and 5.5). Each of the main fact finding techniques has advantages and disadvantages and is appropriate for different situations. Stakeholders is the term used for the people who have an interest in the new system and whose needs must be considered (Section 5.4). UML provides a diagramming technique that can be used to document the stakeholders' requirements. This is the use case diagram, a relatively simple diagram that is supported by written information in the form of a use case description (Section 5.6).

5.2 User Requirements

The aim of developing a new information system must be to produce something that meets the needs of the people who will be using it. In order to do this, we must have a clear understanding both of the overall objectives of the business and of what it is that the individual users of the system are trying to achieve in their jobs. Unless you are in the rare position of developing a system for a new organization, you will need to understand how the business is operating at present and how people are working now. Many aspects of the current system will need to be carried forward into the new system, so it is important that information about what people are doing is gathered and documented. These are the requirements that are derived from the 'current system'. The motivation for the development of a new information system is usually problems with and inadequacies of the current system, so it is also essential to capture what it is that the users require of the new system that they cannot do with their existing system. These are the 'new requirements'.

5.2.1 Current system

The existing system may be a manual one, based on paper documents, forms and files; it may already be computerized; or it may be a combination of both manual and computerized elements. Whichever it is, it is reasonably certain that large parts of the existing system meet the needs of the people who use it, that it has to some extent evolved over time to meet business needs and that users are familiar and comfortable with it. It is almost equally certain that there are sections of the system that no longer meet the needs of the business, and that there are aspects of the business which are not dealt with in the existing system.

It is important that the analyst, gathering information as one of the first steps in developing a new system, gains a clear understanding of how the existing system works: parts of the existing system will be carried forward into the new one. It is also important, because the existing system will have shortcomings and defects which must be avoided or overcome in the new system. It is not always easy or possible to replace existing systems. So-called *legacy systems* may have been developed some time ago and may contain millions of lines of program code which have been added to and amended over a period of time. One approach to dealing with such systems is to create new front-ends, typically using modern graphical user interfaces and object-oriented languages, and *wrap* the legacy systems up in new software. If this is the case, then it is also necessary to understand the interfaces to the legacy systems which the new *wrappers* will have to communicate with.

It is not always possible to leave legacy systems as they are and simply wrap them in new code. It has not been possible to ignore the problems which have faced companies at the turn of the century when it was realized that many systems were in danger of catastrophic collapse as a result of the decision to use two decimal digits to store the year. However, the process of changing the program code in such systems is a matter of understanding the internal working of existing systems rather than gathering information about the way the organization works and the way that people do their jobs.

Not everyone agrees that a detailed understanding of the current system is necessary. Ed Yourdon (1989) argues that it is a waste of time to model the current system in great detail. Yourdon points out that so much time can be spent investigating and modelling the current system that the analysts lose sight of their objective, and impatient users cancel the project. He makes the case for concentrating on the behaviour that is required of the new system. The opposite position is taken by SSADM (Structured Systems Analysis and Design Method) which expends considerable time on the investigation and modelling of the current system. This is done in order to refine it to its logical essence and to be able to merge it with the new requirements to produce a model of the required system that includes the essentials of the existing system.

We believe that a case can be made for investigating the existing system.
- Some of the functionality of the existing system will be required in the new system.
- Some of the data in the existing system is of value and must be migrated into the new system.
- Technical documentation of existing computer systems may provide details of processing algorithms that will be needed in the new system.
- The existing system may have defects which we should avoid in the new system.

- Studying the existing system will help us to understand the organization in general.
- Parts of the existing system may be retained. Information systems projects are now rarely 'green field' projects in which manual systems are replaced by new computerized systems; more often there will be existing systems with which interfaces must be established.
- We are seeking to understand how people do their jobs at present in order to characterize people who will be users of the new system.
- We may need to gather baseline information against which we can set and measure performance targets for the new system.

For all these reasons, an understanding of the current system should be part of the analysis process. However, the analyst should not lose sight of the objective of developing a new system. In the sections on functional, non-functional and usability requirements below, we shall explain what kind of information we are gathering.

5.2.2 New requirements

Most organizations now operate in an environment that is rapidly changing. The relative strength of national economies around the world can change dramatically and at short notice; the fortunes of large companies, which may be an organization's suppliers, customers or competitors, can be transformed overnight; new technologies are introduced which change production processes, distribution networks and the relationship with the consumer; governments and (particularly in Europe) supra-governmental organizations introduce legislation that has an impact on the way that business is conducted. Some authors make the case for developing business strategies to cope with this turmoil. Tom Peters in 'Thriving on Chaos' (1988) argues that we must learn to love change and develop flexible and responsive organizations to cope with the dynamic business environment. A clear result of responding to a dynamic environment is that organizations change their products and services and change the way they do business. The effect of this is to change their need for information. Even in less responsive organizations, information systems become outdated and need enhancing and extending. Mergers and demergers create the need for systems to be replaced. The process of replacement offers an opportunity to extend the capabilities of systems to take advantage of new technological developments, or to enhance their usefulness to management and workforce. Many organizations are driven by internal factors to grow and change the ways in which they operate, and this too provides a motivation for the development of new information systems.

Whether you are investigating the working of the existing system or the requirements for the new system, the information you gather will fall into one of three categories: 'functional requirements', 'non-functional requirements' and 'usability requirements'. Functional and non-functional requirements are conventional categories in systems analysis and design, while usability is often ignored in systems development projects. In many university courses, issues surrounding the usability of systems are taught under the separate heading of Human Factors or Human–Computer Interaction, or are only considered in the design stage of the development process. However, the lesson of human factors research is that usability considerations should be integral to the systems development life cycle, and so they are included here.

Functional requirements

Functional requirements describe what a system does or is expected to do, often referred to as its *functionality*. In the object-oriented approach which we are taking here, we shall initially employ use cases to document the functionality of the system. As we progress into the analysis stage, the detail of the functionality will be recorded in the data that we hold about objects, their attributes and operations. In structured methods, such as SSADM, the function may be the unit around which we structure the system, and the function is described in progressively greater detail as we move through analysis into design and implementation.

At this stage, we are setting out to establish what the system must do, and functional requirements include the following.

- Descriptions of the processing which the system will be required to carry out.
- Details of the inputs into the system from paper forms and documents, from interactions between people, such as telephone calls, and from other systems.
- Details of the outputs that are expected from the system in the form of printed documents and reports, screen displays and transfers to other systems.
- Details of data that must be held in the system.

Non-functional requirements

Non-functional requirements are those which describe aspects of the system that are concerned with how well it provides the functional requirements. These include the following.

- Performance criteria such as desired response times for updating data in the system or retrieving data from the system.
- Anticipated volumes of data, either in terms of throughput or of what must be stored.
- Security considerations.

Usability requirements

Usability requirements are those that will enable us to ensure that there is a good match between the system that is developed and both the users of that system and the tasks that they will undertake when using it. The International Standards Organization (ISO) has defined the usability of a product as 'the degree to which specific users can achieve specific goals within a particular environment; effectively, efficiently, comfortably and in an acceptable manner'. Usability can be specified in terms of measurable objectives, and these are covered in more detail in Chapter 15 on Human–Computer Interaction. In order to build usability into the system from the outset, we need to gather the following types of information.

- Characteristics of the users who will use the system.
- The tasks which the users undertake, including the goals which they are trying to achieve.
- Situational factors which describe the situations which could arise during system use.
- Acceptance criteria by which the user will judge the delivered system.

Paul Booth (1989) describes the issues surrounding system usability in more detail.

Review Question

5.1 Read the following description of a requirement for FoodCo, and decide which parts of it are functional requirements and which are non-functional requirements.

The allocation of staff to production lines should be mostly automated. A process will be run once a week to carry out the allocation based on the skills and experience of operatives. Details of holidays and sick leave will also be taken into account. A first draft Allocation List will be printed off by 12.00 noon on Friday for the following week. Only staff in Production Planning will be able to amend the automatic allocation to fine-tune the list. Once the amendments have been made, the final Allocation List must be printed out by 5.00 pm. The system must be able to handle allocation of 100 operatives at present, and should be capable of expansion to handle double that number.

5.3 Fact Finding Techniques

There are five main fact finding techniques that are used by analysts to investigate requirements. Here we describe each of them in the order that they might be applied in a system development project, and for each one explain the kind of information that you would expect to gain from its use, its advantages and disadvantages, and the situations in which it is appropriate to use it.

5.3.1 Background reading

If an analyst is employed within the organization that is the subject of the fact gathering exercise, then it is likely that he or she will already have a good understanding of the organization and its business objectives. If, however, he or she is going in as an outside consultant, then one of the first tasks is to try to gain an understanding of the organization. Background reading or research is part of that process. The kind of documents that are suitable sources of information include the following:
- company reports,
- organization charts,
- policy manuals,
- job descriptions,
- reports and
- documentation of existing systems.

Although reading company reports may provide the analyst with information about the organization's mission, and so possibly some indication of future requirements, this technique mainly provides information about the current system.

Advantages and disadvantages

+ Background reading helps the analyst to get an understanding of the organization before meeting the people who work there.

+ It also allows the analyst to prepare for other types of fact finding, for example, by being aware of the business objectives of the organization.
+ Documentation on the existing system may provide formally defined information requirements for the current system.
− Written documents often do not match up to reality; they may be out of date or they may reflect the official policy on matters that are dealt with differently in practice.

Appropriate situations

Background reading is appropriate for projects where the analyst is not familiar with the organization being investigated. It is useful in the initial stages of investigation.

5.3.2 Interviewing

Interviewing is probably the most widely used fact finding technique; it is also the one that requires the most skill and sensitivity. Because of this, we have included a set of guidelines on interviewing that includes some suggestions about etiquette in Box 5.1.

A systems analysis interview is a structured meeting between the analyst and an interviewee who is usually a member of staff of the organization being investigated. The interview may be one of a series of interviews that range across different areas of the interviewee's work or that probe in progressively greater depth about the tasks undertaken by the interviewee. The degree of structure may vary: some interviews are planned with a fixed set of questions that the interviewer works through, while others are designed to cover certain topics but will be open-ended enough to allow the interviewer to pursue interesting facts as they emerge. The ability to respond flexibly to the interviewee's responses is one of the reasons why interviews are so widely used.

Interviews can be used to gather information from management about their objectives for the organization and for the new information system, from staff about their existing jobs and their information needs, and from customers and members of the public as possible users of systems. While conducting an interview, the analyst can also use the opportunity to gather documents which the interviewee uses in his or her work.

It is usually assumed that questionnaires are used as a substitute for interviews when potential interviewees are geographically dispersed in branches and offices around the world. The widespread use of desktop video conferencing may change this and make it possible to interview staff wherever they are. Even then, questionnaires can reach more people.

Interviewing different potential users of a system separately can mean that the analyst is given different information by different people. Resolving these differences later can be difficult and time-consuming. One alternative is to use group interviews in order to get the users to reach a consensus on issues. Dynamic Systems Development Method (DSDM) is a method of carrying out systems development in which group discussions are used (Stapleton, 1997). These discussions are run as workshops for knowledgeable users with a facilitator who aims to get the users to pool their knowledge and to reach a consensus on the priorities of the development project.

Box 5.1 Guidelines on Interviewing

Conducting an interview requires good planning, good interpersonal skills and an alert and responsive frame of mind. These guidelines cover the points you should bear in mind when planning and conducting an interview.

Before the interview

You should always make appointments for interviews in advance. You should give the interviewee information about the likely duration of the interview and the subject of the interview.

Being interviewed takes people away from their normal work. Make sure that they feel that it is time well spent.

It is conventional to obtain permission from an interviewee's line manager before interviewing them. Often the analyst interviews the manager first and uses the opportunity to get this permission.

In large projects, an interview schedule should be drawn up showing who is to be interviewed, how often and for how long. Initially this will be in terms of the job roles of interviewees rather than named individuals. It may be the manager who decides which individual you interview in a particular role.

Have a clear set of objectives for the interview. Plan your questions and write them down. Some people write the questions with space between them for the replies.

Make sure your questions are relevant to the interviewee and his or her job.

At the start of the interview

Introduce yourself and the purpose of the interview.

Arrive on time for interviews and stick to the planned timetable—do not over-run.

Ask the interviewee if he or she minds you taking notes or tape-recording the interview. Even if you tape-record an interview, you are advised to take notes. Machines can fail! Your notes also allow you to refer back to what has been said during the course of the interview and follow up points of interest.

Remember that people can be suspicious of outside consultants who come in with clipboards and stopwatches. The cost-benefit analyses of many information systems justify the investment in terms of savings in jobs!

During the interview

Take responsibility for the agenda. You should control the direction of the interview. This should be done in a sensitive way. If the interviewee is getting away from the subject, bring them back to the point. If what they are telling you is important, then say that you will come back to it later and make a note to remind yourself to do so.

Use different kinds of question to get different types of information. Questions can be open-ended—'Can you explain how you complete a timesheet?'—or closed—'How many staff use this system?'. Do not, however, ask very open-ended questions such as 'Could you tell me what you do?'

Listen to what the interviewee says and encourage him or her to expand on key points. Keep the focus positive if possible. Make sure you have understood answers by summarizing them back to the interviewee. Avoid allowing the interview to degenerate into a session in which the interviewee complains about everyone and everything.

You may be aware of possible problems in the existing system, but you should avoid prejudging issues by asking questions that focus too much on problems. Gather facts.

Be sensitive about how you use information from other interviews that you or your colleagues have already conducted, particularly if comments were negative or critical.

Use the opportunity to collect examples of documents that people use in their work, ask if they mind you having samples of blank forms and photocopies of completed paperwork.

After the interview

Thank the interviewee for their time. Make an appointment for a further interview if it is necessary. Offer to provide them with a copy of your notes of the interview for them to check that you have accurately recorded what they told you.

Transcribe your tape or write up your notes as soon as possible after the interview while the content is still fresh in your mind.

If you said that you would provide a copy of your notes for checking then send it to the interviewee as soon as possible. Update your notes to reflect their comments.

Advantages and disadvantages

+ Personal contact allows the analyst to be responsive and adapt to what the user says. Because of this, interviews produce high quality information.
+ The analyst can probe in greater depth about the person's work than can be achieved with other methods.
+ If the interviewee has nothing to say, the interview can be terminated.
− Interviews are time-consuming and can be the most costly form of fact gathering.
− Interview results require the analyst to work on them after the interview: the transcription of tape recordings or writing up of notes.
− Interviews can be subject to bias if the interviewer has a closed mind about the problem.
− If different interviewees provide conflicting information, it can be difficult to resolve later.

Appropriate situations

Interviews are appropriate in most projects. They can provide information in depth about the existing system and about people's requirements from a new system.

5.3.3 Observation

Watching people carrying out their work in a natural setting can provide the analyst with a better understanding of the job than interviews, in which the interviewee will often concentrate on the normal aspects of the job and forget the exceptional situations and interruptions which occur and which the system will need to cope with. Observation also allows the analyst to see what information people use to carry out their job. This can tell you about the documents they refer to, whether they have to get up from their desks to get information, how well the existing system handles their needs. One of the authors has observed staff using a tele-sales system where there was no link between the enquiry screens for checking the availability of stock and the data entry screens for entering an order. These tele-sales staff kept a pad of scrap paper on the desk and wrote down the product codes for all the items they had looked up on the enquiry screens so that they could enter them into the order-processing screens. This kind of information does not always emerge from interviews.

People are not good at estimating quantitative data, such as how long they take to deal with certain tasks, and observation with a stopwatch can give the analyst plentiful quantitative data, not just about typical times to perform a task but also about the statistical distribution of those times.

In some cases where information or items are moving through a system and being dealt with by many people along the way, observation can allow the analyst to follow the entire process through from start to finish. This type of observation might be used in an organization where orders are taken over the telephone, passed to a warehouse for picking, packed and despatched to the customer. The analyst may want to follow a series of transactions through the system to obtain an overview of the processes involved.

Observation can be an open-ended process in which the analyst simply sets out to observe what happens and to note it down, or it can be a closed process in which the analyst wishes to observe specific aspects of the job and draws up an observation schedule or form on which to record data. This can include the time it takes to carry out a task, the types of task the person is performing or factors such as the number of errors they make in using the existing system as a baseline for usability design.

Advantages and disadvantages

+ Observation of people at work provides firsthand experience of the way that the current system operates.
+ Data are collected in real time and can have a high level of validity if care is taken in how the technique is used.
+ Observation can be used to verify information from other sources or to look for exceptions to the standard procedure.
+ Baseline data about the performance of the existing system and of users can be collected.
− Most people do not like being observed and are likely to behave differently from the way in which they would normally behave. This can distort findings and affect the validity.
− Observation requires a trained and skilled observer for it to be most effective.
− There may be logistical problems for the analyst, for example, if the staff to be observed work shifts or travel long distances in order to do their job.
− There may also be ethical problems if the person being observed deals with sensitive private or personal data or directly with members of the public, for example in a doctor's surgery.

Appropriate situations

Observation is essential for gathering quantitative data about people's jobs. It can verify or disprove assertions made by interviewees, and is often useful in situations where different interviewees have provided conflicting information about the way the system works. Observation may be the best way to follow items through some kind of process from start to finish.

5.3.4 Document sampling

Document sampling can be used in two different ways. Firstly, the analyst will collect copies of blank and completed documents during the course of interviews and observation sessions. These will be used to determine the information that is used by people in their work, and the inputs to and outputs from processes which they carry out, either manually or using an existing computer system. Ideally, where there is an existing system, screen shots should also be collected in order to understand the inputs and outputs of the existing system. Figure 5.1 shows a sample document collected from Agate, our case study company.

Secondly, the analyst may carry out a statistical analysis of documents in order to find out about patterns of data. For example, many documents such as order forms contain a header section and a number of lines of detail. (The sample document in Figure 5.1 shows this kind of structure.) The analyst may want to know the distribution of the number of lines in an order. This will help later in estimating volumes of data to be held in the system and in deciding how many lines should be displayed on screen at one time. While this kind of statistical sampling can give a picture of data volumes, the analyst should be alert to seasonal patterns of activity which may mean that there are peaks and troughs in the amount of data being processed.

Advantages and disadvantages

+ Can be used to gather quantitative data, such as the average number of lines on an invoice.
+ Can be used to find out about error rates in paper documents.
− If the system is going to change dramatically, existing documents may not reflect how it will be in future.

Appropriate situations

The first type of document sampling is almost always appropriate. Paper-based documents give a good idea of what is happening in the current system. They also provide supporting evidence for the information gathered from interviews or observation.

The statistical approach is appropriate in situations where large volumes of data are being processed, and particularly where error rates are high, and a reduction in errors is one of the criteria for usability.

5.3.5 Questionnaires

Questionnaires are a research instrument that can be applied to fact finding in system development projects. They consist of a series of written questions. The questionnaire designer usually limits the range of replies that the respondent can make by giving them a choice of options (see Figure 5.2). shows some of the types of question. YES/NO questions only give the respondent two options. If there are more options, the multiple choice type of question is often used when the answer is factual, whereas scaled questions are used if the answer involves an element of subjectivity. Some questions do not have a fixed number of responses, and must be left open-ended for the respondent to enter what they like. Where the respondent has a limited number of choices, these are usually coded with a number which speeds up data entry if the responses are to be analysed by computer software.

Agate
Campaign Summary

Date   23rd February 1999

Client   Yellow Partridge
                Park Road Workshops
                Park Road
                Jewellery Quarter
                Birmingham B2 3DT
                U.K.

Campaign   Spring Collection 1999

Billing Currency   GB £

Item	Curr	Amount	Rate	Billing amount
Advert preparation: photography, artwork, layout etc.	GB £	15,000.00	1	15,000.00
Placement French Vogue	FFr.	47 000,00	11.35	4,140.97
Placement UK Vogue	GB £	5,000.00	1	5,000.00
Placement US Vogue	US $	15,000.00	2.47	6.072.87
Total				30,213.84

This is not a VAT Invoice. A detailed VAT Invoice will be provided separately.

210-212 Carstairs Street, Birmingham B1 5TG
Tel.0121-111-1234 Fax.0121-111-1245
Email.agate@agateltd.co.uk

Figure 5.1 *Sample document from the AGATE case study.*

YES/NO Questions

Do you print reports from the existing system?   YES   NO   10
(Please ring the appropriate answer.)

Multiple Choice Questions

How many new clients do you obtain in a year?   a) 1–10 ☐   11
(Please tick one box only.)   b) 11–20 ☐
                                                     c) 21–30 ☐
                                                     d) 31 +   ☐

Scaled Questions

How satisfied are you with the response time of the stock update?
(Please ring one option.)

1. Very   2. Satisfied   3. Dissatisfied   4. Very   12
   satisfied   dissatisfied

Open-ended Questions

What additional reports would you require from the system?

_____

_____

_____

Figure 5.2 *Types of question used in questionnaires.*

Advantages and disadvantages

+ An economical way of gathering data from a large number of people.
+ If the questionnaire is well designed, then the results can be analysed easily, possibly by computer.
− Good questionnaires are difficult to construct.
− There is no automatic mechanism for follow up or probing more deeply, although it is possible to follow up with an interview by telephone or in person if necessary.
− Postal questionnaires suffer from low response rates.

Appropriate situations

Questionnaires are most useful when the views or knowledge of a large number of people need to be obtained or when the people are geographically dispersed, for example, in a company with many branches or offices around the country or around the world. Questionnaires are also appropriate for information systems which will be used by the general public, and where the analyst needs to get a picture of the types of user and usage that the system will need to handle.

5.3.6 Remembering the techniques

For those who like mnemonics, these techniques are sometimes referred to as SQIRO— Sampling, Questionnaires, Interviewing, Reading (or Research) and Observation.   This order has been chosen to make it possible to pronounce the mnemonic. However, this is not the order in which they are most likely to be used. This will depend on the situation and the organization in which the techniques are being used.

5.3.7 Other techniques

Some kinds of system require special fact finding techniques. *Expert systems* are computer systems which are designed to embody the expertise of a human expert in solving problems. Examples include systems for medical diagnosis, stock market trading and geological analysis for mineral prospecting. The process of capturing the knowledge of the expert is called *knowledge acquisition* and, as it differs from establishing the requirements for a conventional information system, a number of specific techniques are applied. Some of these are used in conjunction with computer-based tools.

5.4 User Involvement

The success of a systems development project depends not just on the skills of the team of analysts, designers and programmers who work on it, or on the project management skills of the project manager, but on the effective involvement of users in the project at various stages of the life cycle. The term *stakeholders* was introduced in Chapter 2 to describe all those people who have an interest in the successful development of the system. Stakeholders include all people who stand to gain (or lose) from the implementation of the new system and managers who control the budget and require the system for their organization. Analysts deal with people at all levels of the organization. In large projects it is likely that a steering committee with delegated powers will be set up to manage the project from the users' side. This will include the following categories of people:

- senior management—with overall responsibility for running the organization,
- financial managers with budgetary control over the project,
- managers of the user department(s) and
- representatives of users.

Users will be involved in different roles during the course of the project as:

- subjects of interviews to establish requirements,
- representatives on project committees,
- those involved in evaluating prototypes,
- those involved in testing,
- subjects of training courses and
- end-users of the new system.

Case Study Example

The section that follows applies what has been covered in this chapter so far to the case study.

One of the first tasks in fact finding is to draw up a plan that outlines what information is being sought, which techniques will be used, who is involved and how long the fact finding will take. A draft plan for fact finding at Agate is shown below. The jobs of the subjects are those shown in the list in Figure A1.1 in the Agate case study.

Objective	Technique	Subject(s)	Time commitment
To get background on the company and the advertising industry.	Background reading	Company reports, trade journals.	0.5 day
To establish business objectives. Agree likely scope of new system. Check out involvement of non-UK offices.	Interview	Two directors	2 x 1 hour each
To gain understanding of roles of each department. Check out lIne management and team structure in the Creative department. To agree likely interviewees among staff.	Interview	Department heads (only 1 account manager)	2 x 1 hour each
To find out how the core business operates.	Interview	1 account manager 1 graphic designer 1 copy writer 1 editor	1.5 hours each
To follow up development of business understanding.	Observation	2 creative staff	0.5 day each
To determine role of support/admin staff and relationship to core business.	Interview	2 admin staff (based on experience with the company)	1.5 hours each
To establish what records and resources are kept.	Interview/ document sampling	Filing clerk Resource librarian	2 x 1 hour each
To determine what use is made of current computer system. To determine functionality of current system.	Interview	Computer manager	2 x 1 hour
To establish additional requirements for new system.	Interview	2 account managers 3 staff from Creative Department	3 x 1 hour each
To establish accounting requirements for new system.	Interview	Accountant Credit controller 1 purchasing assistant 1 accounts clerk	1.5 hours each

Review Questions

5.2 Name the five main fact finding techniques and list one advantage and one disadvantage of each.

5.3 Imagine that you will be interviewing one of the three staff in Production Planning at FoodCo. Draw up ten questions that you would want to ask him or her.

5.5 Documentation Standards

Information systems professionals need to record facts about the organization they are studying and its requirements. As soon as the analysts start gathering facts, they will need some means of documenting them. In the past the emphasis was on paper forms, but now it is rare for a large-scale project to depend on paper-based documentation. As you will see, systems analysts and designers maintain information about the new system in a mixture of diagrams and text. The important thing to bear in mind is that within a project some set of standards should be adhered to. These may be the agreed standards of the organization carrying out the analysis and design project or they may be a requirement of the organization that is having the work done. For example, in the UK, government projects have been expected to use SSADM, while in the US military projects must conform to a number of Department of Defense (DoD) standards.

Standards are important as they promote communication in the same way as a common language. They enable communication between members of the development team if they all document the information in the same standard formats. They promote communication over time, as other people come to work on the system, even several years after it has been implemented in order to carry out maintenance. They also promote communication of good practice, as experience of what should be recorded and how best to do that recording builds up over time and is reflected in the techniques that are used.

We have chosen to use the Unified Modeling Language (UML) in this book, as we believe that it is developing into an industry standard. UML consists mainly of a graphical language to represent the concepts that we require in the development of an object-oriented information system. However, there will also be other kinds of documents, not all of which fit into the UML framework. In large-scale projects a librarian may be required to keep track of these documents and ensure that they are stored safely and in a way that enables them to be retrieved when required. Such documents include the following:

- records of interviews and observations,
- details of problems,
- copies of existing documents and where they are used,
- details of requirements,
- details of users and
- minutes of meetings.

Even in smaller projects which cannot justify a librarian, a filing system with an agreed set of conventions on how material is to be filed, and for recording who has taken items from the filing system is good practice.

The diagrams in a modelling language such as UML also conform to agreed standards. Diagrammatical models are used extensively by systems analysts and designers in order to:

- communicate ideas,

- generate new ideas and possibilities,
- test ideas and make predictions,
- understand structures and relationships.

Computer Aided Software Engineering (CASE) tools are used to draw these diagrams and to maintain the associated data about the various things that are shown in the diagrams. In Chapter 3 we discussed the use of CASE tools in more detail. It is a platitude to state that 'A picture is worth a thousand words', but for most people, the visual representation of information allows them to take in a range of facts and relationships in a glance. This is illustrated in Figure 5.3 which shows the information about Agate that is listed in Figure A1.1 using an organization chart rather than a textual list. Modelling techniques should aid (and enforce):

- simplicity of representation   only showing what needs to be shown,
- internal consistency—within a set of diagrams,
- completeness—showing all that needs to be shown,
- hierarchical representation—breaking the system down and showing more detail at lower levels.

The next section introduces the first of the UML diagramming techniques to be explained in detail. We shall check it against these criteria at the end of the section.

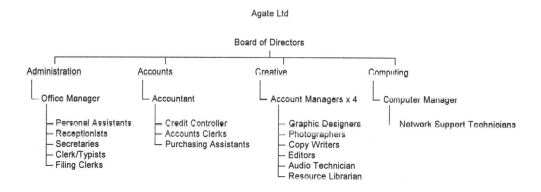

Figure 5.3 *Organization chart for Agate Ltd in the UK.*

5.6 Use Cases

Use cases are textual descriptions of the functionality of the system from the users' perspective. Use case diagrams are used to show the functionality that the system will provide and which users will communicate with the system in some way when it provides that functionality. Figure 5.4 shows an example of a use case diagram. This is a relatively simple diagramming technique, and its notation is explained below in Section 5.6.2.

Use case diagrams were developed by Jacobson et al. (1992), indeed the subtitle of the book in which they are presented is 'A Use Case Driven Approach', although Jacobson and his co-authors offer a complete approach to the development of object-oriented software systems.

5.6.1 Purpose

The use case model is part of what Jacobson calls the requirements model; he also includes a problem domain object model and user interface descriptions in this requirements model. Use cases specify the functionality which the system will offer from the users' perspective. They are used to document the scope of the system and the developer's understanding of what it is that the users require.

The textual *use case descriptions* provide a description of the interaction between the users of the system, termed *actors*, and the high level functions within the system, the use cases. These descriptions can be in summary form or in a more detailed form in which the interaction between actor and use case is described in a step-by-step way. Whichever approach is used, it should be remembered that the use case describes the interaction as the user sees it, and is not a definition of the internal processes within the system, or some kind of program specification.

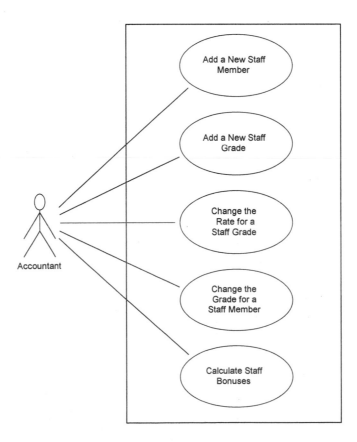

Figure 5.4 *Example use case diagram.*

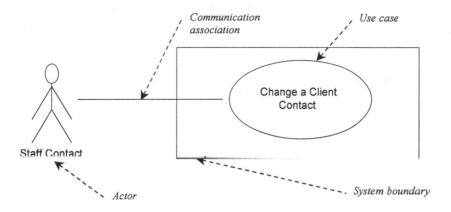

Figure 5.5 *The notation of the use case diagram.*

5.6.2 Notation

Use case diagrams show three aspects of the system: actors, use cases and the system itself. Figure 5.5 shows the elements of the notation.

Actors represent the roles that people or other systems take on when communicating with the particular use cases in the system. In Agate, there is no job title Staff Contact: a director, an account manager or a member of the creative team can take on the role of being staff contact for a particular client company, so one role can represent several people or job titles. Equally, a particular person or job title may be represented by more than one actor on use case diagrams. This is shown in Figure 5.6. A director or an account manager may be the Campaign Manager for a particular client campaign, as well as being the Staff Contact for one or more clients.

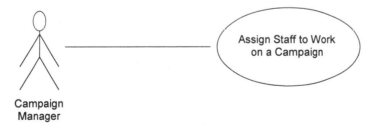

Figure 5.6 *Use case showing* Campaign Manager *actor.*

The use case description associated with each use case can be brief:

> *The campaign manager selects a particular campaign. A list of staff not already working on that campaign is displayed, and he or she selects those to be assigned to this campaign.*

Alternatively, it can provide a step-by-step breakdown of the interaction between the user and the system for the particular use case. An example of this extended approach is provided below.

Actor Action	System Response
1. The campaign manager enters the client name.	2. Lists all campaigns for that client.
3. Selects the relevant campaign.	4. Displays a list of all staff members not already allocated to this campaign.
5. Highlights the staff members to be assigned to this campaign.	6. Presents a message confirming that staff have been allocated.

Alternative Courses
Steps 1–3. The campaign manager knows the campaign name and enters it directly.

Constantine (1997) makes the distinction between *essential* and *real* use cases. Essential use cases describe the 'essence' of the use case in terms that are free of any technological or implementation details, whereas real use cases describe the concrete detail of the use case in terms of its design. During the analysis stage, use cases are almost always essential, as the design has not yet been decided upon. In a real use case, Step 2 in the use case description for `Assign Staff to Work on a Campaign` could be described as 'Lists all campaigns for the client in a list box, sorted into alphabetical order by campaign title.'

Each use case description represents the usual way in which the actor will go through the particular transaction or function from end to end. Possible major alternative routes that could be taken are listed as *alternative courses*. The term *scenario* is used to describe use cases in which an alternative course is worked through in detail, including possible responses to errors. The use case represents the generic case, while the scenarios represent specific paths through the use case.

As well as the description of the use case itself, the documentation should include the purpose or intent of the use case, that is to say details of the task which the user is trying to achieve through the means of this use case, for example:

> *The campaign manager wishes to record which staff are working on a particular campaign. This information is used to validate timesheets and to calculate staff year-end bonuses.*

Two further kinds of relationships can be shown on the use case diagram itself. These are the «extends» and «uses» relationships. The two are easy to confuse. «Extends» is used when you wish to show that a use case provides additional functionality that may be required in another use case. In Figure 5.7, the use case `Print Campaign Summary` extends `Check Campaign Budget`. This means that at a particular point in `Check Campaign Budget` the user can optionally invoke the behaviour of `Print Campaign Summary`, which does something over and above what is done in `Check Campaign Budget` (print out the information in this case). There may be more than one way of extending a particular use case, and these possibilities may represent significant variations on the way the user uses the system. Rather than trying to capture all these variations in one use case, you would document the core functionality in one and then extend it in others. Extension points can be shown in the diagram, as in `Check Campaign Budget` in Figure 5.7. These show where in the textual use case description the extension occurs.

«Uses» applies when there is a sequence of behaviour that is used frequently in a number of use cases, and you want to avoid copying the same description of it into each use case in which it is used. Figure 5.7 shows that the use case `Assign Staff to Work on a Campaign` «uses» `Find Campaign`. This means that when an actor uses `Assign Staff` to Work on a Campaign the behaviour of Find Campaign will also be used in order to select the relevant Campaign.

As well as describing the use cases, it is worth describing who the actors are in terms of job titles. Although at this stage in the life cycle, we are concentrating on requirements, later we shall need to know who the actual users are for each high level function that is represented by a use case. This may help in specifying the security for different functions or in assessing the usability of the functions.

Bear in mind that actors need not be human users of the system. They can also be other systems that communicate with the one that is the subject of the systems development project, for example, other computers or automated machinery or equipment.

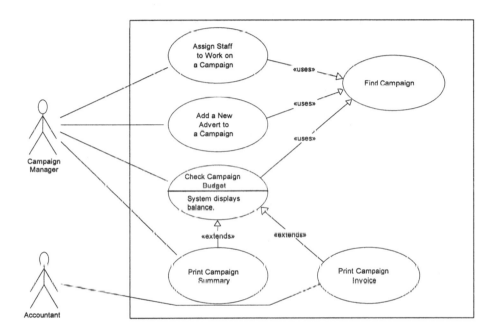

Figure 5.7 *Use case showing «extends» and «uses».*

In Section 5.5 we introduced four criteria by which to judge a modelling technique; now we can assess use cases against these criteria.

Simplicity of representation—only showing what needs to be shown. Use cases show only the user perspective of the system; the diagram is not unnecessarily cluttered with design or implementation details.

Internal consistency—within a set of diagrams. The same names should be used for the same actors across diagrams. The use of the «uses» relationship helps to enforce the discipline of reusing elements of use cases that will apply in several places.

Completeness—showing all that needs to be shown. A full set of use case diagrams should show all the functionality of the system from the users' point of view.

Hierarchical representation—breaking the system down and showing more detail at lower levels. Although we have not covered it yet, the internal interaction of each use case can be represented in a sequence diagram. This will be explained in Chapter 8.

5.6.3 CASE tool support

Drawing any diagram and maintaining the associated documentation is made easier by a CASE tool, as described in Chapter 3. All the use case diagrams in this chapter have been produced using a CASE tool. A CASE tool should also provide facilities to maintain the data dictionary associated with the diagram elements, and to produce reports. Automatically generated reports can be merged into documents that are produced for the client organization. A summary description of each use case forms part of the requirements model or requirements specification which it is necessary to get the client to agree to.

SELECT Enterprise®, the CASE tool that we have used, maintains the data dictionary entries through a series of tabbed property dialogues. Figure 5.8 shows the Use Case Editor dialogue for the use case Check Campaign Budget shown in Figure 5.7. Figure 5.9 shows the automatically generated report output for this use case.

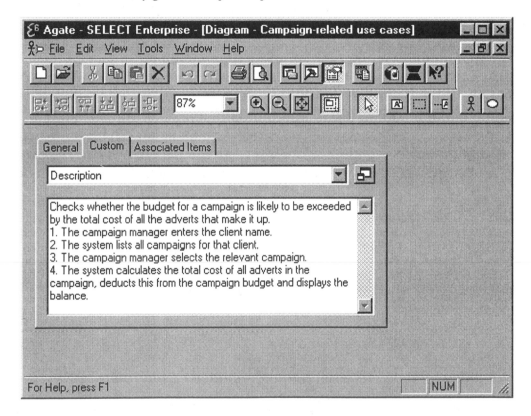

Figure 5.8 *Use Case Editor dialogue in SELECT Enterprise.*

5.7 Summary

Analysts investigating an organization's requirements for a new information system use five main fact finding techniques—background reading, interviews, observation, document sampling and questionnaires. They use these to gain an understanding of the current system and its operation, of the enhancements the users require to the current system and of the new requirements that users have for the new system.

Using agreed standards to document requirements allows the analysts to communicate these requirements to other professionals and to the users. Use case diagrams are one diagramming technique which is used to summarize the users' requirements in a high level overview of the way that the new system will be used.

```
Model: Agate
Date: Monday, March 01, 1999
Time: 11:53:29


Name: Check Campaign Budget
Type: Use Case
Alternate Courses: Steps 1-3   The campaign manager knows
the campaign name and enters it directly.
Description: Checks whether the budget for a campaign is
likely to be exceeded by the total cost of all the adverts
that make it up.
1. The campaign manager enters the client name.
2. The system lists all campaigns for that client.
3. The campaign manager selects the relevant campaign.
4. The system calculates the total cost of all adverts in
the campaign, deducts this from the campaign budget and
displays the balance.
Extension Points: When system displays balance:
      Print Campaign Summary
      Print Campaign Invoice


This item is used on the following diagrams:
      Campaign-related use cases
      Check Campaign Budget sequence diagram
```

Figure 5.9 *Extracts from automatically generated reports produced by SELECT Enterprise.*

Case Study Example

You have already seen several examples from the case study in this section. The use cases are determined by the analyst from the documentation that is gathered from the fact finding process. What follows is a short excerpt from an interview transcript, which has been annotated to show the points which the analyst would pick up on and use to draw the use case diagrams and produce the associated documentation. The interview is between Dave Harris, a systems analyst, and Peter Bywater, an Account manager at Agate. It is from one of the interviews with the objective 'To establish additional requirements for new system' in the fact finding plan in the earlier case study section in this chapter.

Dave Harris: You were telling me about concept notes. What do you mean by this?

Peter Bywater: At present, when we come up with an idea for a campaign we use a word-processor to create what we call a concept note. We keep all the note files in one directory for a particular campaign, but it's often difficult to go back and find a particular one.

DH: So is this something you'd want in the new system?

PB: Yes. We need some means to enter a concept note and to find it again.

(This sounds like two possible use cases. Who are the actors?)

DH: So who would you want to be able to do this?

PB: I guess that the staff working on a campaign should be able to create a new note in the system.

DH: Only them?

(Any other actors?)

PB: Yes, only the staff actually working on a campaign.

DH: What about finding them again? Is this just to view them or could people modify them?

PB: Well, we don't change them now. We just add to them. It's important to see how a concept has developed. So we would only want to view them. But we need some easy way of browsing through them until we find the right one.

(Who are the actors for this?)

DH: Can anyone read the concept notes?

PB: Yes, any of the staff might need to have a look.

DH: Would you need any other information apart from the text of the concept itself?

(Thinking ahead to Chapter 6!)

PB: Yes. It would be good to be able to give each one a title. Could we use the titles then when we browse through them? Oh, and the date, time and whoever created that concept note.

DH: Right, so you'd want to select a campaign and then see all the titles of notes that are associated with that campaign, so you could select one to view it?

(Thinking about the interaction between the user and the system.)

PB: Yes, that sounds about right.

...

From this information, Dave Harris is going to be able to develop the use case descriptions for two use cases:

> Create concept note,
> View concept note.

The use case diagram is shown in Figure 5.10. The use case descriptions will be as follows.

Create concept note.
A member of staff working on a campaign can create a concept note which records ideas, concepts and themes that will be used in an advertising campaign. The note is in text form. Each note has a title. The person who created the note, the date and time are also recorded.

Browse concept notes.

Any member of staff may view concept notes for a campaign. The campaign must be selected first. The titles of all notes associated with that campaign will be displayed. The user will be able to select a note and view the text on screen. Having viewed one note, others can be selected and viewed.

The interaction here is quite straightforward, so we shall not need a more detailed breakdown of the interaction between user and system.

Note that in Figure 5.10, because Campaign Staff is included in the general category of Staff, we do not need to show a communication association between the Campaign Staff actor and the Browse concept notes use case.

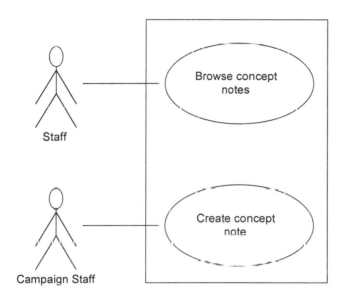

Figure 5.10 *Use cases for concept notes.*

Review Questions

5.4   What is the purpose of producing use cases?

5.5   Describe in your own words the difference between the «extends» and «uses» relationships in use case diagrams.

5.6   What is the difference between an 'essential' and a 'real' use case?

5.7   Write a use case description in the extended form, used for the Assign Staff to a Campaign example in Section 5.6.2, for either Create Concept Note or Browse Concept Note.

5.8   Think of the different possible uses you could make of a library computer system and draw a use case diagram to represent these use cases.

Case Study Work, Exercises and Projects

5.A Refer to the material for the second case study—FoodCo (introduced in Case Study B1). Draw up your initial fact finding plan along the lines of the plan given above.

5.B Read the following excerpt from a transcript of an interview with one of the production planners at FoodCo. Draw a use case diagram and create use case descriptions for the use cases that you can find in this information.

Ken Ong: So what happens when you start planning the next week's allocation?

Rik Sharma: Well, the first thing to do is to check which staff are unavailable.

KO: Would that be because they are on holiday?

RS: Yes, they could be on holiday, or they could be off sick. Because staff are handling raw food, we have to be very careful with any illness. So factory staff often have to stay off work longer than they would if they were office workers.

KO: So how do you know who's off sick and who's on holiday?

RS: They have to complete a holiday form if they want a holiday. They send it to the Factory Manager, who authorizes it and sends it to us. We take a copy, and enter the details into our system. We then return the form to the member of staff.

KO: What details do you enter?

RS: Who it is, the start date of the holiday and the first date they are available for work again.

KO: What about illness?

RS: The first day someone is off sick they have to ring in and notify us. We have to find someone to fill in for them for that day if we can.

KO: Right. Let's come back to that in a minute. How do you record the fact that they're off sick for your next week's production plan?

RS: We make an entry in the system. We record which member of staff it is, when they went off sick, the reason and an estimate of how many days they're likely to be off.

KO: Right, so how do you get at that information when you come to plan next week's allocation?

RS: Well, we run off three lists. We enter Monday's date, and it prints us off one list showing who is available all week, a second list showing who is not available all week, and a third list showing who is likely to be available for part of the week.

KO: Then what?

RS: Then we start with the people who are available all week and work round them. We pull each operative's record up on the screen and look at two main factors— firstly their skills and experience, and secondly which line they're working on at the moment and how long they've been on that line. Then we allocate them to a line and a session in one of the three factories.

KO: So you can allocate them to any one of the three factories. Do you enter the same data for each one?

RS: No, there are slight variations in the allocation screen for each of the factories—mainly for historical reasons.

...

Further Reading

Booth (1989) Chapter 5 describes the issues surrounding the usability of systems in more detail than we can here, and explains the process of Task Analysis.

Hart (1997) gives a detailed explanation of the techniques that are specific to the development of expert systems.

Roberts (1989) addresses the role of users in a systems development project. This book is one of a series of guides written for civil servants in the UK government service, and is relatively bureaucratic in its outlook. However it ranges widely over the issues that users may face. Yourdon (1989) discusses users and their roles in Chapter 3.

Jacobson et al. (1992) present the original ideas behind use cases as an analysis technique.

6

Requirements Analysis

Objectives
In this chapter you will learn:
- why we analyse requirements;
- the role of class diagrams in requirements analysis;
- the technical terms used when working with class diagrams;
- how to use the UML class diagram to build a model of user requirements.

6.1 Introduction

In any organized activity that aims at producing useful artefacts, the main focus of both analysis and design activities is on models. This is equally true for projects to build highways, space shuttles, television sets or software systems. In the days before computer modelling, aircraft designers built wooden or metal scale models of new aircraft to test their characteristics in a wind tunnel. A skilled furniture designer may use a mental model, visualizing a new piece of furniture without drawing a single line.

In IS development, models are usually both abstract and visible. On the one hand many of the products are themselves abstract in nature. Most software is not very tangible to the user. On the other hand, software is usually constructed by teams of people who need to see each other's models. However, even in the case of a single developer working alone, it is still advisable to construct visible models. Software development is a complex activity, and it is extremely difficult to carry all the necessary details in one person's memory.

In Chapter 5 we saw how use case diagrams are used to build a model of users' requirements for a new system. In this chapter we discuss what a requirements model is, and what it needs to do (Section 6.2). We then explain how the class diagram helps to analyse in more detail the requirements that are expressed by use cases. We show how to elaborate the class definitions using attributes and operations, and how to analyse their logical structure using associations (Section 6.3). We then discuss the role and contents of the data dictionary (Section 6.4). Finally, we give practical guidance on carrying out the steps described in the earlier sections of the chapter (Section 6.5).

6.2 What Must a Requirements Model Do?

The most influential factor for the success of an IS project is whether the software product fulfils its users' requirements. Models constructed from an analysis perspective aim at determining these requirements.   This means much more than just gathering and documenting facts and requests. The use case technique introduced in Chapter 5 does not capture enough detail to show all inter-relationships among requirements.   Nor does it show enough of the structure of the problem situation for us to be sure we understand what is really happening.   We must go beneath the surface and analyse the information collected. We must examine the way in which different, possibly conflicting, requirements affect each other.   Then we must communicate our understanding clearly and unambiguously to those who will design and build the software.   We do all of this by building further models, and these must meet several objectives.

- They must contain an overall description of functions.
- They must represent any people, physical things and concepts that are important to the analyst's understanding of what is going on in the application domain.
- They must show connections and interactions among these people, things and concepts.
- They must show the business situation in enough detail to evaluate possible designs.
- Ideally, they should also be organized in such a way that they will be useful later for designing the software.

In fact, we can reverse the last point to give a general measure of usefulness for a requirements model.   The software developed to help meet a need should have a structure that reflects that of the situation in which the need arises.   Thus our description of the application domain must include its conceptual structure.   The UML class diagram is designed to do just this.   It communicates an analysis of requirements in such a way that the model's structure can later be directly translated into software components.

6.2.1 What is a model?

Like any map, models represent something else.   They are useful in several different ways, precisely because they differ from the things that they represent.

- A model is quicker and easier to build.
- A model can be used in simulations, to learn more about the thing it represents.
- A model can evolve as we learn more about a task or problem.
- We can choose which details to represent in a model, and which to ignore.
- A model can represent real or imaginary things from any domain.

Many different kinds of thing can be modelled.   Civil engineers model bridges, city planners model traffic flow, economists model the effects of government policy and composers model their music.   This book is a model of the activity of object-oriented analysis and design.

A useful model has just the right amount of detail and structure, and represents only what is important for the task at hand.   This point was not well understood by at least one character in *The Restaurant at the End of the Universe* by Douglas Adams (1980).   Here, a group of space colonists are trying to re-invent things they need after crash-landing on a strange planet.

'And the wheel,' said the Captain, 'What about this wheel thingy?   It sounds a terribly interesting project.'
'Ah,' said the marketing girl, 'Well, we're having a little difficulty there.'
'Difficulty?' exclaimed Ford.   'Difficulty?   What do you mean, difficulty?   It's

the single simplest machine in the entire Universe!'
The marketing girl soured him with a look.
'Alright, Mr. Wiseguy,' she said, 'you're so clever, you tell us what colour it should be.'

Real projects do get bogged down in this kind of unnecessary detail, if insufficient care is taken to exclude irrelevant considerations (though this example is a little extreme).

What IS developers must usually model is a complex situation, usually within a human activity system. We may need to model what different stakeholders think about the situation, so our models need to be rich in meaning. We must represent functional and non-functional requirements (see Section 5.2.2). The whole requirements model must be accurate, complete and unambiguous. Without this, the work of designers and programmers later in the project would be much more difficult. At the same time, it must not include premature decisions about how the new system is going to fulfil its users' requests, otherwise designers and programmers may later find their freedom of action too restricted. Most systems development models today are in the form of diagrams, with supporting textual descriptions and logical or mathematical specifications of processes and data[1].

Each use case for the Agate case study is a model of a system function seen from a user's perspective, for example `Calculate Staff Bonuses`. If information is also available about a non-functional requirement related to a use case, this can be recorded in the use case description. For example, staff bonuses must be calculated each month before the deadline for providing inputs to the payroll system. This illustrates one way in which requirements may be inter-related.

6.3 Fundamental Techniques

6.3.1 Identifying classes

The class diagram is fundamental to object-oriented analysis. Through successive iterations, it provides both a high-level basis for systems architecture, and a low-level basis for the allocation of data and behaviour to individual classes and object instances, and ultimately for the design of the program code that implements the system. It is important to identify classes correctly. However, given the cyclic nature of the object-oriented life cycle, it is not essential to get this right on the first attempt.

On a first pass, classes can be identified from the use case descriptions in the data dictionary. Depending on how these were written, and what was known at the time, they may contain a mixture of classes and instances. Sometimes the analyst can easily recognize a class before being aware of any particular instances. At other times a class is identified only through recognition of similarities and differences among object instances. Both

---

[1] Some approaches rely primarily on formal logic techniques and rigorous mathematical specification. These are most often applied to real-time and safety-critical systems, such as those that control aircraft in flight or manage nuclear power plants, and are not covered in this book.

approaches are acceptable. In fact, the UML notation for an instance is similar to that for a class, and classes and instances can appear in the same diagram.

Some things and concepts important to understanding an application domain can be found quite easily by identifying all nouns and noun phrases in the use case descriptions. Let us apply this to the Agate case study. The descriptions for two use cases introduced in Section 5.6.2 are repeated below, with nouns underlined.

Use case description for `Check Campaign Budget`.

Checks whether the <u>budget</u> for a <u>campaign</u> is likely to be exceeded by the <u>total cost</u> of all the <u>adverts</u> that make it up.

Use case description for `Assign Staff to Work on a Campaign`.

The <u>Campaign Manager</u> selects a particular <u>campaign</u>. <u>Staff</u> not already working on that campaign are displayed, and he or she selects those to be added to a <u>list of staff</u> working on the <u>campaign</u>.

Though far from a complete analysis, this is already a fair start at identifying some important elements of a requirements class model. Many (perhaps not all) of these will be represented by classes in the final system. In the longer term, classes are only retained if they help to explain the nature or structure of the application domain, so not everything in this list will necessarily remain a class. We will see later how to distinguish classes from other model elements. But we can say confidently that all elements in this list are necessary to an understanding of the work of a Campaign Manager. Figure 6.1 shows the provisional classes entered into SELECT Enterprise.

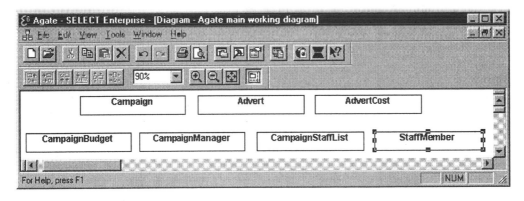

Figure 6.1 *Beginning a class diagram for Agate in SELECT Enterprise.*

Each class is shown as a rectangular box. Class names are written in the singular, although we know there are many staff, campaigns, adverts, etc. This convention reinforces the view of a class as a *descriptor* for a collection of objects, rather than the collection itself. Another convention (derived from object-oriented programming style) is that most names are written in lower case, although classes are capitalized at the beginning of the name. Multiple words are run together, punctuated by upper case letters at the start of each new word to improve readability, for example, `CampaignStaffList`. Some names have been clarified, for example, `AdvertCost` instead of just 'cost'.

In this simplest form, each class has only a name, with all other details to be added later. The diagram will eventually grow quite complex, but at this stage it is little more concise or unambiguous than the text description. Nevertheless it is a start at abstracting the useful details from a description that may be cluttered with many irrelevant facts. As more detail is added, a class diagram will soon be much easier to absorb and less ambiguous than the corresponding text—subject, of course, to the modeller's skill and comprehension. In real-life, even first-cut models are often more complicated than this, and a great deal more effort is needed to arrive at a preliminary understanding.

6.3.2 Relative stability of classes and instances

Classes often represent the more permanent aspects of an application domain. For example, as long as Agate continues to operate in the advertising business, its business activities are likely to involve campaigns, budgets and adverts, creative staff and campaign managers. The description of each class (its data and behaviour—i.e. what it knows, and what it can do) is also relatively stable, and will probably not change frequently.

By contrast, object instances often change frequently, reflecting the need for the system to maintain an up-to-date picture of a dynamic business environment. Instances are subject to three main types of change during system execution. Firstly, they are created. For example, when Agate undertakes a new campaign, details are stored in a new `Campaign` object. When a new member of staff is recruited, a corresponding `StaffMember` object is created.

Secondly, they can be destroyed. For example, after a campaign is completed and all invoices are paid, eventually there comes a time when it is no longer interesting to the company[2]. All information relating to the campaign is then deleted by destroying the relevant `Campaign` instance.

Finally, they can be updated, which means a change to the recorded values of one or more characteristics. This is typically done so to keep each object in step with the thing that it represents. For example, a client may increase the budget for a campaign, in order to cover a longer run of a TV commercial than was originally planned. To reflect this, the budget value set in the corresponding `Campaign` object must also be changed.

6.3.3 Attributes

Attributes are part of the essential description of a class. They belong to the class, but not in the same way that objects do. At the level of a class, attributes are the common structure of 'what a member of the class can know'. Each object has its own, possibly unique, value for each attribute.

In Chapter 4, we identified some possible attributes of `Client` and `StaffMember`.

---

[2] For example, in the UK some financial records must be kept for seven years after the end of the tax year to which they apply, before they may safely be destroyed.

These are shown in Figure 6.2. Note that the symbol for each class is now sub-divided into three compartments. The top compartment contains the class name while the second contains the attribute names (which begin with a lower case letter). The third compartment remains empty for the moment.

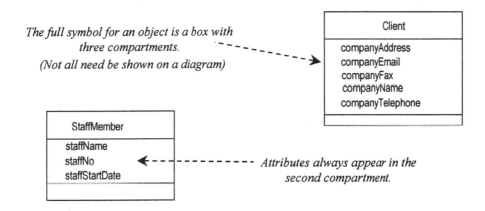

Figure 6.2 *Including attributes on a class diagram.*

In much the same way that a class is partly defined by its attribute structure, instances are described by the values of their attributes. To describe a client we give an appropriate value to each attribute. 'FoodCo' is the value given to the company name attribute for the instance of Client that represents the real-world client FoodCo. To describe an instance completely, we give a value to all its attributes, as in Figure 6.3.

Figure 6.3 *Including attribute values on an instance diagram.*

Some attribute values change during the life of an object. For example, FoodCo may change their email address, their telephone number or their address. They may even decide to change the company's name. In each case, the corresponding attribute value should be updated in the object that represents FoodCo. Other attribute values do not change. For example, a client reference number may be assigned to each client. As long as this is correctly recorded in the first place, there is probably no need for its value ever to change.

6.3.4 Attributes and state

We are now in a position to introduce the idea of object state. The current state of an object is partly described by the instance values of its attributes. When an attribute value changes, the object itself changes state. Some state changes are not significant, which simply means they do not affect the overall behaviour of the system (i.e. the way that it responds to events). Others can have important implications for system behaviour. An example that may be familiar (to some people!) is when your cash card will not allow further cash withdrawals because you have reached your weekly limit. We can understand this in terms of object states, as follows.

Within the bank's computer system, an object `yourAccount` has an attribute `withdrawnDuringWeek` that stores your total withdrawals over the last week, and states `CanWithdraw` and `WithdrawalBarred` (this is a greatly simplified picture—a real bank system is much more complex). Each week, the value of `withdrawnDuringWeek` is reset to zero. The object is in its `CanWithdraw` state. Each time you withdraw cash, `withdrawnDuringWeek` is updated to take account of the transaction. Once the value of `withdrawnDuringWeek` reaches a critical threshold set by the bank, `yourAccount` changes state to `WithdrawalBarred`. Figure 6.4 shows this as a UML state transition diagram.

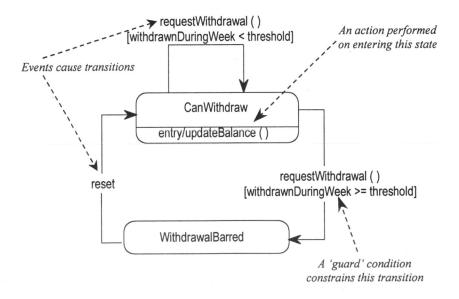

Figure 6.4 *Some states and transitions for a bank account.*

This is not intended as a presentation of state transition diagrams (which are introduced in Chapter 10), but the notation is useful to illustrate what can happen when an object changes state. At the instance level this is merely an update to one attribute value, yet the consequences extend beyond the boundaries of the software system into the user's daily life—as some bank customers know from bitter experience.

6.3.5 Link and association

The concepts of link and association, like those of object and class, are very closely related. A link is a logical connection between two or more objects[3]. In requirements analysis, this expresses a logical relationship in the application domain. An example for Agate is the connection between FoodCo and the 'World Tradition' TV campaign, described by: 'FoodCo is the client for the World Tradition campaign.' This is shown in Figure 6.5.

Linked instances may be from different classes (as with :Client[4] and :Campaign) or from the same class. An example of the latter might be the link supervises between a manager and another staff member who are both instances of StaffMember. Although this is uncommon, a link can even connect an instance to itself. An example might arise for the captain and players in a hockey team where the captain selects the players. Assuming that the captain is also a player, she may have a link to herself—for example 'the captain selects herself for the team'.

Just as a link connects two instances, an association connects two classes, and just as a class describes a set of similar instances, an association describes a set of similar links (links are called 'association instances' by some authors). The similarity is that those links that exist are all between objects of the same classes. An association is a connection between two classes that represents the possibility of their instances participating in a link.

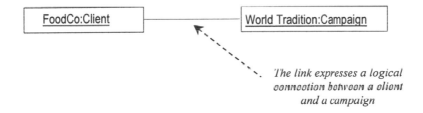

Figure 6.5 *A link between instances.*

Some associations can be recognized easily without first being aware of any specific links. For example, it may be obvious that there is an association between Client and Campaign. Staff at Agate need only record information about clients because they have won (or hope to win) business in the form of campaigns, and it is very unlikely that a campaign would be undertaken except on behalf of a specific client. Other associations are identified only through the existence of their links. There is no clear example of this in the Agate case study, but the rule is that wherever a link exists between two objects, there must always be a corresponding association between their classes.

Like objects, links are not often modelled explicitly at this stage. But they can be useful to check that a model correctly represents users' understanding of the application domain. For example, an analyst may ask for the names of specific staff who are contacts for specific clients in order to confirm some of the characteristics of an association.

---

[3] In most modelling situations, a link connects only two instances, as here. Rarely, a link connects three or more instances. This is not described here.

[4] This form of the notation refers to an anonymous instance of Client—i.e. *any* client.

6.3.6 Significance of associations

While a link between two objects represents a real-world connection, an association between two classes represents the possibility of links. For example, at Agate a member of staff is assigned to each client as a staff contact. This can be mirrored by links between each `Client` and a corresponding `StaffMember`, but a link is modelled only if it supports a specific requirement. We know from the Agate use cases that Campaign Managers need to be able to assign and change a client contact. Therefore the requirements model must permit this link, otherwise it will not be possible to design software that meets these needs.

Figure 6.6 shows some links, but this is not a very economic way of modelling this. To show every link would be unnecessarily complex. There may be many staff and hundreds of clients, not to mention many thousands of links between instances of other classes.

Figure 6.6 *Links between instances of* `StaffMember` *and* `Client`.

Figure 6.7 shows an association that includes all possible `staffContact` links. Rather than modelling every `staffContact` link, this models a general ability to identify a staff contact for each client[5].

Figure 6.7 *Association between* `StaffMember` *and* `Client`.

---

[5] But knowing that two classes have an association does not tell us which instances are linked, if any. An association is abstract and general, not particular.

An association therefore represents constraints on business interactions.

6.3.7 Associations and state

We can now return to our discussion of object state. In Section 6.3.4 we saw that state is partly defined by an object's current attribute values. The second aspect of an object's state is defined by its current set of links to other objects. Whenever a link to another object is created or destroyed, the object changes state. As with state changes that depend on attribute updates, some of these are critical, and some are less so. An example of link creation is when a Campaign Manager assigns a member of staff to be staff contact for a new client. If, instead, an existing staff contact is changed, this involves destroying one link and creating another in its place.

6.3.8 Multiplicity

Associations represent the possible existence of links between objects. It is often important to define a limit on the number of links with objects of another specific class in which a single object can participate. Multiplicity is the term used to describe constraints on the number of participating objects. Multiplicity reflects *enterprise* (or *business*) *rules*, which are constraints on the way that business activities can take place.

A familiar example is the number of people allowed to be the designated account-holder for a bank account. A sole account has *one and only one* account-holder, a joint account may have *exactly two* account-holders and a business partnership account may have *two or more* account-holders. It is important to model these constraints correctly, as they may determine which operations will be permitted by the software. A badly specified system might incorrectly allow an unauthorized second person to draw money from a sole account. Alternatively it might prevent a legitimate customer from being able to draw money from a joint account. In practice, the multiplicity of an association defines upper and lower limits on the number of other instances to which any one object may be linked.

As we saw in the previous section, it is not usually necessary for a requirements model to show the exact names of the members of staff who are staff contacts for every client. This is inappropriate, not only because it may make the diagram so complex it would become unreadable, but also because links may change frequently. By the time we reached the stage of designing our software, such a detailed model would already be out of date. But a requirements model should be able to tell us something about the variety of ways that individual instances can be linked, and in particular any constraints that apply.

We already know that in the Agate case study, each client has one member of staff assigned as staffContact, while each member of staff may be assigned to one or more clients, or none at all. This is modelled by adding a little more information to the simple class diagram in Figure 6.7.

Figure 6.8 shows this, with each end of the association now qualified by its multiplicity. Multiplicity must be read separately in the opposite directions for each association. Thus the possible number of clients allocated to a staff member ranges from 'zero' to 'any number', while the possible number of staff members allocated to a client must be exactly one[6]. To be useful, the notation should be able to show all multiplicities, not just 'exactly

---

[6] This kind of knowledge always emerges, directly or indirectly, from users.

one' and 'zero or more', and Figures 6.9–6.11 show some alternatives.

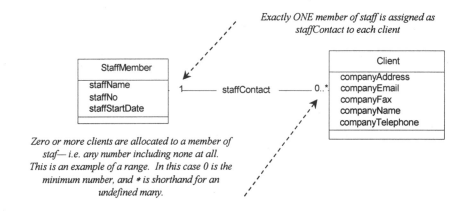

Figure 6.8 *Multiplicity of the association between client and staff.*

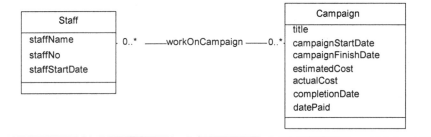

Figure 6.9 *A staff member may work on zero, one or more than one campaign, and a campaign may have zero, one or more staff assigned to work on it.*

Figure 6.10 *Every* StaffMember *must be the contact for exactly one* Client, *and a* Client *may have no contact, or one or more.*

Figure 6.11 *Every* `StaffMember` *must be the contact for one or more* `Clients`, *and a* `Client` *may have no contact, or one or more.*

These examples are not exhaustive, and there are many variations on the form, allowing any range of values to be specified, for example 0..3, 1..5, 2..10, 3..*, or discrete values, such as 3, 5 or 19, or combinations of the two, for example '1,3,7..*'.

To summarize, association multiplicity conveys important information about the structure of the problem domain. Different assumptions about multiplicity have significant effects on the software design. If association multiplicities are modelled incorrectly, this may later make it impossible for the software to do things that users want it to do.

6.3.9 Operations

Operations are the elements of common behaviour shared by all instances of a class. They are actions that can be carried out by, or on, an object. The classes modelled during requirements analysis represent real-world things and concepts, so their operations can be said to represent aspects of the behaviour of the same things and concepts. However, as the basic idea of an object-oriented system is that it should consist of independent, collaborating objects, it is probably easier to understand operations as aspects of behaviour required to *simulate* the way that the application domain works (see Box 6.1). Another way of putting this is that operations are services that objects may be asked to perform by other objects. For example, in the Agate case study, `StaffMember` has an operation that calculates the amount of bonus pay due for a staff member. And, since staff bonus is partly based on the profit of each campaign a member of staff has worked on, `Campaign` has an operation to calculate the profit for each campaign.

Box 6.1 Operations as responsibilities

There is a world of difference between saying, on the one hand, that a`StaffMember` has an operation to calculate its own bonus, and, on the other hand, saying that a real member of staff has responsibility for calculating her own bonus. The former is an appropriate way of simulating the real-world operation, but there is certainly no implication of identity. Nor do we wish to imply that real (but abstract) campaigns are capable of calculating their own profit—or anything else, for that matter. What *is* implied is, firstly, that the ability to carry out these tasks is a requirement of the system, and secondly, that we must make a judgement about where to locate these operations within the model, and our judgement should be based on the idea of object responsibilities. Chapter 8 introduces CRC cards and interaction diagrams, the techniques that are most widely used for allocating responsibilities to classes in a coherent manner.

An operation is a specification for some aspect of the behaviour of a class. Operations are thus defined for the whole class, and are valid for every instance of the class. Figure 6.12 shows some examples of operations for the Agate case study. At this stage we do not need to be concerned with the details of how each operation will work, nor with when they are permitted to work or constrained from working. But we do need to know which operations to include in the requirements model, and we need to make a first guess at their appropriate location within classes. Operation names are always placed in the third compartment of the rectangular class symbol. As for attributes, operations are written beginning with a lower case letter. There is no separate notation for showing the operations of an object instance. In this respect, operations are unlike attributes and associations, in that they have exactly the same significance for instances as they do for classes.

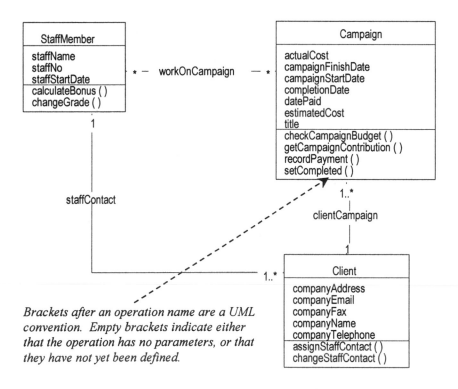

Figure 6.12 *Partial class diagram for Agate, with some attributes and operations.*

Operations are eventually implemented by *methods*, and what a method actually does on any given occasion may be constrained by the value of object attributes and links when the method is invoked. In the bank cash card example described earlier, yourAccount has operations makeWithdrawal() and the attribute withdrawnThisWeek (written briefly as yourAccount.makeWithdrawal() and yourAccount.withdrawnThisWeek). The value of yourAccount.withdrawnThisWeek determines how the operation responds. If your limit has not yet been reached, the withdrawal is made. Otherwise, the operation may cause a message to be returned that says you have exceeded your limit. Alternative courses of action, and the basis for choosing between them, are included in the detailed definition of an operation (we describe this in Chapter 7).

The effects that an operation can have include changing the characteristics of an object, for example, updating an attribute value. Another effect may be to change an object's links with other objects, for example assigning :StaffMember to work on :Campaign. The creation of this particular link is necessary for that member of staff to earn a bonus from the campaign profits. Some operations delegate part of their work to other objects: aStaffMember.calculateBonus() provides an example, as this operation must call aCampaign.getCampaignContribution() for each associated campaign before it can arrive at an answer. Thus each individual operation may represent only a small part of a larger task that is distributed among many objects. A relatively even distribution of computational effort is regarded as highly desirable—it is a large step on the way to building a well-modularized system.

6.3.10 Operations and state

Perhaps it is already evident that there is a relationship between an object's operations and its states. An object can only change its state through the execution of an operation. In fact, we are saying no more here than that attributes cannot store or update their own values, and links cannot make or break themselves. But this illustrates an important point about the way that object encapsulation assists modularity in a system. The services that an object provides can only be accessed through the object's interface, which consists entirely of operation signatures defined at the class level. In order to get an object to do anything at all—change its data, create or destroy links, even respond to simple queries, another object must send a message that includes a valid call on an operation.

6.4 The Data Dictionary

The arguments for the use of diagrams in requirements analysis include the abstraction of essential information for the task at hand and the avoidance of ambiguity. However, some information 'hidden' by the analyst (in the sense that it does not appear in her diagrams) because it does not assist with understanding the initial requirements at an appropriate level of detail, will nevertheless be useful for other tasks. A common example of the kind of information usually hidden at this stage is the detailed logic of an operation. This may include the sequence of its sub-steps, the criteria for any choices or decisions made, and the circumstances in which either the whole operation or part of it may be repeated. While not shown on a class diagram, this information becomes vital during the design and programming stages of the project. In the event that this information is captured during requirements analysis (which it often is) we need a place to keep it safely until it is needed.

It is also often the case that no diagram can adequately remove all ambiguity from the description of the requirements. An example from the Agate case study might be how to distinguish a designated staff contact from another member of staff who happens to have some contact with the client. There may be circumstances where a precise definition of the role of a staff contact may be important, particularly where there are legal, contractual or financial implications. We need a place where this precise definition can be stored for later reference.

Both these needs are met by the use of a data dictionary. A data dictionary contains a precise, authoritative definition of each element in the models developed during an entire project life cycle. These elements will include most of the model components described in the preceding sections of this chapter. In particular, classes, associations, attributes and operations will all be defined in the data dictionary. Normally an entry is created for a new element as soon as it is added to a model, although the details are often left blank until later.

Nowadays most data dictionaries are maintained automatically by a CASE tool. Figure 6.13 illustrates this, and shows three views of the operation properties tabbed dialogue in SELECT Enterprise, the CASE tool used to prepare some of the diagrams for this book.

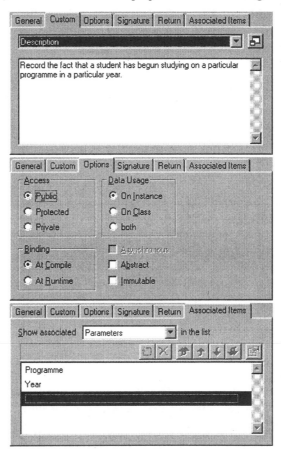

Figure 6.13 *Views of the operation properties tabbed dialogue in SELECT Enterprise.*

SELECT Enterprise provides typical data dictionary features. When a new element is added to a diagram, it automatically creates a new (empty) data dictionary entry for that element. It also automatically records all logical connections between elements—for example, the class to which an attribute or operation belongs—and gives each type of element its own type of data dictionary entry, allowing the analyst to browse, say, all attribute definitions or all class definitions together.

A CASE tool monitors modifications to stored models, automatically adjusting the data dictionary entries in line with changes to the diagrams. In many cases a CASE tool can warn the analyst that a change (perhaps the removal of an element from a diagram) may introduce an inconsistency into a set of diagrams, although it will not usually prevent this, as a model that is under development may contain many temporary inconsistencies. The analyst's part in this is to record a range of optional information for each element. Note that much of this is 'optional' only in the sense that it may not be required now, but will definitely be needed at a later stage.

If no information is available at the time, the CASE tool allows the analyst to leave this entry in the data dictionary completely blank, other than the operation name, which is obligatory. But anything that is recorded will be in a format useful for other tasks later on, such as designing, programming or testing the operation. In this way the data dictionary helps ensure that necessary information is recorded in an appropriate format, whenever it becomes available.

Similar Editors are available for classes, attributes and associations. These control the structure of information that can be recorded for each element. For most, the minimum information required is a unique name. The CASE tool also records the class to which attributes and operations belong, deducing this automatically from the diagram so that it need not be entered by the analyst.

The data dictionary provides a single shared resource where virtually all project information of any importance for the developers can be stored, for retrieval at any time, by any member of the project team authorized to work with that information. And since modern CASE tools allow project information to be accessed over a network, team members do not necessarily need to be working at the same site.

6.5 Drawing a Class Diagram

In this section we look at some practical aspects of drawing a class diagram. This involves suggesting where to look for the necessary information, and also a recommended sequence for carrying out the various tasks. It should be stressed that neither the information sources nor the sequence of tasks are intended to be prescriptive. Nor do we cover every eventuality on every project. In practice, one project differs greatly from another, and some steps may be omitted altogether or carried out at a completely different stage of the life cycle. Experienced analysts will always use their own judgement on how to proceed in a given situation.

6.5.1 Finding objects from use cases

The use cases are usually the first place to look for objects. You should search each use case name and description for all nouns and noun-phrases. Figure 6.14 suggests some categories to look for, together with examples of each category. These are a checklist to help you think of all the different classes you may need to model. Note that many examples appear in more than one category. This illustrates that it is not important how you interpret a category. What counts is that you find all the relevant classes for the system you are modelling.

To start with, it is best to write or type your potential classes as a rough list. You will probably want to add and cross off many items before the list becomes reasonably stable. Once you have an initial list, write a brief description for each item. This becomes part of your data dictionary once you have refined the list. Check through your list carefully. Even

Category	Examples
People.	Mr Harmsworth (Marketing Manager at FoodCo), Dilip (an Agate copywriter).
Organizations	Jones & Co (a forklift truck distributor), the Soong Motor Company, Agate's Creative Department.
Structures.	Team, project, campaign, assembly.
Physical things.	Fork-lift truck, electric drill, tube of toothpaste.
Abstractions of people.	Employee, supervisor, customer, client.
Abstractions of physical things.	Wheeled vehicle, hand tool, retail goods.
Conceptual things	Campaign, employee, rule, team, project, customer, qualification.
Enduring relationships between members of other categories.	Sale, purchase, contract, campaign, agreement, assembly, employment.

Figure 6.14 *Looking for objects.*

the most experienced analyst will probably have included some potential classes that might cause confusion later if they are retained in the model.

Beyond the scope of the system?

You may have included people, things or concepts that are not strictly necessary to describe the application domain that you are investigating. Remove any that you identify from your list. These may become clear from the use case descriptions, but don't worry if the odd one slips through. There will be lots of opportunities to catch these later on. Remember, too, that only the users can finally set the system boundary.

A mistake often made by beginners is to include classes to represent the people who operate the current system. This may happen because their names or job titles appear in a use case description. It is rarely necessary to model the operators of the system. The exception is when they are themselves the subjects of one of the business activities. An example of this might be an office worker handling a company's pension scheme, who is also a member of the scheme. In this case, you may need to model them as a member of the scheme (i.e. a potential object), as well as an operator of the system (i.e. an actor). On the whole, however, actors should not be represented as objects.

Refers to the system as a whole?

You may have included an item that refers to the system you are modelling, or to the department or organization that contains it. It is not usually necessary for a model to contain classes that represent the entire system.

Duplicates another class?

You may have included two items that are really synonyms. If you are not sure, check with your users exactly what they understand by each item on the list. This should become clearer as you write the descriptions for each class.

Too vague?

Eliminate any potential classes for which you are unable to write a clear description, unless you are sure this is only because of a temporary lack of information.

Too specific?

It is usually better to model object classes, rather than instances. Think carefully about any items on the list that are unique. For example, a company may currently have only one supplier, tempting you to model the specific supplier. But the supplier might be replaced tomorrow for business reasons. A class named `Supplier` would be unaffected by this, whereas one named after the particular company would require modification.

Too tied up with physical inputs and outputs?

Avoid modelling things that depend closely on the physical way system inputs and outputs are currently handled. For example, printed order forms and phone enquiries may be how the current system operates, but it is much too soon to make a decision on whether they will play the same role in the new system. Try to think of names that express a logical meaning rather than a physical implementation: `Order` and `Enquiry` would be acceptable alternatives.

On the other hand, physical objects that are an essential part of the business activity should be included. This can depend a lot on context: `Truck` may be an acceptable class in a system to co-ordinate vehicles used for parcel deliveries, but irrelevant in another system that records customer payments for the deliveries, even though invoices and payments might travel on the same truck.

Really an attribute?

An attribute is a characteristic of a class. What makes this a problem is that an item that is an attribute in one domain may be a class in another, depending on the requirements. So some items on your potential class list may be better modelled as attributes. The primary test is this: does the item only have meaning as a description or qualification of another item? To illustrate this, we will look at examples that show how the significance of a date can vary between two different application domains.

In the Agate case study, the significance of a staff member's start date is to allow appropriate salary, bonus and grading calculations to be carried out. It would therefore be appropriate to model `staffStartDate` as a single attribute of `StaffMember`. But now consider a weather forecasting agency, keeping daily records of atmospheric conditions, and producing analyses for different weeks, months and years. Each date may be described by many other variables, e.g. maximum, minimum and average temperature, hours of sunshine, total precipitation, average windspeed, etc. These analyses might also require separate attributes for day of the week, month and year. We might then choose to model a `Date` class, with the other variables as its attributes.

Really an operation?

An operation is an action, a responsibility of a class. This can also be confusing, as some actions can be better modelled as classes. It is particularly easy to confuse the two if the use case descriptions are ambiguous. For an example of an action that can be considered as a class, recall the sale transaction described in Section 6.5.1. There are two considerations that might make a sale transaction a class rather than an operation.

- A sale may have characteristics of its own, that would be best modelled as attributes of a class, e.g. value, date, etc.
- There may be a requirement for the system to remember particular sales over a period of time, e.g. in order to respond to warranty claims, or to audit the shop's accounts.

If there is no requirement to record a history made up of individual sales, or to describe sales in terms of their value, date, etc. it may make more sense to model them as an operation of another class, perhaps as `StockItem.sell()`. For each action on your preliminary class list, consider whether these criteria apply. If they do not, it may be an operation rather than a class.

Really an association?

An association is an interaction of some kind between two classes. But this too can be confusing as we may prefer to represent some interactions as classes. The sales transaction can also be counted as an example of this (a sale is both an action and an interaction). How do we decide which interactions to represent as associations, and which as classes? This can sometimes be a difficult and complex problem. You can apply a similar test to those described above for attributes and operations. If an association is something we need to describe in terms of further characteristics—if it is apparent that it has attributes of its own—then it should be modelled as a class. If it only has meaning as a relationship between two classes, leave it as an association.

But the best answer at this stage is not to spend too long on making the distinction. The important thing during requirements analysis is to make sure all significant interactions are modelled, whether as classes or associations. We can review our judgements later when we understand more about the situation. Indeed, following the transition from requirements modelling to software design, it is often the case that certain types of association may be changed into classes, or further classes may be added to help implement the association effectively (this is covered in Chapter 13).

6.5.2 Other sources of objects

Classes may be added to the class diagram at almost any stage of the project life cycle. However, most additional classes will be required by software design, and are not a necessary part of understanding the problem domain. Nevertheless, during the requirements modelling stage, there are other sources of classes for your model.

It is worth reviewing any background documentation gathered during the fact finding stage. A second reading, after an initial attempt at the class diagram, is very likely to find more required classes, due to the fact that you now have a better understanding of the problem.

Ideally, user representatives will be closely involved in discussing and developing the class diagram. Nowadays users often work alongside professional analysts as part of the project team. Most projects are a learning experience for everyone involved, so it is not unusual for users' understanding of their own business activity to grow and develop, and it is likely that users will identify a number of additional classes that were not apparent at first.

Your own intuition is another useful source, together with that of colleagues. And you can look for analysis patterns, but this is a more advanced technique that will not be covered in this chapter.

With experience these can all give guidance, but always check your intuitions with someone who knows the business intimately. If you let yourself be overwhelmed by similarities to other projects, you may overlook important differences. As an analyst, you should remember at all times that users are the experts on the business and what the software system should do. Your role is to make users aware of the possibilities offered by modern information systems, and to help translate their requests into a system that meets as many of their needs as possible.

6.5.3 Adding and locating attributes

An attribute is a characteristic that describes a class. Many attributes will already appear in the use case descriptions. Others will become apparent as you think about your model in more detail. The simple rule is that attributes should be placed in the class they describe. This usually presents few problems. For example, the attributes staffNumber, staffName, and staffStartDate all clearly describe a member of staff, so would be placed in the Staff class.

Sometimes it is more difficult to identify the correct class for an attribute. The attribute may not properly belong to any of the classes you have already identified. An example will help to illustrate this. Consider this extract from an interview with Amarjeet Grewal (Agate Finance Director):

Amarjeet Grewal: Agate's pay structure is based on distinct grades. Directors and managers negotiate their own salaries, but other staff are placed on a grade that determines their basic salary. You can only change grade as a result of an appraisal with your line manager.

A member of staff has one grade at a time, but it sounds like they may have several previous grades, and several members of staff may be on the same grade at the same time. Staff and Grade are probably classes with an association between them.

The basic salary for each grade is fixed, usually for a year at a time. Every year after the final accounts are closed, I review the grade rates with the Managing Director, and we increase them roughly in line with inflation.

A grade has only one rate at a time, though it can change, and each rate has a money value. Grade may have a rate attribute.

If the company has performed well, we increase the rates by more than the rate of inflation. In case there are any queries, either from an employee or from the Tax Office, it is most important that we keep accurate records of every employee's grades; that is, the rates for all present and all past grades, and the dates these came into force.

There's quite a lot in this bit. A grade may have several previous rates, which suggests either that Grade has multiple rate attributes, or that Rate and Grade are distinct classes. If the latter, then `Rate` must have a date attribute, since we need to know when it took effect. We must also record when a member of staff changes to a grade, and possibly also when they change from a grade, which suggests one or two more date attributes. Each grade has a date it came into force—another attribute.

It's actually quite complicated, because you can have an employee who changes to several different grades, one after the other, and then the rate for each grade also changes each year. So, for each employee, I have to be able to tell exactly what grade they were on for every day they have worked for the company, and also what the rate for each grade was when they were on it. This is all quite separate from bonus, which is calculated independently each month. For creative staff, bonus is based on the profits from each campaign they have worked on, and for other staff we use an average profit figure for all campaigns.

This is necessarily tentative, but a preliminary analysis yields the following list of classes and attributes:
- classes: `Staff`, `Grade`, `Rate`;
- attributes: `gradeStartDate`, `gradeFinishDate`, `rateStartDate`, `rateFinishDate`, `rateValue`.

In order to be reasonably lifelike, we can assume some other attributes not given above, such as `employeeName` and `gradeDescription`, and also some other operations, such as `changeGrade` and `setFinishDate`. An initial, though incomplete, class diagram might then look like the one in Figure 6.15.

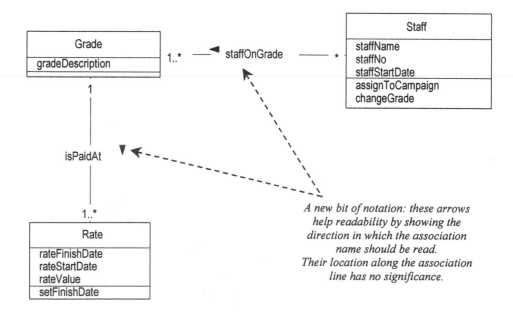

Figure 6.15 *Partially completed Agate class diagram.*

One problem is where to put the attributes `gradeStartDate` and `gradeFinishDate`. These could be placed in `Grade`, but this would commit it to recording multiple start and finish dates. There may be also many members of staff associated with a grade. The computer system must be able to identify the member of staff to which each date applies, so the structure of dates that might need to be stored could grow quite complex. A similar problem occurs if date attributes are placed in `Staff`. The explanation for this difficulty is that these attributes do not describe either a member of staff or a grade in isolation. They only have meaning as a description of the link between a specific member of staff and a specific grade. Thus, the clearest answer is to create an additional class (called an association class) specifically to provide these attributes with a 'home'. This is shown in Figure 6.16.

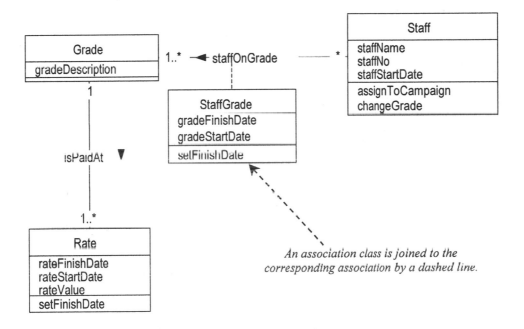

Figure 6.16 *An association class gives a home to attributes that properly belong to a link between two objects.*

Some readers may be familiar with the relational database technique known as normalization, a technique that provides a rigorous guide to placing attributes in tables (or relations) and ensures minimum redundancy of data. The case illustrated is an example of normalization in practice, but a full treatment of the underlying theory is beyond the scope of this book. Normalization is used in object-oriented design (this is discussed in Chapter 17) but on the whole object-oriented approaches concentrate on capturing the structure of the world as perceived by the system's users. Un-normalized relations are therefore acceptable in an object model, provided that they correspond accurately to users' intuitions about how their business activities are organized.

6.5.4 Adding associations

Find associations by considering logical relationships among the classes in the model. Associations may be found in use case descriptions, and other text descriptions of the application domain, as stative verbs (which express a permanent or enduring relationship), or as actions that need to be remembered by the system. 'Customers *are responsible for* the conduct of their account' is an example of the first, while 'purchasers *place* orders' is an example of the second.

But this is not a very reliable way of finding associations. Some will not be mentioned at all, while others may be too easily confused with classes, attributes or operations. A full understanding of the associations in a class model can only be reached later by analysing the interaction between different classes. With practice, the most important ones will be found fairly easily, and for the moment it is not important if some are missed.

6.5.5 Determining multiplicity

Since association multiplicities represent constraints on the way users carry out their business activities, it is important to get these right, and the only way to do this is to question users about each association in turn. This is true even when the existence and character of the association have been inferred from user documents. An analyst should always check what is said in documents, in case it is ambiguous, erroneous, or out of date.

Rosanne Martel: So let me be clear about this. A client must have exactly one staff contact, but a member of staff can be contact for no clients, one client or several clients. Is there an upper limit on that? A campaign must have one client, but a client can have many campaigns. Can you have a client with no campaigns—say, a new client who hasn't given you any business yet?

6.5.6 Finding operations

Operations are really a more detailed breakdown of the high-level system responsibilities already modelled as use cases. An operation can be thought of as a small contribution of one class to achieving the larger task represented by a whole use case. They are sometimes found as action verbs in use case descriptions, but this picture is likely to be fairly incomplete until the interaction between classes has been understood in more depth. Chapter 8 describes how to model class interaction. So, as with associations, do not worry if your first attempt has gaps.

6.5.7 Preliminary allocation of operations

Before attempting to allocate operations to specific classes, it is worth remembering that each class is only a representation of something in the application domain. As an analyst, you are trying to build a logical model that helps to understand the domain, not necessarily a replica that is perfect in every detail. Two guidelines help in deciding which class to locate each operation in, but there is not a single answer—only a satisfactory fit.

(1) Imagine each class as an independent actor, responsible for doing or knowing certain things. For example, we might ask 'What does a staff member need to know or need to be able to do in this system?'

(2) Locate each operation in the same class as the data it needs to update or access. However, this is often problematic, as you may not have identified all the attributes yet.

As a general comment on this stage, the most important thing is not to expect to get things right at the first attempt. You will always need to revise your assumptions and models as your understanding grows.

6.6 Summary

In this chapter we have seen how to build an initial version of the class diagram for a requirements model, basing this mainly on information that can be found within use case descriptions. The important elements of the model at this stage are classes, with their attributes and operations, and associations, which show the relationships between classes, together with any multiplicity constraints. Once this task has been completed to a reasonable first-cut standard, the model embodies a good understanding of the main functional requirements for the system, in terms of its responsibilities for providing services. It has also defined a logical architecture that is a basis for the design work that follows. However, before design can begin, the next stage is to refine the requirements model by identifying any latent generalization that can simplify the structure, and that may suggest opportunities for reusing some of the requirements modelling work that has already been carried out on other projects.

Review Questions

6.1 What are the advantages of models?

6.2 Why are requirements models in IS development neither entirely graphical nor entirely textual?

6.3 Distinguish between attribute and value.

6.4 In what sense are classes generally more stable than their instances, and why is this usually the case?

6.5 Distinguish between link and association.

6.7 What is multiplicity, and why can it be called a constraint?

6.8 What is an operation?

6.9 How are operations related to messages?

6.10 What is an attribute?

6.11 Explain why Section 6.3.7 makes no mention of updating a link when it is changed, but instead only discusses the creation and destruction of links.

Case Study Work, Exercises and Projects

The following transcript gives the first part of an interview that Rosanne Martel conducted with Hari Patel, the Factory Manager in charge of FoodCo's Beechfield Factory. Read this through carefully, and then carry out the exercises that follow.

Rosanne Martel: Hari, for the benefit of the tape, I'd be grateful if you could confirm that you're the manager responsible for all production at Beechfield.

Hari Patel: Yes, that's right.

RM: Good. Now the purpose of this interview is for me to find out about operations on the production lines. Can you tell me how this is organized?

HP: Sure. How much detail do you want?

RM: Can we start with those aspects that are common to all lines? That will give me a general feel, then if there are differences we can go into more detail later.

HP: OK, there are quite a few similarities. First, there are two main grades of shop-floor staff: operatives and supervisors. Different operatives have a range of skills, of course, but that doesn't affect the way the line works.

RM: How many operatives work on a line, and what do they actually do?

HP: There might be anything from around six operatives to over twenty, depending on the product. They really do all the actual work on the line, either by hand or operating a machine. This could be a semi-skilled labourer feeding in the different kinds of lettuce for salad packs, or a more skilled operator running one of the automatic mixing machines. In this factory, unlike Coppice and Watermead, the work is mostly quite unskilled.

RM: How many supervisors are there to each line?

HP: Just one. They are on full-time supervision duties, and they each look after one production line.

RM: Always the same line?

Rosanne is trying to find out what possible classes there are. What else do you think her questions seek to discover?

HP: Well, let's just say nobody has changed line in the last couple of years.

RM: How about the operatives—are they always on the same line too?

HP: No, we swap them around quite a bit. But it doesn't really matter what line an operative works on. They get paid piecework rates depending on the production run, and the rates are based on the job numbers that appear on their timesheets. There's a separate job number for each run.

RM: I'd like a copy of a timesheet please—preferably a real one with some data, if that's all right. We can blot out the name and staff number on the copy for confidentiality.

A sensible request. Real documents with live data are an invaluable source of information. Figure B1.4 shows the timesheet that Rosanne collected.

HP: Sure. Remind me when we finish, and I'll get you one.

RM: Thanks. Now, does one line always produce the same product?

HP: No, that changes from one day to the next. The production planners produce a new schedule every Friday, and this lists all the production runs for each line for the following week.

RM: I'll take a copy of a production schedule too please. So the supervisor finds out on Friday what their line is working on over the next week?

Here Rosanne is checking where the inputs come from, as well as what they contain.

HP: That's right.

RM: Good, I think I've got that clear. Now let's talk about what happens when people come in to work. Do all the lines start up first thing in the morning?

HP: Usually. Production runs generally last for a whole day if possible, or sometimes a half-day. Production Planning try to keep the change-overs simple, so they tend to schedule changes during breaks to avoid wasting productive time.

RM: The lines don't keep running all the time?

HP: No, they stop for coffee and meal breaks.

RM: What role does the line supervisor play in this?

HP: Well, they make sure the lines have enough raw materials, and they deal with minor emergencies. They also monitor output, liaise with production control, keep track of employee absences, and so on.

RM: Can we go through what a supervisor does on a typical run, please, step-by-step?

Another sensible request. Asking someone to go over things again in more detail will often reveal aspects of the situation that are not obvious from a brief description.

HP: First, they make sure everything is ready before the run starts. They check the storage area to see there is enough of each ingredient. If a long run is planned, you don't need all the ingredients ready at the beginning, but there has to be enough to keep the line running smoothly until the next supply drop. They also have to check if the staff allocated to that run have turned up. A line can usually run for a little while with one or two staff missing, but it's best to have everyone there from the start.

RM: How does a supervisor know what ingredients are required, and how many staff?

A good analyst always probes to find out how, what, why, when, where and who.

HP: Every run has a job card, with this information on it. The warehouse gets a copy of the job card too, so in theory they know what supplies to deliver, to which line and when they will be needed.

RM: Does that usually work?

HP: (Laughs). Sometimes!

RM: What if there aren't enough staff?

HP: Sometimes the supervisor can find a spare body on another line. Or they can run the line slower. You can manage with fewer staff if necessary, but productivity is a lot lower.

RM: Let's say the ingredients are all ready, and all the staff are there waiting to go. What next?

HP: The supervisor switches on the line, and then it's mostly troubleshooting and paperwork.

RM: What does the paperwork involve?

HP: Well, they start by taking the names of all the staff at the start of the run. They copy the job number from the job card to the production record sheet and all the timesheets. If it is the first time that operative has worked that week, then the supervisor makes out a new timesheet. When they start the line, they note the time on the production record sheet. Then they keep a rough note of anyone who leaves the line during a run, and how long they're absent.

RM: What kind of problems does the supervisor deal with?

HP: The main problem is if something goes wrong with the run. Say the line breaks down. They would have to call in maintenance, record the downtime while the line's not running, and try to find useful things for the staff to do while they're waiting for it to be repaired. If an ingredient runs out this could also halt the line, and might mean chasing the warehouse, or contacting the farm or an outside supplier. Sometimes people go missing, or leave early because they're sick. The supervisor has to find a replacement as quickly as possible.

RM: Right, now let's go to the end of a run. What information is formally recorded, and by whom?

HP: First the supervisor notes the finish time on the production record sheet.

RM: I'll have one of those too please.

HP: OK, no problem.

Figure B1.3 shows a blank production record sheet.

HP: Next the supervisor phones for someone to come over from Production Control to verify the quantity produced and note this on the production record sheet. Then the supervisor totals all the absences, because if anyone has more than 15 minutes absence, it's deducted from their total unless they have a good reason, say a medical certificate. Then they work out the total hours for each operative. If someone joined the line in mid-session they might not have a timesheet. First the supervisor notes the finish time on the production record sheet., so one is made out now and their hours are added in. By the time all that has been done, Production Control has usually checked out the total quantity produced, and this goes on the production record sheet. After that, it is just returning unused ingredients to the warehouse, tidying up the line ready for the next run, that kind of thing.

RM: Thanks, that was really helpful. Now I'd like to ask about how the piecework formula works. Can you tell me what the calculation is?

HP: To be honest, I can never remember the exact formula. You'd do better asking a supervisor or someone from payroll...

Now carry out the following exercises, based on the information given in the interview transcript.

6.A   Write descriptions for the following use cases:
- Start Line Run
- Record Employee Joining The Line
- Record Employee Leaving The Line
- Stop Line
- Record Line Problem
- End Line Run

6.B   From your use case descriptions, identify potential classes and associations.

6.C   Produce a draft class diagram, initially showing only classes and associations.

6.D   Add to your class diagram any attributes and operations that you can identify from the interview transcript. Make reasonable assumptions and add others that you think would be justified.

Further Reading

Larman (1998) describes the use of UML in object-oriented requirements modelling. Larman's approach is very different from the one taken in this book.

From a historical perspective, Yourdon (1989) is a good introduction to structured analysis and design. This book was written sufficiently late in the period that the author is able to give a balanced view of the thinking that led to the rise of structured methodologies. In retrospect, it is interesting to note that most of the aims (modularity, etc.) were little different from the contemporary goals of object-oriented approaches. It may also add to the reader's appreciation to know that, while Yourdon was one of the highest-profile 'gurus' of structured analysis and design, he later became a leading proponent of object-oriented analysis and design.

7

Refining the Requirements Model

Objectives
In this chapter you will learn:
- what is meant by a component;
- how generalization and aggregation help develop reusable components;
- how to identify generalization and composition;
- how to model generalization and composition.

7.1 Introduction

Once the requirements for a new system have been established in a basic diagram with classes and associations, the next step is to refine the model, with a particular view to creating the conditions for developing components that can be reused either within the current project, or on future projects. In this chapter we introduce the concept of a component and discuss the role of abstraction in an object-oriented approach to identifying and building reusable components (Section 7.2). The principal abstraction mechanisms used for this purpose are the class hierarchy, which we have already met in Chapter 4, and a very different type of abstraction known as composition, which we encounter for the first time in this chapter. We show how these can be used to encourage reuse (Section 7.3). In practice, most reuse actually occurs later in the life cycle than requirements analysis, so the benefits are often not gained directly by analysts. This is a situation that requires careful management, and we discuss this in later chapters. Here we concentrate on laying the groundwork. Over the last few years, the 'Patterns' movement has offered new ways of capturing and communicating generalizable knowledge at many stages of the life cycle, particularly during the analysis and design activities. We introduce analysis patterns briefly at the end of this chapter (Section 7.4).

7.2 Component-based Development

It is quite hard nowadays to think of an industry where the use of standard components, and even standard designs, is not common. To take a familiar example, an architect may design many different houses, each unique in its own particular ways. But it would be very unusual to design every component from scratch. The architect typically chooses components from a catalogue, each house being assembled from a standard set of bricks, roofing timbers, tiles, doors, window frames, electrical circuits, water pipes, carpet tiles, paint, etc. Thus, while one house may be completely unlike another in overall appearance, floorplan and number of rooms, the differences between them lie in the way standard components have been assembled.

In this way a house can be designed to meet the needs of an individual customer, without the architect needing to think about 'low-level' details, such as which clay to use for the bricks, how long they should be fired in the kiln, or at what temperature. These problems have already been understood and solved by the brick manufacturer and can therefore be ignored. The architect need concentrate only on those aspects of the bricks relevant to the overall design of the house, such as colour, texture and size. In fact, many aspects of a brick's design, such as its precise ingredients and details of the manufacturing process, will not be described at all in the catalogue. In effect, these details are hidden from the architect—in other words, abstracted out of the description. Other aspects relevant to the architect, such as colour and size, are described in the catalogue.

7.2.1 Software components: the problems

It sometimes seems that many information systems development organizations believe that the analysis of requirements should begin from scratch on every new project. In one sense, this is necessary, since at first we know nothing about what the requirements for a new system are. It is also advantageous if it encourages analysts to take account of the unique characteristics of the proposed system and its environment. But starting from a position of knowing nothing can also have significant disadvantages. Effort may be wasted on finding new solutions to old problems that have already been adequately solved by others in the past. While it is important that real differences between two projects should not become blurred, this should not prevent the team from capitalizing on successful past work, provided that it is relevant to the current problem.

In the majority of cases, those analysts, designers and programmers who seem unnecessarily to reinvent the wheel do not do so deliberately. Good professionals have always tried to learn as much as possible from experience, both their own and that of their colleagues. An example of this is the long-established use of code libraries such as .DLL (Dynamic Link Library) files in Microsoft Windows. For many years, programmers have built up extensive libraries that range from personal collections of useful sub-routines, to commercially distributed products that contain large numbers of industry-standard components. So why do some software developers carry on reinventing so many wheels? One reason for this is the NIH ('Not Invented Here') syndrome, which mainly seems to afflict programmers. Another reason is the functionally-based decomposition of the modelling techniques in structured analysis, which primarily affects analysts, and to a lesser extent designers.

The NIH syndrome

In spite of all the library resources available today, some professional programmers (even occasionally whole departments) still fall prey to the NIH syndrome. This is the attitude of one who thinks: 'I don't trust other people's widgets, even those that appear to work, suit my purpose and are affordable—I want to invent my own widgets anyway.' This is understandable in someone who enjoys a technical challenge, or has reasons not to trust the work of others, but it usually does not make good commercial sense. After all, if you want a new lightbulb for your room, it does not seem very sensible to invent and build your own. Even if the knowledge and equipment are available to do so, the cost would be prohibitive.

One remedy can be found in object-orientation, partly because of the different attitude to program development it engenders, but also partly because object-orientation actually makes it easier to use library components.

Model organization

Analysts suffer from the NIH syndrome too, but the biggest obstacle to reuse of successful analysis work has been the way that structured models are organized. It is difficult enough to create a model that is useful at other stages of one single project. It is even harder to create a structured model that is useful on a completely different project. This is partly because structured models are partitioned according to functions, which are a particularly volatile aspect of a business domain. There is also little in structured analysis models such as data flow diagrams that explicitly encourages encapsulation. This is an area where object-orientation can make a distinctive contribution to the reuse of requirements analysis.

7.2.2 How object-orientation contributes to reuse

Object-oriented software development tackles the problem of achieving reuse in ways that resemble the practice of other industries that use standard components. The aim is to develop components that are easy to use in systems for which they were not specifically developed. Ideally, software analysts should be in the position described for an architect, free to think about how their client intends to use the system, without needing to worrying about how individual components are built. The productivity gains are potentially enormous.

One of the keys is the encapsulation of internal details of components, so that other components requesting their service need not know how the request will be met. This allows different parts of the software to be effectively isolated in operation and greatly reduces the problems in getting different sub-systems to interact with each other, even when the sub-systems have been developed at different times or in different languages, and even when they execute on different hardware platforms. Sub-systems that have been constructed in this way are said to be *de-coupled* from each other. Later in the book we describe an application of this approach that would enable one sub-system to interact with others over a network (the Agate security sub-system described in Section 19.4).

The effect can be scaled up to the level of complex sub-systems, by applying the same principle of encapsulation to larger groups of objects. Any part of a software system—or, by extension, a model of one—can be considered for reuse in other contexts, provided certain criteria are met.

- A component should meet a clear-cut, but general need (in other words, it delivers a coherent service).
- A component should have a simple, well-defined external interface.

In theory, reusable components can be designed within any development approach, but object-orientation is particularly suited to the task. Well-chosen objects meet both of the criteria above, since an object requesting a service need only know the message protocol, and the identity of an object that can provide it[1]. Another important aspect of object-orientation is the way that models, and hence also code, are organized. For example, Coleman et al. (1994) point out that generalization hierarchies are a very practical way of organizing a catalogue of components. This is because they encourage the searcher to begin first with a general category, then progressively refine their search to more and more specialized levels. This mirrors the way that the Dewey decimal system is used to catalogue books in a library.

Requirements reuse

When we write of a 'reusable requirement', we really mean a reusable *model* of a requirement. To date, this is one of the least developed areas of software reuse. In any case, only parts of any model are likely to be reusable. But the key to reuse in requirements modelling is that models are organized so that they abstract out (hide) those features of a requirement that are not necessary for a valid comparison with a similar requirement on another project. Secondly, the whole point of reuse is to save work, so it should also not be necessary to develop a full model of the second requirement in order to make the comparison. Finally, any relevant differences between the two requirements being compared should be clearly visible—and it should not be necessary to develop a full model of the second requirement in order to see these either.

Generalization

As a simple example of generalization that enables reuse of a component, consider the load-bearing characteristics of a common brick. Depending on the number of floors, type of roofing and other features of the design, different houses place very different loads on the bricks of which their structural walls are built. Yet, within broad parameters, an architect need not be over-precise in calculating the anticipated load, since standard bricks are suitable for use under a range of different loads. This aspect of the design of a brick, which makes it suitable for use in many different situations, rather than being restricted to one narrow set of circumstances, is an example of abstraction by generalization.

Object-oriented developers have one significant advantage over the architect. Using inheritance, a 'software architect' has a way of spawning new products from old ones with minimal effort. There is nothing analogous to this in house-building. For example, each time a new kind of brick is made, it takes much the same effort as any other brick. But inheritance provides a way of designing and building the larger part of a new software component in advance, leaving only the specialized details to be completed at a later stage. This is because, in a class hierarchy, those characteristics that are shared by subclasses are maintained at the superclass level, and are instantly available to any subclass when required.

Composition

Composition is a type of abstraction that encapsulates groups of classes that collectively have the capacity to be a reusable sub-assembly. Unlike generalization, the relationship is that between a whole and its parts. The essential idea is that a complex whole is made of

---

[1] Some recent writers have suggested that useful components are actually at a larger scale than individual objects. This point is discussed further in Chapter 19.

simpler components. These, while less complex than the whole, may themselves be made of still less complex sub-assemblies, elementary components or a mixture of the two.

A simple example of the usefulness of the idea can be seen in a house-builder fitting a window frame to a new house. Like many other house components, window frames are delivered to site as ready-assembled units. All internal details of the sub-assembly are 'hidden' from both architect and builder, in the sense that they do not need to think about them. In some other part of the building industry, specialist designers tackle the problems of what materials, structure, etc. make a good window-frame, and specialist constructors are busy building window-frames to the designer's specification. For this all to happen, it is first necessary that the window-frame be identified as a component that can be encapsulated, and thus treated by the house-builder as a single, simple thing, even though really it is not.

7.2.3 Composition in UML

Composition is based on the rather less precise concept of *aggregation*, which is a feature of many object-oriented programming languages. The UML Notation Guide (Rational, 1997) says the following about the relationship between the aggregation and composition.

> Aggregation is a 'special form of association that specifies a whole–part relationship between the aggregate (whole) and a component part'
> Composition is a 'form of aggregation with strong ownership and coincident lifetime of part with the whole. The multiplicity of the aggregate end may not exceed one (it is unshared) ...'

In addition to this, according to the UML Glossary (Rational, 1997), those parts of a composition that have:

> ... non-fixed multiplicity may be created after the composite itself, but once created they live and die with it ... Such parts can also be explicitly removed before the death of the composite. Composition may be recursive.

Although composition and aggregation may be identified during requirements analysis, their main application is during design and implementation activities, where they can be used to encapsulate a group of objects as a potentially re-usable sub-assembly. This is more than just a matter of labelling the group of objects with a single name. The fact that they are encapsulated is much more important. The external interface for a composition is actually the interface of the single object at the 'whole' end of the association. Details of the internal structure of the composition—that is, what other objects and associations it contains—remain hidden from the client.

On a practical level, composition is familiar to users of most common computer draw packages. For example, many of the drawings in this text were prepared or edited using a widely used drawing package. Within this application, several drawing objects can be selected and grouped. They then behave exactly like a single object, and can be sized, rotated, copied, moved or deleted with a single mouse command. Figures 7.1(a) and (b) show this type of composition as both objects and classes. Notice that this particular example is nested—the composition itself contains a further composition. In just the same way that a composite drawing object can only be handled as a single drawing component, the 'part' objects in a composition structure can usually not be directly accessed, and the whole presents a single interface to other parts of the system.

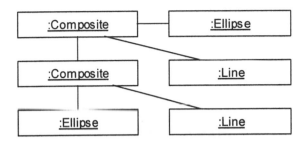

Figure 7.1(a) *Objects in a computer drawing package.*

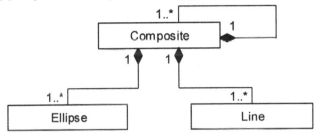

Figure 7.1(b) *Composition used in class diagram to represent composite objects.*

This notation is similar to a simple association, but with a diamond at the 'whole' end. The diamond is filled with solid colour to indicate composition, and left unfilled for aggregation.

Use of composition and aggregation in business-oriented applications is more problematic than in this example, but it is still worth modelling where it conveys useful information about the structure of the business domain. An example from the Agate case study is used later to illustrate aggregation in Section 7.3.2.

7.3 Adding Further Structure

This section shows how to add structure to the class diagram that will help with reuse at later stages of development. First, in Section 7.3.1 we concentrate on generalization, since it is the more useful of the two concepts for this purpose. Then in Section 7.3.2 we show how to model a structure that combines generalization and aggregation.

7.3.1 Modelling generalization

Figure 7.2 shows a note of an interview carried out by Mandy Botnick, an analyst in the Agate case study. Her main objective was to understand more about types of staff and the differences between them.

17 March - brief interview with Amarjeet Grewal (Finance Director)
Purpose - clarification of points from last Thursday's interview

Asked about staff types
*   - only two types seem relevant to system -*
*   creative staff (C) and admin staff (A)*
How do they differ?
*   - main difference is bonus payment...*
*   1. (C) bonus calculated on basis of campaign earnings*
*   (only those campaigns they worked on)*
*   2. (A) paid standard rate for Co, set at annual review*
Any other diffs? Amarjeet says -
*   - C qualifications need to be recorded*
*   - C can be assigned as contact for a client*
*   - C can add notes to campaign*
No other significant differences.
*   (NOTE - at next interview, get details of both algorithms)*

Figure 7.2 *Mandy Botnick's note of the differences between Agate staff types.*

In her haste, Mandy gathered only a handful of facts, but these highlight some useful information that must be modelled appropriately:

- there are two types of staff;
- bonuses are calculated differently;
- different data should be recorded for each type of staff.

Figure 7.3 shows a fragmentary class diagram that corresponds to this (for clarity, only affected classes are shown). The diagram includes a generalization association between StaffMember, AdminStaff and CreativeStaff. Of these, StaffMember is the superclass, while AdminStaff and CreativeStaff are subclasses. The generalization symbol states that all characteristics of the superclass StaffMember (its attributes, operations and associations) are automatically inherited by AdminStaff and CreativeStaff. There is no need to repeat superclass characteristics within a subclass definition.

For example, as StaffMember has an attribute staffName, we can take it for granted that AdminStaff and CreativeStaff also have this attribute. Since StaffMember has an operation getGrade(), AdminStaff and CreativeStaff inherit it too. For reasons of simplification, StaffMember is not shown in Figure 7.3 as having any assocations. Figure 7.4 extends the structure with an association between StaffMember and Grade. The generalization symbol means that this diagram should be interpreted as stating that both AdminStaff and CreativeStaff also have an association with Grade.

From a subclass perspective, inherited features are actually features that belong to the subclasses but have been removed to a higher level of abstraction. These are the shared characteristics that justify the generalization hierarchy. Generalization saves the analyst from the need to show these characteristics explicitly for each subclass to which they apply. Common attributes, operations and associations thus may be shared by several subclasses, but need be shown only once, in the superclass. This aids the general consistency of the model, and can also considerably reduce its complexity.

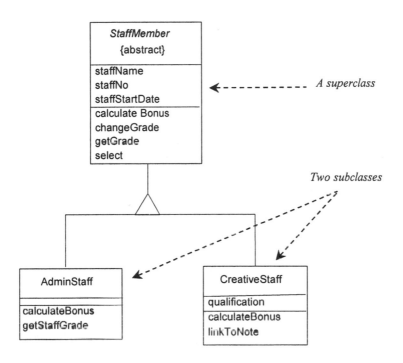

Figure 7.3 *The relationship among Agate staff types described as a generalization hierarchy.*

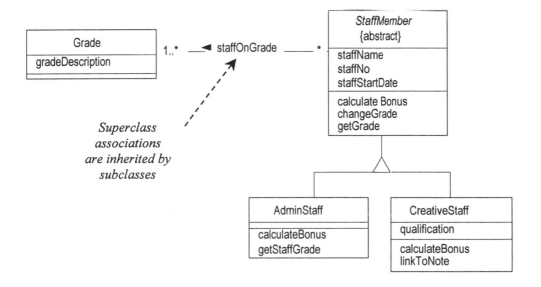

Figure 7.4 *Associations are also inherited.*

Each subclass also requires one or more characteristics that differentiate it from other subclasses in the hierarchy. For example, CreativeStaff has the attribute qualification and the operation linkToNote(), neither of which is shared by AdminStaff, while AdminStaff has the operation getStaffGrade, which is not shared by CreativeStaff.

Redefined operations

What are we to make of the fact that all three classes involved in the hierarchy in Figures 7.3 and 7.4 have an operation calculateBonus()? Has the analyst made a mistake? Or has she failed to take advantage of the economy of representation offered by the generalization notation?

The explanation lies in the fact that, while both AdminStaff and CreativeStaff require an operation calculateBonus(), it works differently in each case. Since the precise logic for the calculation differs between these two groups of staff, the two operations will need to be treated separately later when each algorithm is designed, and also when programme code is written to implement the algorithm. This difference in logic justifies the separate appearance of a superficially similar operation in both subclasses. This sort of fine distinction is not always recognized during analysis.

Why, then, has the operation calculateBonus() been included in the superclass StaffMember? The answer is that it is an attempt at 'future-proofing'. One of the consequences of identifying a superclass is that it may later acquire other subclasses, that are as yet unknown. In this case the analyst has recognized—or assumed—that objects belonging to *all* subclasses of StaffMember are likely to need an operation of some kind to calculate bonus. For this reason, at least a 'skeleton' of the operation is included in the superclass. This may consist of no more than the signature, but since the interface is all that matters to other classes, this alone justifies its inclusion in the superclass definition. Even if the superclass operation is defined in full, some subclasses may choose not to use it because the logic for their version of the operation is different. For all these cases, the operation is said to be *redefined* or *overridden* in the subclass.

Abstract and concrete classes

We now turn to the {abstract} annotation below the StaffMember class name in Figures 7.3 and 7.4. This may at first be puzzling if we recall learning in Chapter 6 that *all* classes are abstract, in the sense that a class is an abstraction of its members. However, StaffMember is abstract in the still more compelling sense that it has no instances. This is one example of a UML *stereotype*, a term used to group together similar elements that can not otherwise be distinguished within the notation.

The {abstract} stereotype can only be applied to a superclass in a generalization hierarchy. All other classes have at least one instance, and are said to be *concrete* or *instantiated*. Applying the {abstract} stereotype to StaffMember means that no staff exist at Agate who are 'general' members of staff, and not members of a particular sub-group. All staff members encountered so far (among those that are relevant to the model) are defined as either AdminStaff or CreativeStaff. Should we later discover another group of staff that is distinct in behaviour, data or associations, and if we need to model this new group, it should be represented in the diagram by a new subclass. The whole reason for the existence of a superclass is that it sits at a higher level of abstraction than its subclasses. This generality allows it to be adapted for use in other systems. While in itself this is not always enough for it to be declared as an abstract class, this is usually the case, and for the moment we can safely ignore exceptions to the rule.

The usefulness of generalization

We can now consider the contribution of generalization hierarchies to reuse. Imagine that the Agate system is completed and in regular use. Some time after installation, the Directors decide that they want to reorganize the company, and one of the results is that Account Managers are to be paid bonuses related to campaign profits. Their bonus is to be calculated in a different way from both administrative and other creative staff, and is to include an element from campaigns that they supervise, and an element from the general profitability of the company. This can be seen in Figure 7.5.

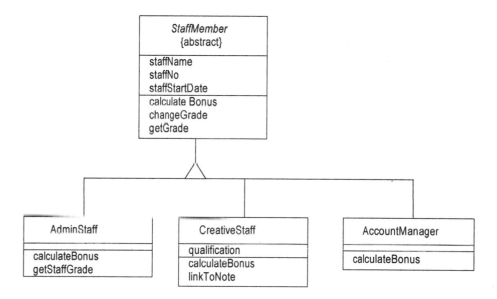

Figure 7.5 *A new subclass is easy to add.*

Adding a new subclass requires relatively little change to the existing class model, essentially just adding a new subclass AccountManager to the staff hierarchy. In practice, some judgement would be needed as to whether this is better modelled as a subclass of StaffMember, or of CreativeStaff, and this would be based on assumptions about difficulty of implementation and likely future benefits. In either case, the impact is minimal.

The reuse lies in the fact that the existing abstract class StaffMember has been used as a basis for AccountManager. The latter can inherit all attributes and operations that are not part of its own specialism—in this case roughly 75% of its entire specification. This is for only one class; over the many classes of a large system, the saving in design and coding effort can clearly be significant. But the opportunity is only available because we have first taken the trouble to identify the generalized aspects of a staff member. This is the main benefit of generalization, that hierarchies can usually be extended without significant effects on existing structures. Were it not for this, there would be little justification for the time taken to organize a class diagram using generalization, when it would be so much quicker to leave it alone once the classes and simple associations are defined.

Identifying generalization (a top-down approach)

It is relatively easy to discover generalization where this exists between classes that have already been identified. The rule of thumb is very straightforward. If an association can be described by the expression *is a* or *is a kind of*, then it can be modelled as generalization. Sometimes this is so obvious, that you may wonder if that is all there is to it, for example, 'administrative staff are a kind of staff'. More often, it is not quite so obvious, but still straightforward. For example, 'a helicopter is a type of aircraft and so is a jumbo-jet' and 'a truck is a type of vehicle and so is a buffalo cart' imply generalizations with similar structures as in Figure 7.6.

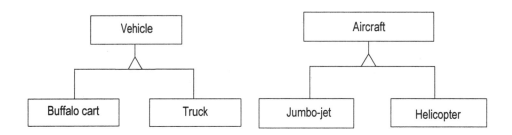

Figure 7.6 *Two similar hierarchies.*

It is not uncommon to find multiple levels of generalization. This simply means that a superclass in one relationship may be a subclass in another. In practice, more than about four or five levels of generalization in a class model is too many (Rumbaugh et al., 1991), but this is primarily for design reasons.

Adding generalization (a bottom-up approach)

An alternative approach to adding generalization is to look for similarities among the classes in your model, and consider whether the model can be 'tidied up' or simplified by introducing superclasses that abstract the similarities. This needs to be done with some care. The purpose of doing this is quite explicitly to increase the level of abstraction of the model, but any further abstraction introduced should still be 'useful'. The guiding principle is still that any new generalization must meet all the tests described in the previous section. Above all, it must be possible to represent the association with the phrase 'is a kind of'.

Some judgement is needed to determine the likely future usefulness of generalization, and this is discussed in the next section.

When not to use generalization

Generalization can be over-used, so it is worth highlighting some implicit judgements made in deciding to model the generalization structure developed in Figures 7.4–7.6. Firstly, as a rule we only model a class as a superclass of another if we are confident that what we know about the former (i.e. its attributes, operations and associations) applies *completely* to the latter. In this example, the analyst had to be reasonably sure that everything she knew about StaffMember applied also to both AdminStaff and CreativeStaff. This is part of the UML definition of generalization: a subclass is 'fully consistent with' its superclass. Even when this is true, if the differences between two potential subclasses are too great, the forced creation of a superclass can give rise to confusion rather than clarity.

We can illustrate this best with an example of inappropriate modelling. At Agate, staff and (some) clients are people (this is perhaps stretching a point, since many clients are in fact companies, but for the sake of illustration we will ignore this and assume all clients are individuals). Following the rule of thumb without reflection, an inexperienced analyst might feel that this justifies the creation of a Person superclass, to contain any common attributes and operations of Client and StaffMember. But it may quickly become apparent that the new class definition contains little but the attribute personName. This is really an attempt to force a generalization hierarchy to include subclasses that are too dissimilar.

Secondly, we should not anticipate subclasses that are not justified by currently known requirements. For example, at Agate AdminStaff and CreativeStaff are distinct classes based on differences in their attributes and operations. We also know about other kinds of staff in the organization, e.g. the directors. But we have no information that suggests a need to model directors as part of the system, and thus we should not create another subclass of StaffMember called DirectorStaff. Even if it were to turn out that we do need to model Directors, there is no reason yet to suppose they must be a distinct class. They might turn out to be members of an existing class. This, too, is part of the UML standard definition: a subclass must 'add additional information' (that, we might add, 'is relevant to the system under study').

There is a tension here. On the one hand, generalization is modelled in order to permit future subclassing in situations which the analyst inevitably cannot fully anticipate. Indeed the ability to take advantage of this is one of the main benefits of constructing a generalization hierarchy. Yet, on the other hand, if generalization is overdone, it just adds needlessly to the complexity of the model for no return. There is at present no simple answer to this problem, save the judgement that comes with experience. But it seems probable that over the next few years, organizations will grow much better at managing the activity of reuse, and part of this will involve the communication of clear guidance to project teams on this tricky point.

7.3.2 Combining generalization with composition or aggregation

At this point it is worth further examining the Agate case study material for further generalization and composition. One possibility is revealed in the statement in Section A1.3 that 'adverts can be one of several types'. For each type, it is certainly true that, for example, a newspaper advert *is a kind of* advert (for simplicity, we assume that 'advert' refers to a *design*, rather than an *insertion*—so an advert that appears five times in one newspaper *is* one advert appearing five times, and not five adverts each appearing once). This suggests that advert could be a superclass, with newspaper advert, etc. as its subclasses. We need to ask if it meets the tests first described in Section 4.2.4. Figure 7.7 shows the partial class model that results from this analysis, but how was this arrived at?

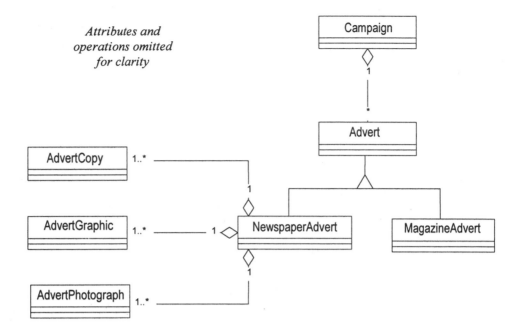

Figure 7.7 *Aggregation and generalization in the Agate model (composition does not appear fully justified in any of these cases).*

The process of thought goes something like this. Firstly, is *everything* that is true of `Advert` also true of `NewspaperAdvert`? The answer appears to be yes. All actual adverts must be one of the specific types, with no such thing in reality as a 'general' advert. `Advert` is thus a sound generalization of common features shared by its specialized subclasses. Secondly, does `NewspaperAdvert` include some details (attributes, operations or associations) that are *not* derived from `Advert`? An authoritative answer would require a detailed examination of attributes and operations for each class, but in general terms the answer appears to be yes. A newspaper advert consists of a particular set of parts. The precise composition of each type of advert is different, and so this structure of associations could not be defined at the superclass level. Finally, there is no reason to suppose that a newspaper advert has any other potential ancestor besides advert, so we do not need to consider this rule at higher levels of recursion.

Next, we can see possible composition in the association between `Campaign` and `Advert`, and in turn between `Advert` and its associated parts. A campaign includes one or more adverts. A newspaper advert includes written copy, graphics and photographs. Magazine adverts, and other types, have similar structure to newspaper adverts, but the detail differs in some respects.

What kind of composition is involved? This can be analysed by reviewing the UML definitions given in Section 7.2.3. Firstly, can an advert belong to more than one campaign? This is not stated in the case study, but it seems unlikely that an advert can simultaneously be part of more than one campaign. Secondly, has each `Advert` a coincident lifetime with its `Campaign`? Again, this is not given explicitly, but a client might wish to use an expensive television advert again on another campaign. This point needs to be clarified, but in the meantime it does not appear justified to model this as composition. Thirdly, can copy, graphics or photographs belong to more than one newspaper advert? This seems unlikely, but should really be clarified. Finally, has each of these components a coincident lifetime with the advert? Probably some do, and some do not. It is hard to imagine that advertising copy would be used again on another advert, but photographs and graphics seem less certain. This is another point to be clarified, but in the meantime composition does not seem justified in any of these cases, and aggregation has therefore been used.

7.4 Analysis Patterns

In this section we offer a brief introduction to the idea of analysis patterns. The use of analysis patterns is an advanced approach that is principally of use to experienced analysts, and we leave the interested reader to follow this up through further reading. They are closely related to design patterns, which we cover in some detail in Chapter 14.

An analysis pattern is essentially a structure of classes and associations that is found to occur over and over again in many different modelling situations. Each pattern can be used to communicate a general understanding about how to model a particular set of requirements, and therefore the model need not be invented from scratch every time a similar situation occurs. Since a pattern may consist of whole structures of classes, the abstraction takes places at a higher level than is normally possible using generalization alone. Fowler (1997b) describes a number of patterns that recur in many business modelling situations, such as accounting, trading and organization structure.

Figure 7.8 shows Fowler's *accountability* pattern as an illustration of an analysis pattern in practice. An accountability structure may be of many kinds, such as management or contract supervision. In the case of Agate, this pattern could represent several different relationships: that between a manager and a member of staff they supervise, that beween a client and a client contact, or that between a client and a campaign manager. Since the details of the relationship itself have been abstracted out as AccountabilityType, this one class structure is sufficiently general to represent any of these relationships, given an appropriate set of attributes and operations. The generalization of Person and Organization as Party similarly allows the pattern to represent relationships between individuals, organizations, or a mixture of the two.

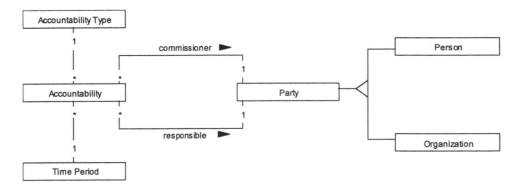

Figure 7.8 *The accountability pattern (adapted from Fowler, 1997).*

7.5 Summary

The primary purpose of refining the requirements model is to maximize opportunities for the reuse of components. These components may be of several kinds. Software components can reduce the effort involved in coding and implementation. Design components can reduce the time required to develop and test a design, and also act as a store of knowledge that embodies best known practice. Analysis components can reduce the time taken to develop a deep understanding of the system. In principle, whatever their form, opportunities for the use of components occur in only three ways. Existing components may be imported from other sources beyond the project boundaries. This requires careful evaluation of the degree of commonality between the requirements of the current project and the features of available components. Secondly, new reusable components may be developed for use on more than one part of the current project. For this to be possible, the requirements model must be at a sufficient level of abstraction for the common features between different aspects of the project to become evident. Finally, new reusable components may be developed for export to other projects. Here, too, it is necessary for the requirements modelling to identify those aspects of the project that can be generalized for use in other contexts.

The abstraction mechanisms of the object-oriented approach, particularly generalization and composition, can make a significant contribution to all three approaches to the reuse of components in systems development. So far, these benefits have been mainly apparent during the design and coding activities, but the starting point for finding reuse opportunities is during requirements analysis.

Review Questions

7.1   What are the advantages of components?
7.2   Explain why the NIH syndrome occurs.
7.3   What does object-orientation offer that helps to create reusable components?
7.4   Distinguish composition from aggregation.
7.5   Why are operations sometimes redefined in a subclass?
7.6   What is an abstract class?
7.7   Why is encapsulation important to creating reusable components?
7.8   Why is generalization important to creating reusable components?
7.9   When should you not use generalization in a model?

Case Study Work, Exercises and Projects

7.A   Find out from your library about the coding system that is used for classifying books, videos, etc. Draw part of the structure in UML notation as a generalization hierarchy. Think up some attributes for 'classes' in your model to show how the lower levels are progressively more specialized.

7.B   Choose an area of commercial activity (business, industry, government agency, etc.) with which you are familiar. Identify some ways in which its products show the use of generalization, and some ways that components used as inputs show the use of generalization.

7.C   Examine the two hierarchies shown in Figure 7.6. Think up some operations and/or attributes that show generalization and specialization for each class at the appropriate levels. Draw new class diagrams that include these operations and attributes. Think of some new classes that specialize each of the 'leaf' classes in your diagram, and add these to the diagram. Add some operations and/or attributes that show clearly the specialisms of each of the new classes. Think of a new class that would be suitable as a superclass of both truck and aircraft, and draw a new diagram that includes this class, showing how it generalizes the other two structures. If necessary, rename some of the given classes in order to clarify the relationships. Check that your structure still makes sense when you compare the 'leaf' classes with the new 'root' class.

7.D   In Section 7.3.2, we established that generalization probably was an appropriate way of modelling the association between `Advert` and `NewspaperAdvert`. Identify the other possible subclasses for `Advert`, and repeat the checks for each. Which of them pass (if any)? Do you think there are really only two levels to the hierarchy? Explain your reasoning. Redraw the Agate class diagram to include all the generalizations that you feel are justified.

7.E   Re-read the case study material for FoodCo and identify possible generalizations or compositions. Add these to the class diagram you drew for Exercises 6.C and 6.D.

Further Reading

All standard texts on object-orientation have a chapter or two that discuss the role of generalization and aggregation in software and requirements reuse.

Rumbaugh et al (1991) gives one of the clearest summaries. This particular book pre-dates the development of UML, but Rumbaugh was later one of the 'Three Amigos' (along with Jacobson and Booch) who founded the UML standard. UML incorporates a great deal of OMT notation, particularly that used for the class diagram.

Fowler (1997b) describes a number of analysis patterns that may be applied at this stage of modelling, although these are not presented in UML. Coad et al. (1997) also gives a number of analysis and design patterns; most are presented in Coad notation, but some are also shown in OMT notation and in Unified notation (an early version of UML).

8

Object Interaction

Objectives
In this chapter you will learn:
- how to identify object collaboration using CRC cards;
- how to develop object collaboration from use cases;
- how to model object collaboration using an interaction sequence diagram;
- how to model object collaboration using an interaction collaboration diagram;
- how to cross-check between interaction diagrams and a class diagram.

8.1 Introduction

Communication and collaboration between objects are fundamental to object-orientation. This mirrors the world in which we live, where most human endeavour involves communication and collaboration between individuals. For example, each employee in a manufacturing organization has specialized tasks. Different employees collaborate with each other in order to satisfy a customer request. This involves communicating to request information, to share information and to request help from each other. In a similar way an object-oriented application comprises a set of autonomous objects, each responsible for a small part of the system's overall behaviour. These objects produce the required behaviour through collaboration, by exchanging messages that request information, that give information or that ask another object to perform some task (Section 8.2).

In order for the collaboration to succeed, two fundamental and interrelated questions must be answered, and the answers to these questions must then be modelled in an appropriate way. Firstly, which object should be responsible for each part of the overall behaviour? Essentially this means allocating operations to appropriate classes. This has been addressed in part by the allocation of attributes, which was used to guide a preliminary allocation of operations (see Chapter 6). However, the attributes that are assigned to a class determine only to some extent the behaviour that is appropriate for it. We must now develop firmer foundations for the allocation of operations. Secondly, what collaborations between objects are required to generate the overall application functionality? This means identifying what messages should be sent between objects and the conditions under which they should be sent. In this chapter we introduce the CRC technique (Section 8.3), which, while not part of UML, is very useful for exploring these issues and modelling the possible answers. UML itself provides two notations for modelling interaction between objects: interaction sequence diagrams and interaction collaboration diagrams. These are introduced in Section 8.4 and Section 8.5, respectively.

8.2 Object Interaction and Collaboration

When an object sends a message to another object an operation is invoked in the receiving object. For example, in the Agate case study there is a requirement to be able to determine the current cost of the advertisements for an advertising campaign. For a particular campaign this might be achieved if the `Campaign` object sends a message to each of its `Advert` objects asking them for their current cost. In a programming language, sending the message `getCost()` to an `Advert` object, might use the following syntax.

```
advertCost = anAdvert.getCost()
```

Note that in this example the advert object is identified by the variable name `anAdvert`.

The cost of the advert returned by the operation `getCost()` is stored in the variable `actualCost` in the sending object. In order to calculate the sum of the costs for all adverts in a campaign the above statement must be executed repeatedly. However, rather than think in terms of operation invocation we use the metaphor of message passing to describe object interaction as this emphasizes that objects are encapsulated and essentially autonomous. Message passing can be represented on an object diagram, as in Figure 8.1 where the message is shown as an arrow. This notation is rather like that of the collaboration diagram, which we discuss later, and the similarity demonstrates that the latter is a natural extension of the class diagram.

Figure 8.1 *Object messaging.*

It can be difficult to determine what messages should be sent by each object. In this case, it is clear that the `getCost()` operation should be located in the `Advert` class. This operation requires data that is stored in the `advertCost` attribute, and this has been located in `Advert`. We can also readily see that an operation that calculates the cost of a `Campaign` must be able to find out the cost of each `Advert` involved. But this is a simple collaboration and the allocation of these operations is largely dictated by the presence of particular attributes in the classes. More complex requirements may involve the performance of complex tasks, such that an object receiving one message must itself send messages that initiate further collaboration with other objects.

It is an aim of object-oriented analysis and design to distribute system functionality evenly among its classes. This does not mean that all classes have exactly equal levels of responsibility but rather that each class should have appropriate responsibilities. Where responsibilities are evenly distributed, each class tends not to be unduly complex and, as a result is easier to develop, to test and to maintain. A class that is relatively small and self-contained has a much greater potential for reuse than one that is large, one that is complex or one that has responsibilities or functionality that are not clearly focused.

An even distribution of responsibility among classes has the important side-effect of producing a system that is more resilient to changes in its requirements. When the users' requirements for a system change it is reasonable to expect that the application will need some modification, but ideally the change in the application should be of no greater magnitude than the change in the requirements. An application that is resilient in this sense costs less to maintain and to extend than one that is not. Figure 8.2 illustrates this.

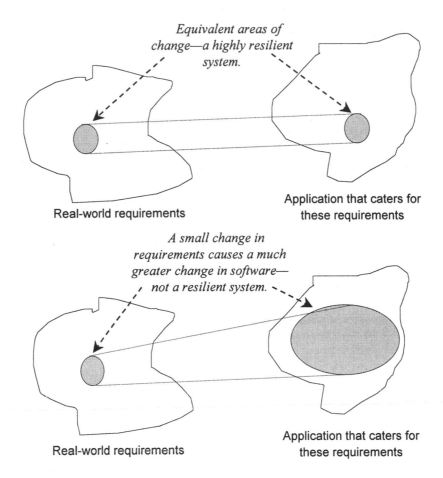

Figure 8.2 *Resilience of a design.*

The focus of a model of object interaction is to determine the most appropriate scheme of messaging between objects in order to support a particular user requirement. As we saw in Chapter 5, user requirements are first documented by use cases. Each use case can be seen as a dialogue between an actor and the system, that results in objects performing tasks so that the system can respond in the way that is required by the actor. For this reason many interaction diagrams explicitly include one or more objects to manage the dialogue, and when such objects are not shown explicitly it can be assumed in most cases that they will need to be identified at a later stage. The identification and specification of dialogue objects is in part an analysis activity and in part a design activity. During analysis our concern is to identify the nature of a dialogue in terms of the user's need for information and his or her access to the system's functionality. The detailed specification of interface objects that can manage the dialogue is primarily a design activity and is discussed in Chapters 15 and 16.

8.3 CRC (Class Responsibility Collaboration) Cards

We have recommended use cases as the starting point for the identification of classes. A further examination of the use cases also helps in identifying operations and the messages that classes need to exchange. However, it is often easier to think first in terms of the overall responsibilities of a class rather than its individual operations. A *responsibility* is a high level description of something a class can do. It reflects the knowledge or information that is available to that class, either stored within its own attributes or requested via collaboration with other classes, and also the services that it can offer to other objects. A responsibility may correspond to one or more operations. It can be difficult to determine the most appropriate choice of responsibilities for each class as there may be many alternatives that all appear to be equally justified.

Class Responsibility Collaboration (CRC) cards provide an effective technique for exploring the possible ways of allocating responsibilities to classes and the collaborations that are necessary to fulfil the responsibilities.[1] CRC cards can be used at several different stages of a project for different purposes. For example, they can be used early in a project to aid the production of an initial class diagram and to develop a shared understanding of user requirements among the members of the team. Here we concentrate on their use in modelling object interaction. The format of a typical CRC card is shown in Figure 8.3.

Class Name:	
Responsibilities	Collaborations
Responsibilities of a class are listed in this section.	*Collaborations with other classes are listed here, together with a brief description of the purpose of the collaboration.*

Figure 8.3 *Format of a CRC card.*

CRC cards are an aid to a group role playing activity that is often fun to do[2]. Index cards are used in preference to pieces of paper due to their robustness and to the limitations that their size (approx. 15cm x 8cm) imposes on the number of responsibilities and collaborations that can be effectively allocated to each class. A class name is entered at the top of each card and responsibilities and collaborations are listed underneath as they become apparent. For the sake of clarity, each collaboration is normally listed next to the corresponding responsibility.

Wirfs-Brock et al. (1990) and others recommend the use of CRC cards to enact a system's response to particular scenarios. From a UML perspective, this corresponds to the

---

[1] CRC cards were invented by Kent Beck and Ward Cunningham (1989) while they were working together on a Smalltalk development project for Tektronix.

[2] A useful spin-off is that this can support team building and help a team identity to emerge.

use of CRC cards in analysing the object interaction that is triggered by a use case. The process of using CRC cards is usually structured as follows.

- Conduct a brainstorming session to identify which objects are involved in the use case.
- Allocate each object to a team member who will role play that object.
- Role play the use case. This involves a series of negotiations among the objects (role played by team members) to explore how responsibility can be allocated and to identify how the objects can collaborate with each other.
- Identify and record any missing or redundant objects.

Before beginning a CRC session it is important that all team members are briefed on the organization of the session. Some authors (Bellin and Simone, 1997) recommend that a CRC session should be preceded by a separate exercise that identifies all the classes for that part of the application to be analysed. The team members to whom these classes are allocated can then prepare for the role-playing exercise by considering in advance a first-cut allocation of responsibilities and identification of collaborations. Others prefer to combine all four steps into a single session and perform them for each use case in turn. Whatever approach is adopted, it is important to ensure that the environment in which the sessions take place is free from interruptions and conducive to the free flow of ideas (Hicks, 1991).

During a CRC card session, there must be an explicit strategy that helps to achieve an appropriate distribution of responsibilities among the classes. One simple but effective approach is to apply the rule that each object (or role-playing team member) should be as lazy as possible, refusing to take on any additional responsibility unless persuaded to do so by its fellow objects (the other role-playing team members). During a session conducted according to this rule, each role-player identifies the object that they feel is the most appropriate to take on each responsibility, and attempts to persuade that object to accept the responsibility. For each responsibility that must be allocated, one object (one of the role-players) is eventually persuaded by the weight of rational argument to accept it.[3] This process can help to highlight missing objects that are not explicitly referred to by the use case description. When responsibilities can be allocated in several different ways it is useful to role-play each allocation separately to determine which is the most appropriate. The aim normally is to minimize the number of messages that must be passed and their complexity, while also producing class definitions that are cohesive and well focused.

We illustrate how a CRC exercise might proceed by considering the use case Add a New Advert to a Campaign. The use description is repeated below for ease of reference.

The Campaign Manager selects the required Campaign for the Client concerned and adds a new Advert to the existing list of adverts for that campaign. The details of the Advert are completed by the Campaign Manager.

This use case involves instances of Client, Campaign and Advert, each role played by a team member. The resulting CRC cards are shown in Figure 8.4, and in the discussion that follows we explain how they can be derived from the use case.

---

[3] An alternative strategy is for each object to be equally keen to take on a responsibility, with the final choice determined by negotiation. Irrespective of the strategy chosen, it is important that all team members understand the need for an effective distribution of responsibilities.

Class Name *Client*	
Responsibilities	Collaborations
Provide client information. *Provide list of campaigns.*	*Campaign provides campaign details.*

Class Name *Campaign*	
Responsibilities	Collaborations
Provide campaign information. *Provide list of adverts.* *Add a new advert.*	*Advert provides advert details.* *Advert constructs new object.*

Class Name *Advert*	
Responsibilities	Collaborations
Provide advert details. *Construct adverts.*	

Figure 8.4 *CRC cards for the use case* Add a New Advert to a Campaign.

The first issue is how to identify which client is involved. In order to find the correct Client the Campaign Manager (an actor and therefore outside the system boundary from the perspective of this use case) needs access to the client's name. Providing a client name and any other details for that client is clearly a responsibility of the Client object.

Next, the Campaign Manager needs a list of the campaigns that are being run for that client. This list should include the title, start date and finish date for each campaign. Although a Campaign object holds details of the campaign, it is not clear which object (and hence which class) should be responsible for providing a list of campaigns for a client. The team member playing the Campaign object argues that although it knows which Client object commissioned it, it does not know which other Campaign objects have been commissioned by the same Client. After some discussion, the Client object is persuaded to accept responsibility for providing a list of its campaigns and the Campaign object is persuaded that it should provide the information for this list. Once the Campaign Manager has obtained details of the campaigns for that client she requests that the Campaign object provide a list of its adverts, to which list the new advert will be added. Since the Campaign object already has responsibility for looking after the list of adverts it is reasonable for it to add the new advert to its list. In order to do this it must collaborate with the Advert class which, by definition, has responsibility for creating a new Advert object. This completes the analysis of the use case interaction, and the new responsibilities and collaborations that have been identified are added to the cards.

During a CRC session, the team can keep track of the relationships between classes by sticking the index cards on a large board and attaching pieces of thread or string to represent collaborations. This is particularly useful when CRC cards are used early in the development cycle to help produce a class diagram. The cards and pieces of thread can be a very effective prototype of the class diagram[4]. CRC cards can also be extended in various ways. For example, superclasses and subclasses can be shown beneath the class name and some users of the technique also like to list attributes on the back of each card.

8.4 Interaction Sequence Diagrams

An *interaction sequence diagram* (or simply a *sequence diagram*) is one of two kinds of UML interaction diagram. The other is the collaboration diagram, which we describe in Section 8.5. In the UML Notation Guide (Rational, 1997) an interaction is defined as:

> [A] behavioural specification that comprises a sequence of message exchanges among a set of objects within a collaboration to accomplish a specific purpose.

A sequence diagram shows an interaction between objects arranged in a time sequence. Sequence diagrams can be drawn at different levels of detail and also to meet different purposes at several stages in the development life cycle. The commonest application of a sequence diagram is to represent the detailed object interaction that occurs for one use case. A sequence diagram of this kind can be seen as an expansion of a use case to the lowest possible level of detail. Those drawn during the analysis phase differ from those drawn during design in two major respects. Analysis sequence diagrams normally do not include design objects nor do they usually specify message signatures in any detail.

8.4.1 Basic concepts and notation

Figure 8.5 shows a sequence diagram for the use case Add a New Advert to a Campaign. The vertical dimension represents time and all objects or classes involved in the interaction are spread horizontally across the diagram (the horizontal ordering of objects or classes is arbitrary and has no modelling significance). Time normally proceeds down the page. However, a sequence diagram may be drawn with a horizontal time axis if required, and in this case, time proceeds from left to right across the page. Each object or class is represented by a vertical dashed line, called a *lifeline*, with an object symbol at the

---

[4] This is sometimes known as *paper CASE* technology!

top. A message is shown by a solid horizontal arrow from one lifeline to another and is labelled with the message name. Each message name may optionally be preceded by a sequence number that represents the sequence in which the messages are sent, but this is not usually necessary on a sequence diagram since the message sequence is already conveyed by their relative positions down each lifeline.

When a message is sent to an object, it invokes an operation of that object. Once a message is received, the operation that has been invoked begins to execute. The period of time during which an operation executes is known as an activation, and is shown on the sequence diagram by a rectangular block laid along the lifeline. The activation period of an operation includes any delay while the operation waits for a response from another operation that it has itself invoked as part of its execution. The message name indicates which particular operation is being invoked.

For example, in Figure 8.5 a Client object receives a listCampaigns() message and a period of activation begins. This is shown by the tall thin rectangle that begins at the message arrowhead. The Client object now sends a message getCampaignDetails() to each Campaign object in turn in order to build up a list of campaigns. Iteration is indicated by a * before the message. The conditions for continuing or ceasing an iteration may be shown beside a message name, although for reasons of space this is not shown in Figure 8.5.

This example of a continuation condition is written as follows.

```
[For all client's campaigns] *getCampaignDetails()
```

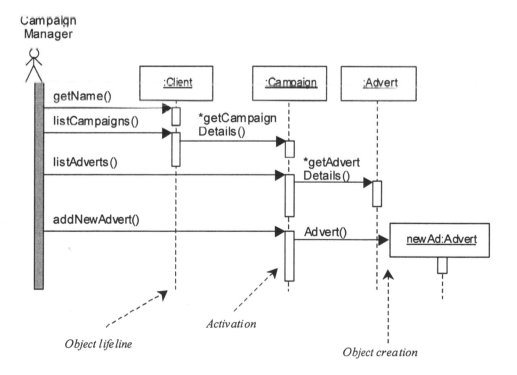

Figure 8.5 *Sequence diagram for the use case* Add a New Advert to a Campaign.

The Campaign Manager next sends a message to a particular `Campaign` object asking it to list its advertisements. The `Campaign` object delegates responsibility for the advertisement title to each `Advert` object although the `Campaign` object retains responsibility for the list as a whole (indicated by the continuation of the activation bar beyond the point where the message is sent).

When an advertisement is added to a campaign an `Advert` object is created. This is shown by the `Advert()` message arrow (this invokes the constructor operation) drawn with its arrowhead pointing directly to the object symbol at the top of the lifeline. Where an object already exists prior to the interaction the first message to that object points to the lifeline below the object symbol. For example, this is the case for the `Campaign` object, which must exist before it can receive an `addNewAdvert()` message. The sequence diagram in Figure 8.5 corresponds directly to the interaction suggested by the CRC cards in Figure 8.4. In effect, the sequence diagram is a more formal representation of the same interaction, but with the messages and the sequence of collaboration both made explicit.

The destruction of an object is indicated by a large X on the lifeline at the point in the interaction when the object is destroyed. An object may either be destroyed when it receives a message (this is shown in Figure 8.6) or it may self-destruct at the end of an activation if this is required by the particular operation that is being executed.

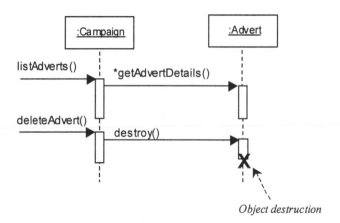

Figure 8.6 *Object destruction.*

An object can send a message to itself. This is known as a *reflexive message* and is shown by a message arrow that starts and finishes at the same object lifeline. The use case `Check Campaign Budget` includes an example of this. For ease of reference the use case description is repeated below.

The campaign budget may be checked to ensure that it has not been exceeded. The current campaign cost is determined by the total cost of all the adverts and the campaign overheads.

The corresponding sequence diagram is shown in Figure 8.7 and this includes a reflexive message `getOverheads()` sent from a `Campaign` object to itself.

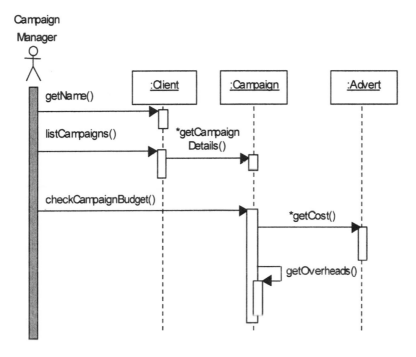

Figure 8.7 *Sequence diagram for the use case* Check Campaign Budget.

In this case the reflexive message invokes a different operation from the operation that sent the message, so a new activation symbol is stacked on the original activation (this is the shorter second rectangle shown offset against the first activation). In certain circumstances an operation invokes itself on the same object; this too can be represented but is not illustrated here[5].

Until this point our discussion has centred on simple use cases and correspondingly simple interactions. These are typical of many modelling situations, but more complex interactions also occur in many systems. It is sometimes also necessary to represent in more detail the synchronization of messages. Figure 8.8 illustrates some variations in the UML notation.

The *focus of control* indicates times during an activation when processing is taking place within that object. Parts of an activation that are not within the focus of control represent periods when, for example, an operation is waiting for a return from another object. The focus of control is shown by shading those parts of the activation rectangle that correspond to active processing by an operation. In Figure 8.8 the Check Campaign Budget use case is redrawn with foci of control shaded. The focus of control for the checkCampaign-Budget() operation is initially with the Campaign object, but is then transferred to the Advert object and the activation rectangle in the Campaign object is now unshaded while the Advert object has the focus of control. The checkCampaignbudget() activation is also unshaded while the getOverheads() operation is activated by the reflexive message getOverheads().

---

[5] If a limit has been set to the depth of recursion this can be recorded in a note.

A *return* is a return of control to the object that originated the message that began the activation. This is not a new message, but is only the conclusion of the invocation of an operation. Returns are shown with a dashed arrow, but it is optional to show them at all since it can be assumed that control is returned to the originating object at the end of the activation in a destination object (asynchronous messages—see below—are exceptions). Returns are often omitted, as in Figure 8.5. Figure 8.8 explicitly shows all returns for the same interaction.

A *return type* is the type of the value that an operation returns to the object that invoked it. These are rarely shown on an analysis sequence diagram, and are discussed further in Chapter 9.

Figure 8.8 *Sequence diagram for* Check Campaign Budget *with foci of control and explicit returns shown.*

A *synchronous message* or *procedural call* is shown with a full arrowhead (see Figure 8.9), and is one that causes the invoking operation to suspend execution until the focus of control has been returned to it. This may be because the invoking operation requires data that is held by the destination object before it can proceed. In Figure 8.8 the Check Campaign Budget use case is shown with procedural calls and explicit returns. Procedural calls are appropriate for the interaction since each operation that invokes another does so in order to obtain data and cannot continue until that data is supplied.

An *asynchronous message*, drawn with a half-arrowhead as in Figure 8.9, does not cause the invoking operation to halt execution while it awaits a return. When an asynchronous message is sent operations in both objects may carry out processing at the same time. Asynchronous messages are frequently used in real-time systems where operations in different objects must execute concurrently, either for reasons of efficiency, or because the system simulates real-world activities that also take place concurrently. It may be necessary for an operation that has been invoked asynchronously to notify the object that invoked it when it has terminated. This is done by explicitly sending a message (known as a *callback*) to the originating object.

An *uncommitted message*, also illustrated in Figure 8.9, is an undetermined message that can represent either synchronous or asynchronous control. It is generally used either where the type of communication is not relevant or where it has not yet been determined.

8.4.2 Time constraints

A sequence diagram can be labelled to document any time constraints that apply to the execution of operations. For example, in Figure 8.9 the labels 'a' and 'b' represent the start and end points of an interaction and the constraint in braces indicates that the elapsed time between 'a' and 'b' must be less than 1 second. Time constraints are frequently used in modelling real-time systems where the application must respond within a certain time, typically for reasons of safety or efficiency. For most other information systems time constraints are not significant and only the sequence of the messages matters.

Figure 8.9 *Sequence diagram showing different message types.*

All message arrows have so far been drawn horizontally across the sequence diagram and at right angles to the object lifelines. This indicates that the time taken to send a message is not significant in comparison to the time taken for operation execution. There is consequently no need to model another activity during the period while a message is in transit. In some applications the length of time taken to send a message is itself significant. For example, in distributed systems messages are sent over a network from an object on one computer to another object on a different computer. If the transit time for a message is significant the message arrow is slanted downwards so that the arrowhead (the arrival of the message) is below (later than) the tail (the origination of the message). The asynchronous message shown in Figure 8.9 illustrates this. When necessary, the vertical time dimension is used to show time constraints or timing requirements. In this case, the annotation indicates that the time taken to send this message has a time constraint: the total period of transit must be less than 5 seconds. For the sake of readability, the start and end times of any one message are usually labelled with the same letter—here we have used c and c' (the latter is said as 'c primed'[6]).

8.4.3 Branching

The interactions that we have considered so far have only one execution path, although some have iterations during their execution. Some interactions have two or more alternative execution pathways. Each reflects a branch in the possible sequence of events for the use case it represents. The notation for branching is illustrated in Fig 8.10.

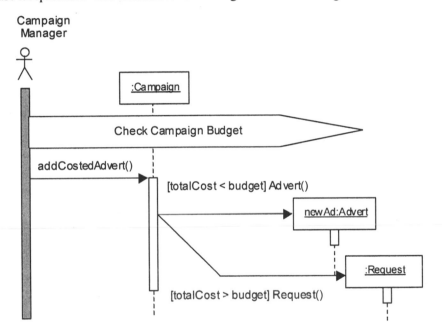

Figure 8.10 *Sequence diagram for* Add a New Advert to a Campaign if within Budget *showing branching.*

This shows a sequence diagram for the use case Add a New Advert to a Campaign if within Budget (an extension of the use case Add a New Advert to a Campaign). The use case description is as follows.

> *A new advertisement is added to a campaign by the campaign manager only if the campaign budget is not exceeded by adding the new advert. If adding the advertisement would cause the budget to be exceeded then a campaign budget extension request is generated. This will be*

---

[6] 'Priming' a letter simply means adding a vertical dash to it—a and a' represent 'a' and 'a primed'.

recorded for later reference. The budget extension is printed and sent to the client at the end of the day.

The first part of this sequence diagram is identical to that for `Check Campaign Budget`, and this is indicated by the annotated bar across the top of the diagram. This is not UML notation—there is not currently any UML notation to show that one sequence diagram is linked to another in this way. However, it is sometimes useful to be able to show this (an alternative way of representing this situation using a *probe* is shown in Figure 16.14). This relationship is also evident from the use case diagram, where `Add a New Advert to a Campaign if within Budget` has a «uses» relationship with `Check Campaign Budget`.

The branching is seen where two messages `Advert()` and `Request()` both start from the same point on the `Campaign` lifeline. Each branch is followed only if the branch condition is true; this is shown in square brackets before the message label. The fact that the `Advert()` message is above the `Request()` message does not imply a time sequence since the two branches are actually alternative execution pathways. Only one branch is followed during any one execution of the use case.

Branching notation can be used at a generic level to create a sequence diagram that represents all possible sequences of interaction for a use case. Looping and branching constructs respectively correspond to iteration and decision points in the use case. At an instance level a sequence diagram shows communication between specific objects. The two kinds of sequence diagram (generic and instance level) are equivalent to one another if the interactions implied by the use case contain no looping or branching constructs.

8.4.4 Dialogue management—interface classes

Most use cases imply at least one interface class that manages the dialogue between the actor and the system. The examples illustrated in this chapter do not include any explicit interface or boundary classes. In Chapter 16 we show a number of sequence diagrams that explicitly include interface classes, but these have been prepared from a design perspective. For example, Figure 16.13 shows a further revision to the sequence diagram for the use case `Check Campaign Budget`. If this is compared with the version shown in Figure 8.8, it can be seen that the management of the interaction between the user and the system is managed explicitly by the `CBWindow` class in Figure 16.13, whereas in Figure 8.8 the dialogue management is not shown at all. Some methodologies (for example Objectory, which is described in Chapter 20) take a stance on this point that requires the identification of interface classes during the analysis stages of the project, while other approaches take the view that this is primarily a feature of design. Larman (1998) approaches the question of interaction modelling from a very different perspective, and suggests that during the analysis activity, system interaction diagrams should be drawn that represent only the interface or boundary object and the messages that are exchanged between the user and the system as a whole.

We believe that it is not usually necessary to show interface classes explicitly on an analysis sequence diagram but sometimes a case can be made for this. One example is when the inclusion of explicit interface classes appears likely to help in establishing the requirements. For example, a sequence diagram could provide a basis for story-boarding a use case, and this in turn may give the client and the potential users a view of how the application might operate and hence help to elicit further requirements for the system.

8.4.5 Modelling real-time systems and concurrency

Real-time systems are broadly characterized by the need to respond to external events within tight time constraints. Partly for this reason, they frequently exhibit concurrent behaviour in the form of simultaneous execution pathways or *threads of control*. An application that has concurrent execution always includes some objects that co-ordinate and initiate threads of control; these are known as *active objects* and belong to active classes. In addition, real-time applications usually include many other objects that work only within a thread of control; these are known as *passive objects* and belong to passive classes. The classes and objects that have been discussed so far in this chapter are passive. Active objects or classes are shown with a heavy border on collaboration diagrams and are frequently composites with embedded parts (the interface or boundary class as a composite is discussed in Chapter 16).

It is important for a sequence diagram of a concurrent system to show clearly which threads of control are active at any time. Control is represented by the synchronous and asynchronous messages introduced earlier in this section. In this usage, a full (synchronous) arrowhead represents the yielding of a thread of control and a half (asynchronous) arrowhead represents a message that is sent without control being yielded.

8.5 Collaboration Diagrams

Collaboration diagrams are the second kind of interaction diagram in the UML notation set. They provide an alternative notation for interactions that also shows links between the collaborating objects. They are more frequently used during design than during analysis, but we discuss them here rather than later in the book because of their strong link with sequence diagrams.

In UML a collaboration is seen as containing a set of modelling elements that interact within a given context for a particular purpose. This is a slightly broader view than the concept of a collaboration employed in the CRC cards technique, in that a UML collaboration can range from the performance of a single operation to the articulation of a whole use case. At one point in the UML Notation Guide (Rational, 1997) it is suggested rather poetically that a collaboration is a 'society of objects'.

8.5.1 Basic concepts and notation

Collaboration diagrams have many similarities to sequence diagrams. They express the same information in a different format, and, like sequence diagrams, they can be drawn at various levels of detail and during different stages in the system development process. Due to their similar content, collaboration diagrams can be used for the automatic generation of sequence diagrams. The most significant difference between the two types of interaction diagram is that a collaboration diagram explicitly shows associations or links between the different class or object *roles*[7] that participate in a collaboration. Unlike sequence

---

[7] The term 'role' indicates that the level of abstraction of a collaboration diagram can range from specific objects within an application to generic object roles that are not specific to one application.

diagrams, there is no explicit time dimension. Figure 8.11 shows an example of a collaboration diagram.

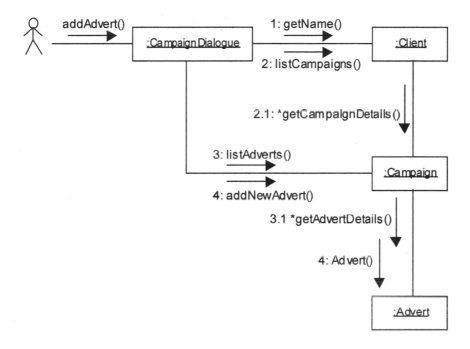

Figure 8.11 *Collaboration diagram for the use case* Add a New Advert to a Campaign.

In a collaboration diagram the interaction is drawn on what is essentially a fragment of a class or object diagram, as can be seen in Figure 8.11. This example is drawn at quite a simple level of detail (but note that it includes a dialogue object :CampaignDialogue). This level of detail is often sufficient to capture the nature of a collaboration. Since the diagram has no time dimension the order in which messages are exchanged is represented by sequence numbers. Some of the sequence numbers are written in a nested style (for example, 2.1 and 3.1) to indicate iterative loops within the interaction that is being modelled. A similar style of numbering is used to indicate branching constructs.

8.5.2 Message labels in collaboration diagrams

Messages on a collaboration diagram are represented by a set of symbols that are the same as those used in a sequence diagram, but with some additional elements to show sequencing and recurrence as these cannot be inferred from the structure of the diagram. Each message label includes the message signature and also a sequence number that reflects call nesting, iteration, branching, concurrency and synchronization within the interaction. The formal message label syntax is as follows:

```
[predecessor]  [guard-condition]  sequence-expression  [return-
value ':='] message-name'('[argument-list]')'
```

(Characters in single quotes such as `':='` are literal values. The other terms will be replaced in actual messages by relevant values. Terms in square brackets [...] are optional. Terms in braces { ... } are repeated zero or more times.)

A *predecessor* is a list of sequence numbers of the messages that must occur before the current message can be enabled. This permits the detailed specification of branching pathways. The message with the immediately preceding sequence number is assumed to be the predecessor by default, so if an interaction has no alternative pathways the predecessor list may be omitted without any ambiguity. The syntax for a predecessor is as follows:

```
sequence-number1 { ',' sequence-number2 } '/'
```

The `'/'` at the end of this expression indicates the end of the list and is only included when an explicit predecessor is shown.

Guard conditions are written in Object Constraint Language (OCL) (see Chapter 9), and are only shown where the enabling of a message is subject to the defined condition. A guard condition may be used to represent the synchronization of different threads of control.

A *sequence-expression* is a list of integers separated by dots (`'.'`) optionally followed by a *name* (a single letter), optionally followed by a *recurrence* term and terminated by a colon. A *sequence-expression* has the following syntax:

```
integer { '.' integer }[name] [recurrence] ':'
```

In this expression *integer* represents the sequential order of the message. This may be nested within a loop or a branch construct, so that, for example, message 4.1.3 occurs after message 4.1.2 and both are contained within the activation of message 4.1. In Figure 8.11 message 3.1 is nested within the activation of message 3. A comparison of this diagram with the corresponding sequence diagram (Figure 8.5) may help the reader to clarify how sequence numbers express a loop construct. The *name* of a sequence-expression is used to differentiate two concurrent messages since these are given the same sequence number. For example, messages 4.2.1a and 4.2.1b are concurrent within the activation of message 4.2. *Recurrence* reflects either iterative or conditional execution and its syntax is as follows:

Branching:	`'['condition-clause']'`
Iteration:	`'*' '['iteration-clause']'`

Some further sample message labels are listed in Figure 8.12.

Type of message	Syntax example
Simple message.	`4: addNewAdvert()`
Nested call with return value. *The return value is placed in the variable* `name`.	`3.1.2: name:= getName()`
Conditional message. *This message is only sent if the condition* `[balance > 0]` *is true.*	`[balance > 0] 5: debit(amount)`
Synchronization with other threads. *Message* `4: playVideo()` *is invoked only once the two concurrent messages* `3.1a` *and* `3.1b` *are completed.*	`3.1a, 3.1b/ 4: playVideo()`

Figure 8.12 *Examples of the syntax for various types of message label.*

Figure 8.13 illustrates the use of a collaboration diagram to show the interaction for a single operation, in this case checkCampaignBudget(), which is one of the operations shown in the sequence diagrams in Figures 8.7 and 8.8.

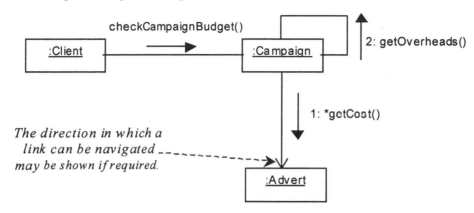

Figure 8.13 *Collaboration diagram for the operation* checkCampaignBudget().

Collaboration diagrams are preferred to sequence diagrams by some developers as they offer a view of object interaction that is easy to relate to a class diagram due to the visibility of object links. This is one of the reasons why they are often used for the documentation of Design Patterns (see Chapter 14). They also show how object roles interact: Design Patterns describe object roles. However, when there are many messages between two objects in one interaction then a collaboration diagram is more difficult to read than the equivalent sequence diagram. Sequence diagrams also offer more explicit visual cues to the duration of each activation. We believe that it is seldom useful to draw both kinds of interaction diagram for the same use case, but some people use sequence diagrams for the generic model of the interaction and collaboration diagrams for specific scenarios. The choice of which is the more appropriate usually depends on the nature of the interaction and the purpose of the diagram. Often neither offers a clear advantage and it should be a matter of organizational policy as to which is normally prepared.

8.6 Model Consistency

The preparation of interaction diagrams (either sequence or collaboration diagrams) involves the allocation of operations to classes. These operations should be listed against the correct classes in the class diagram and if operation signatures have been specified in full these must be consistent. The interaction diagrams and the class diagram(s) should be mutually consistent. A good CASE tool enforces this consistency at a syntactic level, usually by prompting the developer with a list of operations currently allocated to a class when he or she adds a message that is sent to an object of that class. If a corresponding operation has not been defined in the destination class the appropriate operation is added automatically to the class definition.

But to ensure full consistency between a class diagram and a set of related interaction diagrams requires more than the simple syntactic check described above. Every sending object in an interaction diagram must have the ability to send its message to the destination object and this requires it to know the identity or *object reference* of the destination object. There are only two ways that a sending object can know the reference of a destination object. Either it is already known by the sending object via a direct link, which really means that an association exists between the respective classes to which the objects belong. Alternatively, a sending object may obtain the reference it needs indirectly from another object (usually of a different class) that has a link with the destination object. The representation and placement of the object references that represent associations is a design issue that we discuss in detail in Chapter 13. At this stage it is sufficient to ensure that there is some possible pathway via object links (deduced from associations on the class diagram) that connects a sending object to a destination object. Any inconsistency between an interaction diagram and the corresponding class diagram indicates a flaw in one or the other. There may be a necessary association missing from the class diagram, for instance. Note that the existence of an association does not by itself guarantee the existence of any particular link. Where the minimum multiplicity of an association is zero there may not be any object that has a link for that association. If the multiplicity is one (or more), then each object must have at least one link. All message pathways should be analysed carefully.

When both sequence and collaboration diagrams are prepared for the same interaction the threads of execution specified in each must be exactly the same. Statechart diagrams (described in Chapter 10) document information about messages from the perspective of an object rather than an interaction, and it is also important to check for consistency between the statechart diagram for a class and all interaction diagrams that involve objects of that class.

8.7 Summary

Object interaction is a crucial feature of the object-oriented approach to systems development. When discussing object interaction we use the metaphor of message passing to describe the mode of collaboration between objects. Developing interaction diagrams requires a careful analysis of the use cases and may involve the use of CRC cards. These enable project team members to role-play the objects involved in the use case to determine the most appropriate allocation of responsibility. UML uses two equivalent modelling techniques to describe interactions—sequence diagrams and collaborations diagrams—that provide a rich notation. An integral part of the process of developing interaction diagrams is ensuring that they and the class diagrams for an application are mutually consistent.

Review Questions

8.1   List two specific features of bad object-oriented modelling that are discouraged by the use of collaboration diagrams. (Hint: What do we try to achieve by the use of CRC cards?)

8.2   What are the benefits of keeping all classes reasonably small and self-contained?

8.3   What is the difference between a responsibility and a collaboration? How are they related?

8.4   What are the main differences between sequence diagrams and collaboration diagrams?

8.5 What are the essential parts of a message label (i) in a sequence diagram and (ii) in a collaboration diagram?

8.6 What is an object lifeline?

8.7 What is meant by the focus of control?

8.8 How do asynchronous messages differ from synchronous messages (i) in their behaviour and (ii) in their notation?

8.9 In what circumstances are sequence numbers in a collaboration diagram written in nested style (e.g. 3.2.1)?

8.10 What consistency checks should be applied to interaction diagrams?

Case Study Work, Exercises and Projects

Exercises 8.A–8.C are based on the use case Start Line Run for the FoodCo case study, which you developed for Exercise 6.A.

8.A Identify the classes involved and prepare and complete CRC cards for each class. Now develop a sequence diagram that documents the interaction.

8.B Develop an alternative allocation of responsibility (you may find it helps to complete another set of CRC cards). Draw a second sequence diagram and compare it to your first one. Which is the more appropriate, and why?

8.C Prepare a collaboration diagram based upon either CRC card analysis. Perform a consistency check between the collaboration diagram, the corresponding sequence diagram and your class diagram for the case study.

8.D Using a CASE tool with which you are familiar, enter a class diagram together with some associated sequence and collaboration diagrams (e.g. the FoodCo models).

8.E Critically evaluate the extent to which the CASE tool supports UML and the consistency check that is necessary between the different diagrams.

Further Reading

Bellin and Simone (1997) provides a thorough introduction to the use of CRC cards. Hicks (1991) offers sound advice about how to use brain-storming effectively. Larman's (1998) use of sequence diagrams during analysis is both interesting and very different from the approach advocated in this book. Rumbaugh et al. (1991) and Booch (1994) discuss their variants of message modelling in detail. Buschmann et al. (1996) provides interesting examples of system sequence diagrams, although this is in a notation that is not quite UML.

9

Specifying Operations

Objectives
In this chapter you will learn:
- why operations are specified;
- the difference between algorithmic and non-algorithmic methods;
- how to interpret different ways of specifying operations;
- how to specify operations using one method.

9.1 Introduction

Operation specifications play a similar role in a data dictionary to that of other entries, such as attribute specifications. They support a graphical model by adding precision so that users can confirm the correctness of the model, and designers can use them as a basis for software development. But they are potentially the most complex of all entries in a data dictionary, since they explain the detailed behaviour of the system.

We consider the need for specifying operations (Section 9.2), and introduce the 'contract' (Section 9.3) as a kind of black box specification. If the behaviour of an operation is simple, a contract that describes only its external interface may be all that is required, and if its behaviour is not yet understood in any detail, a black box specification may be all that is possible. Often there is also a need to describe an operation's logic, or internal behaviour.

The two general ways of doing this are respectively called 'algorithmic' (or 'procedural') and 'non-algorithmic' (or 'declarative'). A non-algorithmic approach is generally preferred in object-oriented development, but in some situations only an algorithmic approach is sufficiently expressive. We review some established techniques, in particular decision tables and pre- and post-condition pairs (Section 9.4.1) and Structured English (Section 9.4.2). None of these are specific either to UML, or to an object-oriented approach, but all can be used to specify operations in a UML model. A full description is beyond the scope of this book, so we introduce them at an overview level.

UML does not require any specific techniques or notations for specifying operations, but the notation set includes a formal language for specifying general constraints on a model, known as the Object Constraint Language (OCL). We introduce OCL (Section 9.5) and show how it can be used in specifying operations (Section 9.6).

A constraint can only be expressed in a formal notation, such as OCL, if it is known with confidence and understood with some precision. Sometimes what the analyst or designer must record is their uncertainty or lack of understanding. We introduce the UML note as a means of recording information that is either informal or uncertain (Section 9.7).

9.2 The Role of Operation Specifications

Each operation specification is a small but necessary step on a path that begins with a user's idea of a business activity, and leads ultimately to a software system made up of collaborating objects with attributes and methods. From an analysis perspective, an operation specification is created at a point when the analyst's understanding of some aspect of an application domain can be fed back to users, ensuring that the proposals meet users' needs. From a design perspective, an operation specification is a framework for a more detailed design specification, that later guides a programmer to a method that is an appropriate implementation of the operation in code. An operation specification can also be used to verify that the method does indeed meet its specification, which in turn describes what the users intended, thus checking that the requirements have been implemented.

Novice programmers often do not appreciate the need to design, still less specify, an operation before beginning to write it in program code. This is partly because beginners are given such simple tasks, e.g. to write a program that can calculate and display the area of a rectangle. More importantly, the student is shielded from the activity of requirements analysis. In effect, the teacher has already carried this out, and the student is presented with its results as a starting point: 'There is a need for a programme to calculate the area of a rectangle'. Why? Any answer given to a student will be couched in educational terms, such as: 'This will help you to develop important basic skills in ...'

Of course, the situation just described is quite artificial, and most students know this perfectly well. But it is only once the complexity or scale of a software system reaches a certain threshold, that the production of code too early becomes extremely inefficient, and very possibly disastrous. To code a relatively small sub-task in a large system requires some understanding of the ways in which that sub-task will interact with other sub-tasks. If this understanding has not yet been achieved, assumptions must be made, and these may later turn out to be inappropriate, even disastrous, for the system as a whole.

Object-oriented programming is generally more immune to this kind of problem than other programming approaches, but it is still important to describe the logical operation of the planned software as early as possible. Modelling object interaction (see Chapter 8) is part of this description, as it helps to determine the distribution of behaviour among the various classes. A detailed description of individual operations must also now be provided.

There are differences of opinion on how much specification should be done. On the one hand, Rumbaugh et al. (1991) suggest that only operations that are 'computationally interesting' or 'non-trivial' need be specified. 'Trivial' operations (e.g. those that create or destroy an instance, and those that get or set the value of an attribute) need not be specified at all. Further, all specifications are simple, consisting only of the operation's signature and a description of its 'transformation' (i.e. its logic). On the other hand, Allen and Frost (1998), recommend the specification of all operations, although the level of detail may vary according to the anticipated needs of the designer. We recommend the latter approach, mainly because of the problems that can arise later in a project if full documentation is not maintained. It is important to keep at least to a minimal documentation standard, even for operations that are very simple.

Each operation has a number of characteristics, which should be specified at the analysis stage. Users must confirm the logic, or rules, of the behaviour. The designer and the programmer responsible for the class will be the main users of the specification, as they need to know what an operation is intended to do: does it perform a calculation, or transform data, or answer a query? Designers and programmers of other parts of the system also need to know about its effects on other classes. For example, if it provides a service to other classes, they need to know its signature. If it calls operations in other classes or updates the values of their attributes, this may establish dependencies that guide how these classes should be packaged during design or implementation.

Defining operations should neither be begun too early, nor left too late. For Allen and Frost (1998), this task should be left until the class diagram has stabilized. In a project where the development activity has been broken down at an early stage to correspond to separate sub-systems, this may refer only to that part of the class diagram which relates to a particular sub-system. But for any given part of the model, it is important to create all operation specifications before moving into the object design activity.

9.3 Contracts

The term 'contract' is a deliberate echo of legal or commercial contracts between people or organizations. Signing (or becoming a party to) a contract involves making a commitment to deliver a defined service to an agreed quality standard. For example, a small ground-care company has a contract to mow the grass on the lawn in front of the Agate headquarters building. The contract stipulates how often the grass must be cut (every two weeks from April to October), the maximum height of the grass immediately after it is cut (no more than 3 centimetres) and how much Agate will pay for the service (£80 per cut). The contract does *not* spell out how the work will be done—for example, what type of mower should be used (electric or petrol, cylinder or rotary), how many staff or mowers should be involved, or in which direction the lawn should be cut.

In the language of system theory, a contract is an interface between two systems. In this example, Agate is a business system and the ground-care company is a system for mowing Agate's grass. The contract defines inputs and outputs, and treats the grass-mowing system to some extent as a black box, with its irrelevant details hidden. Which details are deemed irrelevant is always a matter of choice, and any contract can specify that some details of the implementation should be visible to other systems. For example, Agate's directors might not wish to permit the ground-care contractor to use toxic pesticides or weedkillers. This can be included as a constraint in the contract.

Meyer (1997) was one of the first to draw an analogy between commercial contracts and service relationships between objects. The use of the term has become widespread in object-oriented development since it stresses the encapsulation of classes and sub-systems in a model. Cook and Daniels (1994) use the concept extensively in the Syntropy™ methodology, and Allen and Frost (1998) apply it also in The SELECT Perspective methodology.

One of Meyer's principal reasons for using the analogy of a contract is that design-by-contract helps to achieve a software design that is correct in terms of its requirements. During requirements analysis, we do not yet need the full technical rigour that is required of a design specification, but there is still a clear advantage in adopting an approach that can later be extended seamlessly through design into implementation. Specification by contract means that operations are defined primarily in terms of the services they deliver, and the 'payment' they receive (usually just the operation signature).

Contracts can also be applied at a much higher level of abstraction than individual operations. Larman (1998) describes the use of contracts to define services provided by a system as a whole. Whether written for a single operation, for the behaviour of the system as a whole, or for some intermediate packaged component, the structure of a contract is very similar. A commercial contract usually identifies the parties, the scope (i.e. the context in which it applies), the agreed service, and any performance standards that apply. In just the same way, in object-oriented modelling we identify the nature of the service provided by the server object, and what must be provided by the client object in order to obtain the service. These various aspects can be summarized as follows.

- The intent or purpose of the operation.
- The operation signature including the return type (probably established during interaction modelling).
- An appropriate description of the logic (the following sections present some alternative ways of describing the logic of an operation).
- Other operations called, whether in the same object or in other objects.
- Events transmitted to other objects.
- Attributes set during the operation's execution.
- The response to exceptions (e.g. what should happen if a parameter is invalid).
- Any non-functional requirements that apply.

(Adapted from Larman, 1998 and Allen and Frost, 1998.)

Most of this is self-explanatory, but the critical part of an operation specification is the logic description, and it is to this that we turn in the next section.

Review Question

9.1   What are the two main purposes of an operation specification?

9.4 Describing Operation Logic

Rumbaugh et al. (1991) suggest an informal classification of operations that is a useful starting point in considering the various ways of describing their logic. Firstly, there are operations that have side-effects. Possible side-effects include the creation or destruction of object instances, setting or returning attribute values, forming or breaking links with other objects, carrying out calculations, sending messages or events to other objects or any combination of these. A complex operation may do several of these things, and, where the task is at all complex, an operation may also require the collaboration of several other objects. It is for this reason that we identify the pattern of object collaboration before specifying operations in detail. Secondly, there are operations that do not have side-effects. These are pure queries; they request data but do not change anything within the system.

Like classes, operations may also be stereotyped as {abstract} or {concrete} (although this decision is often the result of design considerations, and is therefore not always made at this stage). Abstract operations have a form that consists at least of a signature, sometimes a full specification, but they will not be given an implementation (i.e. they will not have a method). Typically, abstract operations are located in the abstract superclasses of an inheritance hierarchy. They are always overridden by concrete methods in concrete subclasses.

A specification may be restricted to defining only external and visible effects of an operation, and we may choose either an algorithmic or a non-algorithmic technique for this. A specification may also define internal details, but this is effectively a design activity.

9.4.1 Non-algorithmic approaches

A non-algorithmic approach concentrates on describing the logic of an operation as a black box. In an object-oriented system this is generally preferred for two reasons. Firstly, classes are usually well-encapsulated, and thus only the designers and programmers responsible for a particular class need concern themselves with its internal implementation. Collaboration between different parts of the system is based on public interfaces between classes and sub-systems implemented as operation signatures (or message protocols). As long as the signatures are not changed, a change in the implementation of a class, including the way its operations work, has no effect on other parts of the system[1]. Secondly, the relatively even distribution of effort among the classes of an object-oriented system generally results in operations that are small and single-minded. Since the processing carried out by any one operation is simple, it does not require a complex specification.

Even in non object-oriented approaches, a declarative approach has long been recognized as particularly useful where, for example, a structured decision is made, and the conditions that determine the outcome are readily identified, but the actual sequence of steps in reaching the decision is unimportant. For situations like this, structured methods made use of non-algorithmic techniques such as decision tables and pre- and post-condition pairs (described in the following sections).

Decision tables

A decision table is a matrix that shows the conditions under which a decision is made, and the actions that may result. Decision tables cater best for situations where there are multiple outcomes, or *actions*, each depending on a particular combination of input *conditions*. One common form shows the conditions in the form of questions that can be answered with a simple yes or no. Actions are listed, and check-marks are used to show how they correspond to the conditions. The following is an example of a possible application in the Agate case study. Figure 9.1 shows a corresponding decision table.

When a campaign budget is overspent, this normally requires prior approval from the client, otherwise Agate is unlikely to be able to recover the excess costs. A set of rules has been established to guide Campaign Managers when they identify a possible problem. If the budget is expected to be exceeded by up to 2%, a letter is sent notifying the client of this. If the budget is expected to be exceeded by more than 2%, a letter is sent and the staff contact also telephones the client to invite a representative to a budget review meeting. If the campaign is not thought likely to exceed its budget, no action is taken.

---

[1] In practice, the situation is more complex than this might suggest. Some object-oriented languages, e.g. C++, allow a lot of flexibility as to how much of the internal implementation of a class is visible to other classes.

Conditions and actions	Rule 1	Rule 2	Rule 3
Conditions			
Is budget likely to be overspent?	N	Y	Y
Is overspend likely to exceed 2%?	-	N	Y
Actions			
No action	X		
Send letter		X	X
Set up meeting			X

Figure 9.1 *A decision table with two conditions and three actions, yielding three distinct rules.*

The vertical columns with Y, N and X entries are known as *rules*. Each rule is read vertically downwards, and the arrangement of Ys and Ns indicates which conditions are true for that rule. An X indicates that an action should occur when the corresponding condition is true (i.e. has a Y answer). We can paraphrase the table into text as follows.

Rule 1. If the budget is not overspent (clearly in this case the scale of overspend is irrelevant, indicated by a dash against this condition), no action is required.

Rule 2. If the budget is overspent and the overspend is not likely to exceed 2%, a letter should be sent.

Rule 3. If the budget is overspent and the overspend is likely to exceed 2%, a letter should be sent and a meeting set up.

A single rule may have multiple outcomes that overlap with the outcomes of other rules. Decision tables are very useful for situations that require a non-algorithmic specification of logic, reflecting a range of alternative behaviours. But this is relatively unusual in an object-oriented system, where thorough analysis of object collaboration tends to minimize the complexity of single operations.

Review Question

9.2   To what kinds of situation are decision tables particularly suited?

Pre- and post-conditions

As its name suggests, this technique concentrates on providing answers to the following questions.

- What conditions must be satisfied before an operation can take place?
- What are the conditions that can apply (i.e. what states may the system be in) after an operation is completed?

Let us consider an example from Agate. The operation `Advert.getCost()`[2] was first

---

[2] We have added brackets to the names of operations to distinguish them from attributes. This does not mean that these are operations with no parameters, only that the parameters are not shown.

discussed in Section 8.2. Let us suppose that it has the following signature.

```
Advert.getCost():Money
```

This operation has no pre-condition. (We may note that the, object sending the message must know the identity of the object that contains the operation, but this is not in itself a pre-condition for the operation to execute correctly when invoked). The post-conditions should express the valid results of the operation upon completion. In this case, a money value is returned (for simplicity, we ignore the question of valid values for an advert cost, but we should note that in reality this attribute may be able to take only a limited range of values, depending on business constraints).

Pre-condition: none
Post-condition: a valid money value is returned

More complex examples can easily by constructed from the use case descriptions, or by consulting users if existing descriptions are not sufficiently detailed. Consider the use case `Assign Staff to Work on a Campaign`. This involves calling the operation `Campaign.assignStaff()` for each member of staff assigned. Let us assume that the signature of this operation is as follows.

```
Campaign.assignStaff(creativeStaff.id)
```

This example has one pre-condition: a calling message must supply a valid identity for a `creativeStaffObject`. There is one post-condition: a link must be created between the two objects.

Pre-condition: `creativeStaffObject.id` is valid
Post-condition: a link is created between `campaignObject` and
`creativeStaffObject`

Let us look at one more example from Agate, with more complex conditions. This is taken from the use case `Change the Grade for a Member of Staff` (we assume that the use case is being invoked for a member of creative staff). This use case involves several operations including:

```
CreativeStaff.changeGrade()
StaffGrade.setFinishDate()
StaffGrade() (the constructor operation that creates a new instance of this class)
```

We examine only one of these in detail, `CreativeStaff.changeGrade()`, but our specification must still recognize calls made to other operations during execution. Let us assume that the operation signature is as follows.

```
StaffGrade.changeGrade(creativeStaffObject.id, gradeObject.id,
                       gradeChangeDate)
```

The pre-conditions are straightforward, consisting only of a valid `CreativeStaff.id` and `Grade.id`. The post-conditions are more involved, as once the operation is completed we should expect several effects to have taken place. A new instance of `StaffGrade` is created, and this is linked to the appropriate `creativeStaffObject` and `gradeObject` (by a `StaffOnGrade` link). The new `staffGradeObject` is also linked to the previous `staffGradeObject` (by a `PreviousGrade` link). Attribute values in the new `staffGradeObject` are set by its constructor operation (including `gradeStartDate`, which is set equal to the supplied parameter `gradeChangeDate`). The attribute `StaffGrade.gradeFinishDate` in the previous instance is also set, through a message to invoke the operation `StaffGrade.setFinishDate`. A full logic description is thus as follows.

> *pre-conditions:* `creativeStaffObject.id` is valid
> `gradeObject.id` is valid
> `gradeChangeDate` is a valid date
>
> *post-conditions:* a new `staffGradeObject` exists
> the new `staffGradeObject` is linked to the
> `creativeStaffObject`
> the new `staffGradeObject` is linked to the previous one
> the value of the previous `staffGradeObject.gradeFinishDate` is
> set equal to `gradeChangeDate`

For many operations in an object-oriented model, such a specification would be sufficiently detailed. It must meet the following two tests.

- A user should be able to check that it correctly expresses the business logic.
- A class designer should be able to produce a detailed design of the operation for a programmer to code.

However, while a declarative approach to operation specification usually meets all the needs of object-oriented development, there is still sometimes a case for using an algorithm. One example would be a requirement that involves carrying out a calculation where the sequence of steps is significant, and neither a designer nor a programmer could reasonably be expected to come up with a formula that produces the correct result.

Review Question

9.3   Why is it important to specify both pre- and post-conditions for an operation?

9.4.2 Algorithmic approaches

An algorithm describes the internal logic of a process or decision by breaking it down into small steps (the word derives from al-Kwarazhmi, an Arab mathematician of the 9th century). The level of detail to which this is done varies greatly, depending on the information available at the time and on the reason for defining it. An algorithm also specifies the sequence in which the steps are performed. In the field of computing and information systems, algorithms are used either as a *description* of the way in which a programmable task is currently carried out (this is their purpose in operation specification), or as a *prescription* for a program to automate the task. This dual meaning again reflects the differing perspectives of analysis (understanding a problem and determining what must be done to achieve a solution) and design (the creative act of imagining a system to implement a solution). An algorithmic technique is almost always used during method design, because a designer is concerned with the efficient implementation of requirements, and must therefore select the best algorithm available for the purpose. But algorithms can also be used with an analysis intention. A major difference here is that there is no need for the analyst to worry about efficiency, since the algorithm need only illustrate accurately the results of the operation.

Algorithms are generally organized procedurally, which is to say that they use the fundamental programming control structures of sequence, selection and iteration. We can illustrate this in the Agate case study by considering the operation that calculates the total cost of a campaign. This operation is invoked during the use case Check a Campaign Budget. For ease of reference, the use case description is repeated below.

> *The campaign budget may be checked to ensure that it has not been exceeded. The current campaign cost is determined by the total cost of all the adverts and the campaign overhead costs.*

Let us suppose that there is a precise (though simple) formula for this calculation, based on summing the individual total costs of each advert, and adding the campaign overhead costs. For further simplicity, let us assume that the overhead cost part of the calculation simply involves multiplying the total of all other costs by an overhead rate (this approximates to normal accounting practice). To convey an understanding of the calculation, we can begin by representing it as a mathematical formula.

total_campaign_cost = (sum of all advert_costs) * overhead_rate

This does not explicitly identify all the steps, but a sequence can be deduced. In fact, several possible sequences can be deduced, but any sequence that always produces a correct result will do. One possible sequence, at a very coarse level of detail would include the following steps:

(1) add up all the individual advert costs;

(2) multiply the total by the overhead rate;

(3) the resulting sum is the total campaign cost.

For such a relatively simple calculation as this one, the formula itself would almost certainly serve better as a specification, but some are a lot more complex. When it is necessary to specify the sequence of calculation in more detail, we can use Structured English for this.

Review Questions

9.4 What are the main differences between algorithmic and non-algorithmic approaches to operation specification?

9.5 Why are non-algorithmic (or declarative) approaches generally preferred in object-oriented development?

9.6 Why are operation specifications in an object-oriented project likely to be small?

Structured English

This is a 'dialect' of written English that is about halfway between everyday non-technical language and a formal programming language. When it is necessary to specify an operation procedurally, this is the most useful and versatile technique. Its advantages include the possibility, with a little care, of retaining much of the readability and understandability of everyday English. It also allows the construction of a formal logical structure that is easy to translate into program code. Structured English is very easy to write iteratively, at successively greater levels of detail, and it is easily dismantled into components that can be reassembled in different structures without a lot of reworking. The logical structure is made explicit through the use of keywords and indentation, while the vocabulary is kept as close as possible to everyday usage in the business context. Above all, expressions and keywords that are specific to a particular programming language are avoided. The result ideally is something that a non-technical user is able to understand, alter or approve, as necessary, while it should also be useful to the designer. This means it must be capable of further development into a detailed program design without undue difficulty.

The main principles of Structured English are as follows. A specification is made up of a number of simple sentences, each consisting of a simple imperative statement or equation. Statements may only be combined in restricted ways that correspond to the sequence, selection and iteration control structures of structured programming. The very simplest specifications contain only sequences, and differ little from everyday English except in that they use a more restricted vocabulary and style (many organizations have their own Structured English house-style). Figure 9.2 gives some examples.

```
get client contact name
sale cost = item cost * ( 1 - discount rate )
calculate total bonus
description = new description
```

Figure 9.2 *Typical statements in Structured English.*

Selection structures show alternative courses of action, the choice between them depending on conditions that prevail at the time the selection is made. For example, an *if-then-else* construct, which has only two possible outcomes, is shown in Figure 9.3.

```
if client contact is 'Sushila'
   set discount rate to 5%
else
   set discount rate to 2%
end if
```

Figure 9.3 *An if-then-else construct.*

If the two alternatives are not really different actions, but are rather a choice between doing something and not doing it, the 'else' branch can be omitted. Figure 9.4 shows this simpler form.

```
if client contact is 'Sushila'
   set discount rate to 5%
end if
```

Figure 9.4 An *'else' branch is not always necessary. In this example, if Sushila is not the contact, the rate is not changed.*

Note that in each case the end of the structure is marked by end if. This marker allows the entire structure to be treated logically as an element, as if it were a single statement in a sequence.

Multiple outcomes are handled either by a *case* construct or by a *nested if*. Figure 9.5 illustrates a case structure.

```
begin case
case client contact is 'Sushila'
   set discount rate to 5%
case client contact is 'Wu'
   set discount rate to 10%
case client contact is 'Luis'
   set discount rate to 15%
otherwise
   set discount rate to 2%
end case
```

Figure 9.5 *A case construct can handle many outcomes.*

The 'otherwise' branch of a case construct can be omitted if it is not required, although it is generally good practice to include a catch-all to ensure completeness. Figure 9.6 shows the same selection specified using a *nested-if* construct.

```
if client contact is 'Sushila'
   set discount rate to 5%
else
   if client contact is 'Wu'
      set discount rate to 10%
   else
      if client contact is 'Luis'
         set discount rate to 15%
      else
         set discount rate to 2%
      end if
   end if
end if
```

Figure 9.6 *A nested-if variant of selection between multiple outcomes.*

Figure 9.6 also illustrates how indentation can help the readability of a specification. For each corresponding set of control statements (lines beginning with 'if', 'else' and 'end if'), the indentation from the left margin is the same. This helps to show which sequence statements ('set discount rate to 10%', etc.) belong to each structure.

The third type of control structure is iteration. This is used when a statement, or group of statements, needs to be repeated. Typically this is a way of applying a single operation to a set of objects. Logically, once something has begun to occur repeatedly, there must be a condition for stopping the repetition (unless the repetition is to continue indefinitely). There are two main forms of control of iteration. These differ in whether the condition for ending the repetition is tested before or after the first loop. Figure 9.7 gives examples of both, showing some typical applications of this kind of structure.

```
do while there are more staff in the list
   calculate staff bonus
   store bonus amount
end do
```

Test applied before the loop is entered—if the list is empty, no bonus is calculated.

```
repeat
   allocate member of staff to campaign
   increment count of allocated staff
until count of allocated staff = 10
```

Test applied after the loop is exited—this ensures that the action is processed (or attempted) at least once. In this case the 'until' line acts as an end-of-structure marker.

Figure 9.7 *Alternative forms of iteration.*

Different types of structure can be nested inside each other, as shown in Figure 9.8.

```
do while there are more staff in the list
   calculate staff bonus
   store bonus amount
   begin case
   case bonus > £250
      add name to 'star of the month' list
   case bonus < £25
      print warning letter
   end case
end do
```

Figure 9.8 *Nesting different types of control structure inside each other.*

In this way, a Structured English specification can be made as complex as it needs to be, and it can also be written in an iterative, top-down manner. For example, an initial version of an algorithm is defined at a high level of granularity. Then, provided the overall structure is sound, more detail is easily added progressively. In refining the level of detail, structures can be nested within each other to any degree of complexity, although in practice it is unlikely that even the most complex operation would need more than two to three levels of nesting at most. It is in any case sensible to limit the complexity. One often-quoted guideline is that no Structured English specification should be longer than one page of typed A4, or one screen if it is likely to be read within a CASE tool environment.

The style in all the examples given above is based on that of Yourdon (1989), but this should not be taken as necessarily prescriptive. What passes for acceptable style varies widely from one organization to another, and in practice an analyst should follow the house-style, whatever it happens to be.

Pseudo-code

Pseudo-code differs from Structured English in that it is closer to the vocabulary and syntax of a specific programming language. There are thus many different dialects of pseudo-code, each corresponding to a particular programming language. They differ from each other in vocabulary, in syntax and in style. Structured English avoids language specificity primarily to avoid reaching conclusions about design questions too early. Sometimes there seems no good reason to hold back, for example, because the final implementation language has been decided early in the project. This can be misleading, as it may be desirable at a later stage to redevelop the system in a different programming language. If the operations have been specified in a language-specific pseudo-code, it would then be necessary to rewrite them.

However language-specific it may be, pseudo-code remains only a skeleton of a program, intended only to illustrate its logical structure without including full design and implementation detail. In other words, it is not so much a fully developed program as an outline that can later be developed into program code.

Activity diagrams

Activity diagrams are a part of the UML notation set that can be used to specify the logic of procedurally complex operations. Since they are closely related to the UML statechart, they are not described in this chapter, but instead they are introduced in Section 10.7 following the sections that introduce statechart notation.

Review Questions

9.7   What are the three kinds of control structure in Structured English?
9.8   What is a sensible limit on the size of a Structured English specification?

9.5 Object Constraint Language

In drawing any class diagram, much of the time and effort is spent in working out what constraints apply. For example, the multiplicity of an association represents a constraint on how many objects of one class can be linked to any object of the other class. This particular example can be adequately expressed in the graphical language of the class diagram, but this is not equally so for all constraints. Among those for which it is not true are many of the constraints within operation specifications. For example, many pre- and post-conditions in a contract are constraints on the behaviour of objects that are party to the contract. Sometimes the definition of such constraints can be done in an informal manner (as in the examples in Section 9.4.1), but where greater precision is required, OCL provides a formal language.

OCL expressions are constructed from a collection of pre-defined elements and types, and the language has a precise grammar that enables the construction of unambiguous statements about the properties of model components and their relationships to each other. Many OCL statements consist of the following features:

- A *context* that defines a domain within which the expression is valid. This is underlined for clarity, and is often an instance of a specific type, for example an object in a class diagram. A link (i.e. an instance of an association) may also be the context for an OCL expression.

- A *property* of that instance which is the context for the expression. Properties may include attributes, association-ends and query operations.
- An OCL *operation* that is applied to the property. Operations include (but are not restricted to) the arithmetical operators *, +, - and /, logical operators such as 'and', 'or' and 'implies' and set operators such as 'size', 'isEmpty' and 'select'.

Figure 9.9 gives some examples of expressions in OCL, mainly taken from the OCL Specification Document, which is part of the UML documentation (Rational, 1997). All have an object of some class as their context. The figure shows examples of OCL syntax and an interpretation of the meaning of each.

OCL expression	Interpretation
`Person` `self.age`	In the context of a specific person, the value of the property 'age' of that person—i.e. a person's age.
`Person` `self.income >= 5,000`	The property 'income' of the person under consideration must be greater than or equal to 5,000.

`Person`
    `self.wife->notEmpty implies`
    `self.wife.sex = female`
	If the set 'wife' associated with a person is not empty, then the value of the property 'sex' of the wife must be female. The boldface denotes an OCL keyword, but has no semantic import in itself.

`Company` `self.employee->size <= 50`	The size of the set of the property 'employee' of a company must be less than or equal to 50. That is, a company cannot have more than 50 employees.
`Company` `self.employee->select (age > 50)`	This specifies the set of employees of a company whose age is greater than 50.

Figure 9.9 *Examples of some expressions in OCL.*

OCL can specify many constraints that cannot be expressed directly in diagrammatic notation, and it is also useful as a precise language for pre- and post-conditions. Figure 9.10 shows the general syntax for operation specification.

```
Type::operation(parameter1:type,parameter2:type):return type
    pre: parameter1 > ...
         parameter2 = ...
    post:result = ...
```

Figure 9.10 *Syntax for operation pre- and post-conditions.*

Note that the contextual 'type' is the type which owns the operation as a feature. The 'pre' expressions are functions of operation parameters, while 'post' expressions are functions of self and/or operation parameters. For an example from the Agate case study, we revisit the operation `creativeStaff.changeGrade()`, for which a logic specification was constructed earlier (in Section 9.4.1). Figure 9.11 shows this rewritten in OCL.

```
CreativeStaff::changeGrade(id:String, grade.id:String,
gradeChangeDate:Date)
    pre:    self.id->notEmpty
            self.grade.id->notEmpty
            self.gradeChangeDate >= today (assumes no retrospective changes)
    post:   staffGrade[grade.id]->exists
            self.staffGrade->notEmpty
            self.staffGrade[grade.id].previousGrade->notEmpty
            self.staffGrade.gradeFinishDate = gradeChangeDate :
            Boolean
```

Figure 9.11 *Pre- and post-conditions for* `creativeStaff.changeGrade()`, *rewritten in OCL.*

An *invariant* is a formal expression that is always true for part of a model. If associated with an operation specification, an invariant is a condition that remains true for an object, and which must therefore not be altered by an operation side-effect. Formal definition of invariants is valuable because they provide rigorous tests for execution of the software.

For example, the value of a `campaignObject.estimatedCost` should always equal the sum of all associated `advertObject.estimatedCost` values. In OCL, this can be written as in Figure 9.12.

```
Campaign
    self.estimatedCost = self.Advert.estimatedCost->sum
```

Figure 9.12 *An example of an invariant that should remain true, regardless of operation side-effects.*

Review Questions

9.9   What are the three components of most OCL expressions?

9.10   What is an invariant?

9.6 Creating an Operation Specification

Figure 9.13 shows the sequence diagram for the use case Check a Campaign Budget first introduced in Chapter 8 (the use case description is repeated above in Section 9.4.2). In this particular example the message `checkCampaignBudget` invokes the operation `Campaign.checkCampaignBudget()`.

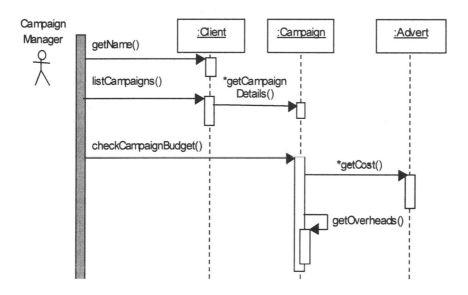

Figure 9.13 *Sequence diagram for the use case* Check Campaign Budget.

A specification for Campaign.checkCampaignBudget() is given below. We have used different fonts to signpost the specification as follows. This font (Arial) labels the **specification structure**, while this (Courier) highlights its content. Comments on the reasoning behind the specification are formatted like this (in Technical).

Operation specification: checkCampaignBudget

Operation intent: return campaign committed expenditure and budget.

The invocation appears not to require any parameters, but does have a return type that includes both the total committed expenditure and the budget, so we can expect it to contain two numerical values. Let us assume that there is a Money type available. The signature is shown below, followed by the pre- and post-conditions..

operation signature: <u>Campaign::checkCampaignBudget()</u>
<u>campaignBudget:Money, committedExpenditure:Money</u>

logic description (pre- and post-conditions):
```
pre: self.id->exists
post: campaignBudget = self.estimatedCost
      committedExpenditure = self.adverts.estimatedCost->sum
```

As can be seen from the ISD, this operation calls two other operations and these must be listed. In a full specification, full signatures would be recorded, but we omit this detail here.

Other operations called: Advert.getCost, self.getOverheads

Events transmitted to other objects: none

The only messages are those required to call the operations just mentioned, whose return values are required by this operation (remember that an 'event' is a message that starts another distinct thread of processing).

Attributes set: none

This is a query operation whose only purpose is to return data already stored within the system.

Response to exceptions: `none defined`

Here we could define how the operation should respond to error conditions, e.g. what kind of error message will be returned if a calling message uses an invalid signature.

Non-functional requirements: `none defined`

Several non-functional requirements may be associated with the operation, but these need to be determined through discussion with users. They may include, for instance, response time under live conditions (enquiries that are made frequently typically require a faster response) or the format of the output (e.g. if there is a house standard that overspent budgets are displayed in red). However, these are really design issues, and would be noted at this stage only if the information happens to be available at the time.

9.7 Use of Notes

There is often more than one appropriate way of modelling a scenario. A textual note is

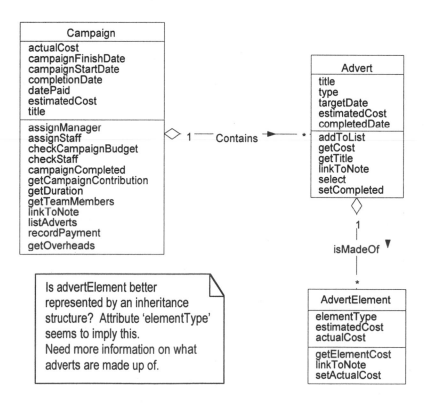

Figure 9.14 *A Note as a comment on a diagram.*

useful in this case, since it is stored with the diagram, and is thus available to any other team member working on that part of the system, but it does not itself change the semantics of the model. Alternatively, an analyst may wish to document their reasons for choosing a particular approach. Notes also provide a location for any semantic information not recorded elsewhere on the diagram or in the associated data dictionary (i.e. they do not apply specifically to any operation or attribute specification, or they belong to a specification that has not yet been created).

Figure 9.14 shows a note on a class diagram, used in this case to record a comment about the model that cannot be captured in the notation of class diagram. Here, the analyst has used the note to record her reservations about the model, because she believes that an aspect of the business is not yet well understood.

9.8 Summary

Operations specifications are the most detailed description of the behaviour of a system model. As such, they are also one of the more significant elements in the data dictionary. In this chapter we introduced the 'contract' as a framework for specifying operations, in terms of the service relationship between classes. We also introduced several techniques for describing operation logic, distinguishing between those that are algorithmic and those that are non-algorithmic. These include decision tables, pre- and post-condition pairs, and Structured English.

OCL was introduced as a formal language for specifying constraints on an object model, including operation pre- and post-conditions and invariants. Finally, we introduced the note as a notation for recording informal aspects of a model or comments on modelling assumptions or decisions.

Review Question

9.11   Why might an analyst wish to create notes on a class diagram?

Case Study Work, Exercises and Projects

9.A   Consider the first sequence diagram you drew for the use case `Start Line Run` from the FoodCo case study (in Exercise 8.A). Choose an operation in one of the classes involved in this ISD and write a contract for it. Make reasonable assumptions where necessary, and use a pre- and post-conditions approach for describing its logic.

9.B   Consider the decision table in Figure 9.1. Suppose you have learned that an extra condition must be taken into account: the rules in the current table actually apply only to campaigns with a total budget of £1,000 or over, but for smaller campaigns the thresholds for each action are different. Thresholds for smaller campaigns are as follows. For an expected overspend of less than 10%, no action is taken. For expected overspends of 10%–19%, a letter is sent. For an expected overspend of

20% or more a letter is sent and a meeting is arranged. How does this change the decision table? Draw a new version that takes into account the additional circumstances.

9.C   Consider the decision table in Figure 9.1. Which of the three control structures are required to convert this into a Structured English specification? Re-write the decision table in Structured English format.

9.D   Find out how a decision tree differs from a decision table (e.g. from one of the books listed in Further Reading). Produce a decision tree that corresponds to the decision table in Figure 9.1. What are the relative advantages and disadvantages of decision trees, decision tables and Structured English?

Further Reading

Most techniques for operation specification were in regular use long before object-oriented analysis became widespread, and many of the best treatments of these are in textbooks that deal with structured approaches to systems analysis.

Yourdon (1989) gives a thorough presentation of Structured English and pre- and post-conditions

Senn (1989) gives good coverage of decision tables.

Larman (1998) describes a contract-based approach to object-oriented aanalysis and design, with examples taken through to Java code.

Meyer (1997) is the definitive text on design-by-contract in object-oriented software engineering. This book is very comprehensive, but note that it is aimed at technically-oriented readers.

Further examples of operation specifications can be found in Case Study Chapter A2.

10

Specifying Control

Objectives
In this chapter you will learn:

- how to identify requirements for control in an application;
- how to model object life cycles using statecharts;
- how to develop statechart diagrams from interaction diagrams;
- how to model concurrent behaviour in an object;
- how to ensure consistency with other UML models.

10.1 Introduction

The various types of UML notation that have been introduced so far enable us to model the static structure of an application (class diagrams) and the way in which objects interact (sequence and collaboration diagrams). Another important aspect of an application that must be modelled is the way that its response to events can vary depending upon the passage of time and the events that have occurred already. For an application such as a real-time system it is easy to understand that the response of the system to an event depends upon its state. For example, an aircraft flight control system should respond differently to events (for example, engine failure) when the aircraft is in flight and when the aircraft is taxiing along a runway. A more mundane example is that of a vending machine, which does not normally dispense goods until an appropriate amount of money has been inserted. This variation in behaviour is determined by the state of the machine—which depends on whether or not sufficient money has been inserted to pay for the item selected. In reality, of course, the situation is more complicated than this. For example, even when the correct amount of money has been inserted, the machine cannot dispense an item that is not in stock. It is important to model state dependent variations in behaviour such as these since they represent constraints on the way that a system should behave.

Objects can have similar variations in their behaviour dependent upon their state. These variations in behaviour represent important constraints on the way that an object behaves and are dictated by the requirements for the system. UML uses statecharts to model states and state dependent behaviour for objects and for interactions. The notation used in UML is based upon work by Harel (1987) and was adopted by OMT and also in the second version of the Booch approach (Booch, 1994).

There is an important link between interaction diagrams and statecharts. A model of state behaviour in a statechart captures all the possible responses of a single object to all the use cases in which it is involved. By contrast, a sequence or a collaboration diagram captures the responses of all the objects that are involved in a single use case. A statechart can be seen as a description of all the possible life cycles that an object of a class may follow. It can also be seen as a more detailed view of a class. Some CASE tools handle statecharts as a child diagram to a class on a class diagram.

It is not only object-oriented approaches that have recognized the importance of modelling the lifetime behaviour of entities. For example, SSADM uses entity life histories for this purpose, and some approaches to the development of real-time systems, such as that of Yourdon, model the state behaviour of a system or sub-system.

The statechart is a versatile technique, and can be used within an object-oriented approach for other purposes than the modelling of object life cycles. In Chapter 16 we show how to use statecharts to build models of human–computer dialogues.

10.2 States and Events

All objects have a state. The current state of an object is a result of the events that have occurred to the object, and is determined by the current value of the object's attributes and the links that it has with other objects. Some attributes and links of an object are significant for the determination of its state while others are not. For example, in the Agate case study staffName and staffNumber attributes of a StaffMember object have no impact upon its state, whereas the date that a staff member started his or her employment at Agate determines when the probationary period of employment ends (after six months, say). The StaffMember object is in the Probationary state for the first six months of employment. While in this state, a staff member has different employment rights and is not eligible for redundancy pay in the event that they are dismissed by the company.

The UML Notation Guide (Rational, 1997) defines a state as follows:

> A state is a condition during the life of an object or an interaction during which it satisfies some condition, performs some action or waits for some event. An object waits in a state for a finite (non-instantaneous) time.

The possible states that an object can occupy are limited by its class. Objects of some classes have only one possible state. For example, in the Agate case study a Grade object either exists or it does not. If it exists it is available to be used, and if it does not exist it is not available. Objects of this class have only one state, which we might name Available. Objects of other classes have more than one possible state. For example, an object of the class GradeRate may be in one of several states. It may be Pending, if the current date is earlier than its start date, Active, if the current date is equal to or later than the start date but earlier than the finish date (we assume that the finish date is later than the start date), or Lapsed, if the current date is later than the finish date for the grade. The current state of a GradeRate object can be determined by examining the values of its two date attributes (alternatively, the GradeRate class might have a single attribute (an enumerated type—that has an integer value for each possible state) with values that indicate the current state of an object). It is important to note that movement from one state to another for a GradeRate object is dependent upon events that occur with the passage of time. Figure 10.1 shows a statechart for GradeRate.

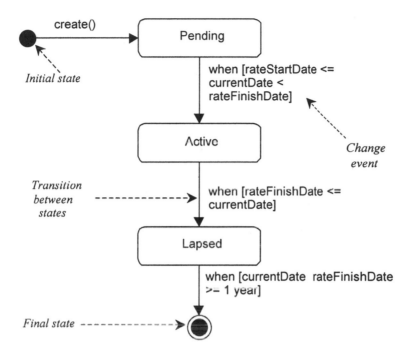

Figure 10.1 *Statechart for the class* GradeRate.

Movement from one state to another is called a *transition*, and is triggered by an *event*. When its triggering event occurs a transition is said to *fire*. A transition is shown as a solid arrow from the source state to the target state. An event is an occurrence that is relevant to the object or to an application, for example, the cancellation of an advert at Agate. Just as a set of objects is defined by the class of which they are all instances, events are defined by an *event type* of which each event is an instance. For example, the cancellation of an advert in the FoodCo coleslaw campaign is one instance of an event, and the cancellation of an advert in the Soong Motor Co Granda campaign is another instance. Both are defined by the event type cancellationOfAdvert(). It is usually event types that are modelled, but we refer to them simply as events. An event can have parameters and a return value, and in an object-oriented system it is implemented by a message.

Events can be grouped into several general types. A *change event* occurs when a condition becomes true. This is usually described as a Boolean expression, which means that it can take only one of two values: true or false. Change events are annotated by the keyword *when* followed by the Boolean expression in parenthesis. This form of conditional event is different from a guard condition that is only evaluated at the moment that its associated event fires.

A *call event* occurs when an object receives a call for one of its operations either from another object or from itself. Call events correspond to the receipt of a call message and are annotated by the signature of the operation as the trigger for the transition.

A *signal event* occurs when an object receives a signal. As with call events the event is annotated with the signature of the operation invoked. There is no syntactic difference between call events and signal events. It is assumed that a naming convention is used to distinguish between them.

A *time event* is caused by the passage of a designated period of time after a specified event (frequently the entry to the current state). Time events are shown by time expressions as triggers for the transitions. Where the time event reflects the passage of time since the entry to the current state the event is annotated by the keyword *after* followed by the amount of time in parentheses.

The basic syntax for a call or signal event is:

```
event-name '(' parameter-list ')'
```

where the parameter-list contains parameters of the form:

```
parameter-name ':' type-expression
```

separated by commas. (Characters in single quotes, such as ‘(’, are *literals* that appear as part of the event.)

10.3 Basic Notation

The *initial state* (in other words the starting point) of a life cycle is indicated by a small solid filled circle. The initial state is a notational convenience, and an object can not remain in its initial state but must immediately move into another named state. In Figure 10.1 the `GradeRate` object enters the `Pending` state immediately on its creation. A transition from the initial state can optionally be labelled with the event that creates the object. The end point of a life cycle (in other words its final state) is shown by a bull's-eye symbol. This too is a notational convenience, and an object cannot leave its final state once it has been entered. All other states are shown as a rectangle with rounded corners and should be labelled with a meaningful name. In this example all transitions except the transition from the initial state are triggered by change events. The statechart for a `GradeRate` object is very simple since it enters each state only once. Some classes have much more complex life cycles that this. For example, a `BookCopy` object in a library system may change many times between the states `OnLoan` and `Available`.

Figure 10.2 illustrates the basic notation for a statechart with two states for the class `Campaign` and one transition between them. A transition should be annotated with a transition string to indicate the event that triggers it. For call and signal events the format of the transition string is as follows:

```
[event-signature]    [guard-condition]    [action-expression]
[send-clause]
```

where the square brackets indicate that these elements are optional. The event signature takes the following form:

```
event-name '(' parameter-list ')'
```

A *guard condition* is a Boolean expression that is evaluated at the time the event fires. The transition only takes place if the condition is true. A guard condition is a function that may involve parameters of the triggering event and also attributes and links of the object that owns the statechart. A guard condition is contained in square brackets – '[…]'. In Figure 10.2 the guard condition is a test on the `contractSigned` attribute in the class `Campaign` and since the attribute is Boolean it could be written as follows:

```
[contractSigned]
```

This expression evaluates to true only if contractSigned is true. A guard condition can also be used to test concurrent states of the current object or the state of some other reachable object. Concurrent states are explained later in Section 10.4.2.

An *action-expression* is executed when an event triggers the transition to fire. Like a guard condition, it may involve parameters of the triggering event and may also involve operations, attributes and links of the owning object. In Figure 10.2 the action-expression begins with the '/' delimiter character and is the execution of the Campaign object's operation setCampaignActive().

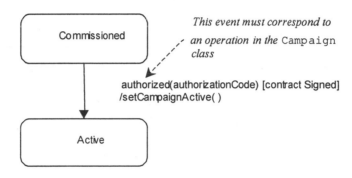

Figure 10.2 *Fragment of the statechart for the class* Campaign.

A *send-clause* is a special case of an action-expression that describes a message sent to another object. The syntax for a send-clause is as follows:

```
'^' destination-expression.destination-message-name
'(' parameter-list ')'
```

A transition may have more than one send-clause each with its own '^' delimiter. The sequence of action-expression statements, operations and send-clauses is significant since it determines the order in which they are executed. Actions are considered to be atomic (that is, they cannot be subdivided) and cannot be interrupted once they have been started. An action will actually take time to execute in a live application but this is dependent only upon internal factors such as the complexity of the program code and the speed of the processor. Unlike the duration of an action, the duration of a state is normally dependent upon external events in the application environment.

So far we have considered only action-expressions that are associated with a transition. It can also be useful to model internal actions that are triggered by events that do not change the state, by entering the state or by exiting the state. In Figure 10.3 the state symbol is shown with two compartments, a name compartment and an internal transitions compartment. Actions or activities that are associated with the state are listed in the latter. Note that an activity may persist for the duration of the state unlike actions which are transitory. An internal transition occurs in response to an event and results in an action but does not cause a change in state. The format for internal transitions is as follows:

```
event-name    '('    parameter-list    ')'    [guard-condition]
[action-expression]
```

Two kinds of internal event have a special notation. These are the *entry event* and the *exit event*, respectively indicated by the key words *entry* and *exit*. These cannot have guard conditions as they are invoked implicitly on entry to the state or exit from the state. Entry or exit *action-expressions* may also involve parameters of incoming transitions (provided that these appear on all incoming transitions) and attributes and links of the owning object. It is important to emphasize that any transition into a state causes the entry event to fire and all transitions out of a state cause the exit event to fire.

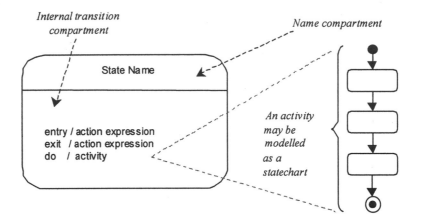

Figure 10.3 *Internal actions and activities for a state*

Activities are preceded by the key word *do* and have the following syntax:

```
'do' '/' activity-name '(' parameter-list ')'
```

Complex activities can be represented by a statechart nested within another state. The nesting of one statechart within another allows the representation of highly complex behaviour. When an activity in a state ends (i.e. the nested statechart reaches its final state) the state is considered completed and the object makes a transition triggered by the completion of this activity. Alternatively an activity may persist as long as the object remains in the state, in which case it does not trigger a transition from the state. The activity will only end when some other specified event triggers a transition from the state. Figure 10.4 shows the Menu Visible state for a DropDownMenu object.

Figure 10.4 Menu Visible *state for a* DropDownMenu *object.*

In this example, the entry action causes the menu to be displayed. While the object remains in the Menu Visible state, the activity causes a sound clip to be played and, if the event itemSelected() occurs, the action highlightItem() is invoked. When this state is exited the menu is hidden.

Figure 10.5 shows a statechart for the class Campaign. The transition from the initial state to the Commissioned state has been labelled only with an action-expression that comprises the operations `assignManager()` and `assignStaff()`. Execution of these operations ensures that when a campaign is created a manager and member(s) of staff are assigned to it[1]. The operations are triggered by the event that creates a `Campaign` object. The transition from the `Completed` state to the `Paid` state has a guard condition that only allows the transition to fire if total amount due (`paymentDue`) for the `Campaign` has been completely paid (note that this guard condition allows a `Campaign` to enter the `Paid` state when the customer overpays).

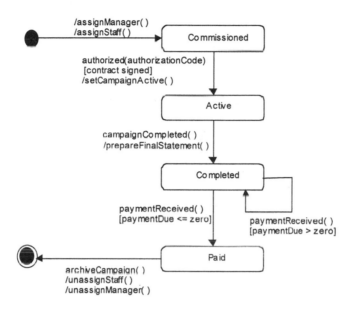

Figure 10.5 *Statechart for the class* `Campaign`.

The recursive transition from the `Completed` state models any payment event that does not reduce the amount due to zero or beyond. Only one of the two transitions from the `Completed` state (one of which is recursive) can be triggered by the `paymentReceived` event since the guard conditions are mutually exclusive. It would be bad practice to construct a statechart where one event can trigger two different transitions from the same state. The UML interpretation of this would be that either transition could fire indeterminately. A life cycle is only unambiguous when all the transitions from each state are mutually exclusive.

If the user requirements were to change, so that an overpayment is now to result in the automatic generation of a refund, the statechart can be changed as follows. Since the action that results from an overpayment is different from the action that results from a payment that reduces `paymentDue` to zero, a new transition is needed from the `Completed` state to the `Paid` state. The guard conditions from the `Completed` state must also be modified. Figure 10.6 shows a statechart that captures this requirement. It is important to appreciate that the statecharts in Figures 10.5 and 10.6 are not equivalent to each other, but capture different versions of the users' requirements.

---

[1] Unless the specifications for these operations permit a null option.

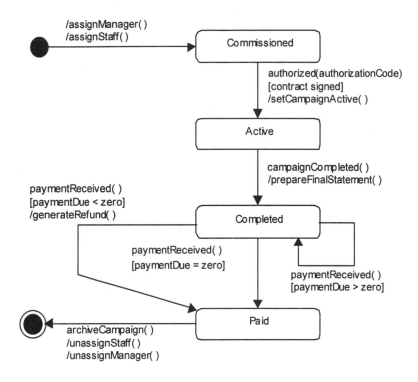

Figure 10.6 *A revised statechart for the class* `Campaign`

10.4 Further Notation

The statechart notation can be used to describe highly complex time-dependent behaviour. Hierarchies of states can be nested and concurrent behaviour can also be represented.

10.4.1 Nested states

When the state behaviour for an object or an interaction is complex it may be necessary to represent it at different levels of detail and to reflect any hierarchy of states that is present in the application.   For example, in the statechart for `Campaign` the state `Active` encompasses several *substates*. These are shown in Figure 10.7 where the `Active` state is seen to comprise three disjoint substates: `Advert Preparation`, `Scheduling` and `Running Adverts`. This diagram now shows a single state which contains within it a nested state diagram. In the nested statechart within the `Active` state, there is an initial state symbol with a transition to the first substate that a `Campaign` object enters when it becomes active. The transition from the initial state symbol to the first substate does not need to be labelled. If it is left unlabelled it is implicitly fired by any transition to the `Active` state. A final state symbol may also be shown on a nested state diagram. A transition to the final state symbol represents the completion of the activity in the enclosing state (i.e. `Active`) and a transition out of this state triggered by the completion event. This transition may also be unlabelled (as long as this does not cause any ambiguity) since the event that triggers it is implied by the completion event.

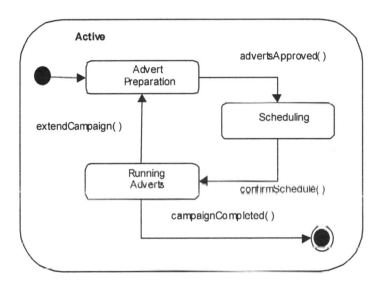

Figure 10.7 *The* `Active` *state of* `Campaign` *showing nested substates.*

When a campaign enters the `Active` state in Figure 10.7 it first enters the `Advert Preparation` substate, then if the adverts are approved it enters the `Scheduling` substate and finally enters the `Running Adverts` substate when the schedule is approved. If the campaign is deemed completed the object leaves the `Running Adverts` substate and also leaves the `Active` enclosing state, moving now to the `Completed` state (see Figure 10.5). If the campaign is extended the `Advert Preparation` substate is re-entered (Figure 10.7). A high level statechart for the class `Campaign` can be drawn to include within the main diagram the detail that is shown in the nested statechart for the `Active` state if so desired.

10.4.2 Concurrent states

Objects can have *concurrent* states. This means that the behaviour of the object can best be explained by regarding it as a product of two distinct sets of substates, each state of which can be entered and exited independently of substates in the other set. Figure 10.8 illustrates this form.

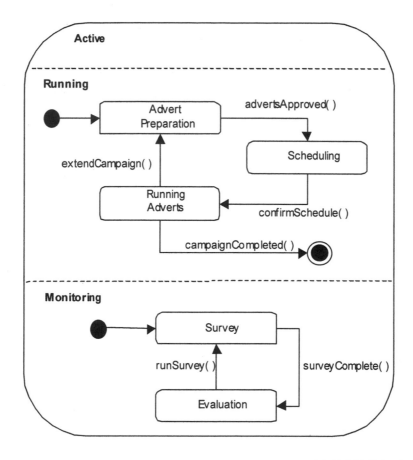

Figure 10.8 *The* Active *state with concurrent substates.*

Suppose that further investigation reveals that at Agate a campaign is surveyed and evaluated while it is also active. A campaign may occupy either the Survey substate or the Evaluation substate when it is in the Active state. Transitions between these two states are not affected by the campaign's current state in relation to the preparing and running of adverts. We model this by splitting the Active state into two concurrent nested statecharts, Running and Monitoring, each in a separate sub-region of the Active statechart. This is shown by dividing the state icon with a dashed line. These concurrent substates for the Active state of the Campaign class are shown in Figure 10.8.

A transition to a complex state such as this one is equivalent to a simultaneous transition to the initial states of each concurrent statechart. An initial state must be specified in both nested statecharts in order to avoid ambiguity about which substate should first be entered in each concurrent region. A transition to the Active state means that the Campaign object simultaneously enters the Advert Preparation and Survey states. A transition may now occur within either concurrent region without having any effect on the state of the object in the other concurrent region. However, a transition out of the Active state applies to all its substates (no matter how deeply nested). In a sense, we can say that the substates inherit the campaignCompleted() transition from the Active state since it applies implicitly to them all. This is equivalent to saying that an event that triggers a transition out of the `Active` state also triggers a transition out of any substates that are currently occupied. Inherited transitions can be masked if a transition with the same trigger is present in one of the nested statecharts (as is the case for the `campaignCompleted()` transition in the `Active` state).

Figure 10.9 shows the use of synchronization bars to show explicitly how an event triggering a transition to a state with nested concurrent states causes specific concurrent substates to be entered and also shows that the super-state is not exited until both concurrent nested statecharts are exited.

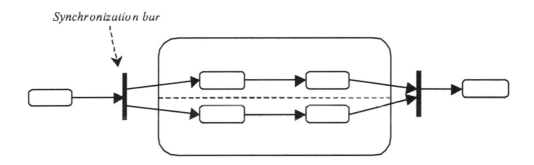

Synchronization bar

Figure 10.9 *Synchronized concurrent threads.*

10.5 Preparing a Statechart

Statecharts can be prepared from various perspectives. The statechart for a class can be seen as a description of the ways that use cases can affect objects of that class. Use cases give rise to interaction diagrams (sequence diagrams or collaboration diagrams) and these can be used as a starting point for the preparation of a statechart.

Interaction diagrams show the messages that an object receives during the execution of a use case. The receipt of a message may correspond to an event for the receiving object. Some messages do not represent events that cause a transition to fire. For example, simple 'get' messages (e.g. `getTitle()`) and query messages (e.g. `listAdverts()`) are not events in this sense. This is because they do not change the values of any of the object's attributes, nor do they alter any of its links with other objects. Some messages change attribute values without changing the state of an object. For example, a message `receivePayment()` to a `Campaign` object will only cause a change of state to `Paid` if it represents payment of the full amount due.

Figure 10.10 shows a sequence diagram for the use case `Set Campaign Completed`. The incoming message `campaignCompleted()` to a `Campaign` object causes the `campaignCompleted()` event to occur and triggers a transition from the `Active` state to the `Completed` state. Incoming messages to an object generally correspond to an event and trigger a state change. Allen and Frost (1998) describe the use of interaction diagrams to develop a statechart as a behavioural approach.

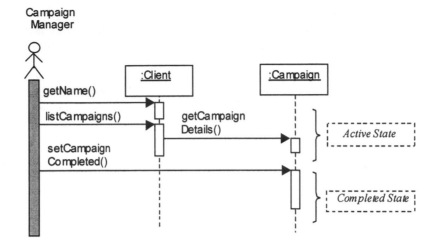

Figure 10.10 *Sequence diagram for use case* Set Campaign Completed.

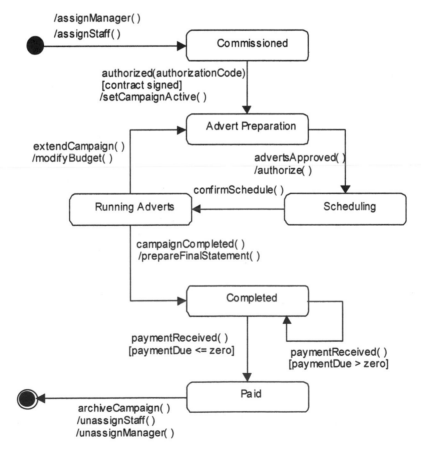

Figure 10.11 *Initial statechart for the* Campaign *class—a behavioural approach.*

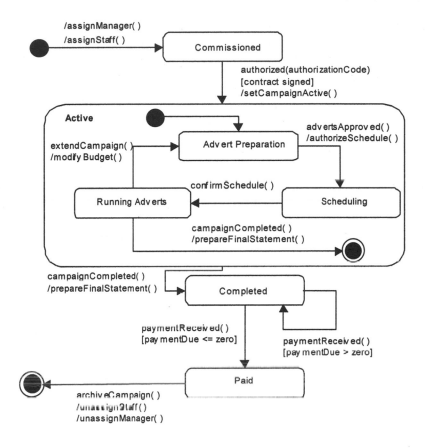

Figure 10.12 *Revised statechart for the* Campaign *class.*

The preparation of a statechart from a set of interaction diagrams has the following sequence of steps.

(1) Examine all interaction diagrams that involve each class that has heavy messaging.
(2) Identify the incoming messages on each interaction diagram that may correspond to events. Also identify the possible resulting states.
(3) Document these events and states on a statechart.
(4) Elaborate the statechart as necessary to cater for additional interactions as these become evident, and add any exceptions.
(5) Develop any nested statecharts (unless this has already been done in an earlier step).
(6) Review the statechart to ensure consistency with use cases. In particular, check that any constraints that are implied by the statechart are appropriate.
(7) Iterate steps 4, 5 and 6 until the statechart captures the necessary level of detail.
(8) Check the consistency of the statechart with the class diagram, with interaction diagrams and with any other statecharts.

The sequence diagram in Figure 10.10 has been annotated to indicate the state change that is triggered by the event campaignCompleted(). In order to identify all incoming messages that may trigger a state change for an object, all interaction diagrams that affect the object should be examined (sequence diagrams are probably easier to use for this purpose than collaboration diagrams, but this is a matter of personal preference). Analysis of the interaction diagrams produces a first-cut list of all events (caused by incoming messages) that trigger state changes, and also a first-cut list of states that the object may enter as a result of these events. If only major interactions have been modelled then the lists will not be complete but they can still provide an effective starting point.

The next step is to prepare a draft statechart for the class. Figure 10.11 illustrates the level of detail that might be shown in a first-cut statechart for the Campaign class. This would need to be expanded in order to reflect any events that have not been identified from the interaction diagrams, and also to include any exceptions. Complex nested states can be refined at this stage. A review of the statechart in Figure 10.11 results in the addition of the Active state to encompass the states Advert Preparation, Scheduling and Running Adverts (shown in the revised statechart in Figure 10.12).

The statechart is then compared to use cases in order to check that the constraints on class behaviour shown in the statechart satisfy the requirements documented in the use case. In this example the states Surveying and Evaluating have not yet been included. These might be identified in a final sweep up to check that the statechart is complete, and could then be added as concurrent states within the Active state.

Let us suppose that further investigation of behaviour that can affect a campaign reveals that in some circumstances a campaign can be cancelled. This is not permitted after a campaign has been completed but it can be cancelled while it is in the Commissioned or in the Active state. In either case cancellation costs are calculated for billing to the client. If the campaign is active then advertisement schedules are also cancelled. A final statechart that includes this additional requirement is shown in Figure 10.13.

An alternative approach to the preparation of statecharts is based on the consideration of life cycles for objects of each class. This approach does not use interaction diagrams as an initial source of possible events and states. Instead, they are identified directly from use cases and from any other requirements documentation that happens to be available. First, the main system wide events are listed (at Agate 'A client commissions a new campaign' might be one of the first to consider). Each event is then examined in order to determine which objects are likely to have a state dependent response to it.

The steps involved in the life cycle approach to state modelling are as follows:

(1) Identify major system-wide events.
(2) Identify each class that is likely to have a state dependent response to these events.
(3) For each of these objects produce a first-cut statechart by considering the typical life cycle of an object.
(4) Examine the statechart and elaborate to encompass more detailed event behaviour.
(5) Enhance the statechart to include exception handling.
(6) Review the statechart to ensure that is consistent with the use cases. In particular, check that the constraints that the statechart implies are appropriate.
(7) Iterate through steps 4, 5 and 6 until the statechart captures the necessary level of detail.
(8) Ensure consistency with class diagram and interaction diagrams and other statecharts.

The life cycle approach is less formal than the behavioural approach in its initial identification of events and relevant objects. It is often helpful to use a combination of the two, since each provides checks on the other. A life cycle approach might produce Figure 10.5 as an initial first-cut statechart for the Campaign class but further elaboration should still result in the statechart shown in Figure 10.13.

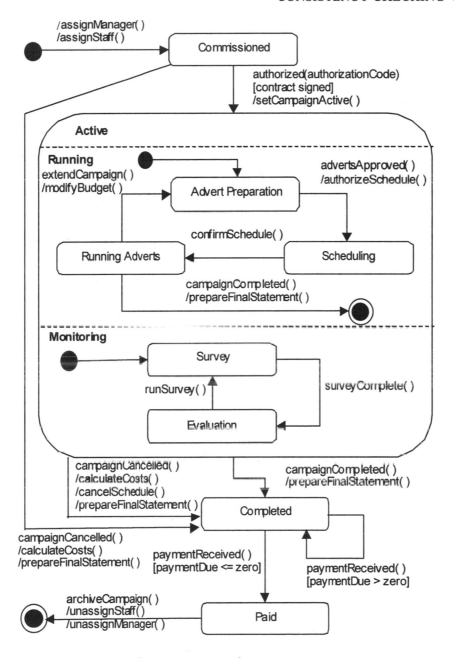

Figure 10.13 *Final version of* Campaign *statechart.*

10.6   Consistency Checking

The need for consistency between different models was discussed in Chapter 8 in relation to interaction diagrams. Statecharts must also be consistent with other models.

- Every event should appear as an incoming message for the appropriate object on an interaction diagram.
- Every action should correspond to the execution of an operation on the appropriate class, and perhaps also to the despatch of a message to another object.
- Every event should correspond to an operation on the appropriate class (but note that not all operations correspond to events).
- Every outgoing message sent from a statechart must correspond to an operation on another class.

Consistency checks are an important task in the preparation of a complete set of models. This process highlights omissions and errors and encourages the clarification of any ambiguity or incompleteness in the requirements.

10.7 Activity Diagrams

An activity diagram is a variation of a statechart that focuses on a flow of activity driven by internal processing within an object rather than by events that are external to it. In an activity diagram most (or all) states are action states (also called *activities*), each of which represent the execution of an operation. The transition out of an action state is normally triggered by completion of the operation. An activity diagram describes changes in state or the changes in activity within a procedure. They are similar to flowcharts (Skidmore, 1994) and are most useful when the events that trigger state changes are internally generated. This usually reflects the flow of control within a procedure.

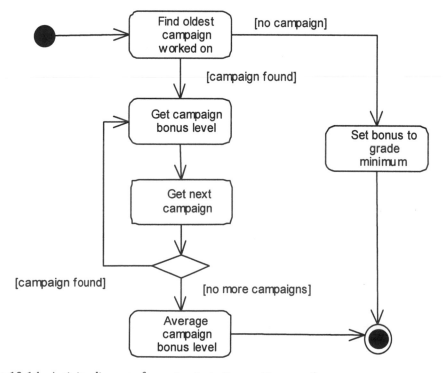

Figure 10.14 *Activity diagram for* calculateBonus() *operation*

Activity diagrams can be used for a number of different purposes:

- to model a human task (in business modelling for instance);
- to describe a system function that is represented by a use case;
- in operation specifications, to describe the logic of an operation.

An action state is a state that has an internal action and at least one outgoing transition that is triggered by the completion of the internal action. In the activity diagram for the operation calculateBonus() shown in Figure 10.14, the state Find oldest campaign worked on is an action state. It has two outgoing transitions each with a guard condition defined so that only one can be triggered by the completion of the action. Note that the activity commences from the initial state symbol and moves to the first action state, Find oldest campaign worked on. The activity in this state identifies all campaigns that a staff member has worked on since the last bonus calculation. If a staff member has not worked on any campaigns in this period then a guarded transition to the Set bonus to grade minimum state fires since its guard condition no campaign is now satisfied.

The use of guard conditions provides a useful mechanism to represent decisions. An alternative approach is the use of a diamond-shaped decision icon to represent each decision point. These have two or more outgoing arrows each labelled with a distinct guard condition and no event trigger. For example in Figure 10.14 the transition from the action state Get next campaign flows to a decision icon. The result of the decision must satisfy one of the guard conditions on the output arrows (and only one, since they are mutually exclusive). A decision icon must have at least one input arrow from an action state. More than one input arrow can be shown if there is more than one action state that leads directly to this decision point. Decision points can be nested to represent complex decisions (Figure 10.15).

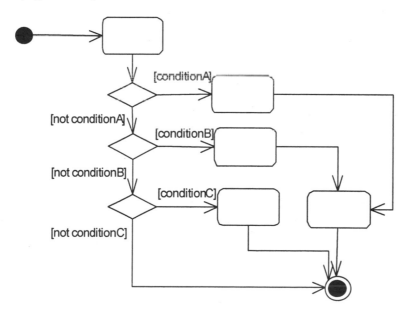

Figure 10.15 *Activity diagram—nested decisions*

Parallel activities can be represented on activity diagrams by the use of synchronization bars. Figure 10.16 shows an activity diagram for the activity Make a cup of tea. Three transitions leave the first synchronization bar which indicates that three threads of activity follow it, namely: Fill kettle with water, Find cup, and Find tea bag. These three activities may be in parallel. The synchronization bar, before the action state Place tea bag in cup indicates that both the action states Find cup and Find tea bag must be completed before the action state Place tea bag in cup can be initiated. The synchronization bar before Add water to cup indicates that the kettle must have boiled and the tea bag must have been placed in the cup before the action state Add water to cup can begin.

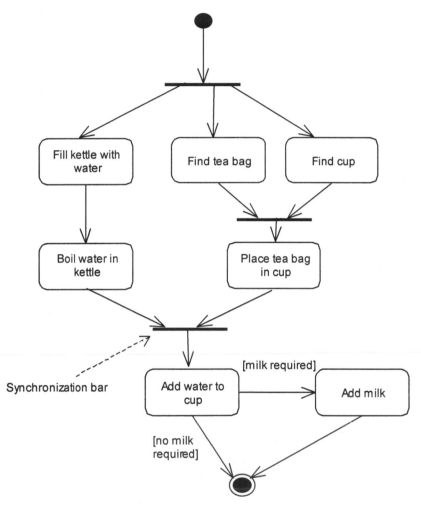

Figure 10.16 *Activity diagram*—Making a cup of tea.

An activity diagram can be divided into *swimlanes*. Each swimlane represents one focus of responsibility in the activity and each may be handled by a distinct operation in one or more objects[2]. The order of the swimlanes has no significance. Each action is assigned to

---

[2] Activity diagrams that use swimlanes are reminiscent of the system flowcharts that were used in the 1970s and 1980s (see Skidmore, 1994 for examples of system flowcharts).

one swimlane. Figure 10.17 shows an activity diagram for running a campaign. Three swimlanes are shown, each bordered by two solid black lines. The first action state is Request Campaign. Within the client swimlane the control flow (represented by a solid arrow) moves from this action to the Pay action. The dashed arrow from the transition represents an object flow that causes the Campaign object to change to the Commissioned state (the state is shown in parentheses). This Campaign object is now an input to the Authorize action that has as its output the Campaign object in the Active state. The object flow to the transition from the Pay action to the Accept Payment action indicates that a completed campaign is an input to the Accept Payment action.

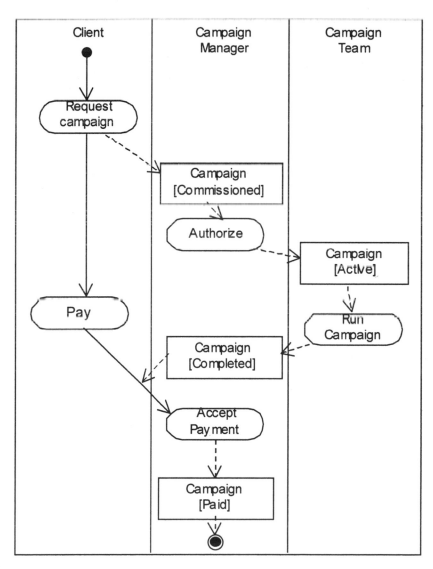

Figure 10.17 *Activity diagram*—Running a campaign.

10.8 Summary

The specification of the control aspects of an application is an important aspect of both analysis and design. The control aspects of an application are described in part by interaction diagrams but these focus only on a use case or an operation. In order to capture fully the control constraints for each class it is necessary to model the impact of events on that class and to model the resulting state changes with their attendant limitations on behaviour. It is only necessary to prepare statecharts for classes that have state dependent variations in behaviour. UML's statechart notation permits the construction of detailed models that may include the nesting of states and the use of concurrent states to capture complex behaviour.

Statecharts must be checked for consistency with their associated class and interaction diagrams and this may highlight the need to make modifications to these other models.

An alternative form of the statechart is the activity diagram which can be used to model procedural flow of control for either use cases or for operations. Activity diagrams are most useful when there is a need to model complex procedural behaviour.

The notations provided by UML are very detailed and should be used with some care. There is no advantage in producing a statechart or an activity diagram that utilizes every UML feature unless this is really necessary for the application that is being modelled. Ideally, state models should be kept as simple as possible but should have sufficient detail to make them unambiguous and informative. The use of multiple nested states does not aid clarity unless the behaviour being described is itself complex.

Review Questions

10.1   Define event, state and transition.
10.2   What is the effect of a guard condition?
10.3   Why should all the guard conditions from a state be mutually exclusive?
10.4   What does it mean to say that an object can be in concurrent states?
10.5   How do nested states differ from concurrent states?
10.6   What is the difference between an action and an activity?
10.7   What UML model element can be regarded as the parent of a statechart?
10.8   Against which other UML diagrams should a statechart be cross-checked?
10.9   What cross-checks should be carried out?
10.10 What are activity diagrams used for?

Case Study Work, Exercises and Projects

10.A   Using the interaction sequence diagrams that you prepared for Exercises 8.A–8.C, list events that affect a `ProductionLine` object and identify appropriate states for this object.
10.B   Prepare a statechart for the class `ProductionLine`.
10.C   List any changes that may have to be made to the class diagram for the FoodCo case study in the light of preparing this statechart.

10.D Prepare an activity diagram for the operation `startProductionRun()` (do not use swimlanes)

10.E Prepare another activity diagram for `startProductionRun()`, this time using swimlanes. List the advantages and disadvantages of swimlanes.

Further Reading

Statecharts have been used widely to model complex control behaviour. Various non-object-oriented approaches have used statecharts very effectively. In particular, the texts by Ward and Mellor (1985, 1986) and Hatley and Pirbhai (1987) provide detailed descriptions of their application in real-time applications. From an object-oriented perspective both Rumbaugh et al. (1991) and Booch (1994) provide useful descriptions of the Harel notation used in UML. Cook and Daniels (1994) give an interesting alternative perspective on the modelling of events and states.

A2

Agate Ltd Case Study—Excerpt from Requirements Specification

Agate Ltd

A2.1 Introduction

In this chapter we bring together some models (diagrams and supporting textual information) from different chapters in the second part of the book. In Chapters 5 to 10 we have introduced the following UML diagrams:
- use case diagram
- class diagram
- behaviour diagrams
  - o statechart diagram
  - o activity diagram
  - o interaction diagrams
    - − sequence diagram
    - − collaboration diagram

This list is from the UML Summary (Rational, 1997) using the UML classification of types of diagrams.

There is not the space in this book to produce a complete Requirements Specification. However, in this chapter we have included a sample of the diagrams and other information. This is done, firstly to illustrate the kind of material that should be brought together in a Requirements Specification at the end of the Analysis phase of a project, and secondly to allow the reader to see more easily how the elements of different models can be cross-referenced and cross-checked with one another.

A2.2 Use Cases and Actors

A use case diagram is included for a group of use cases that are used by the actors Campaign Manager and Accountant (Figure A2.1). Essential use case descriptions are shown for two of the use cases, one of which has a «uses» association with the other. One of these use cases will be developed in Chapter A3—Design Specification. The actor description for the actor Campaign Manager is also provided.

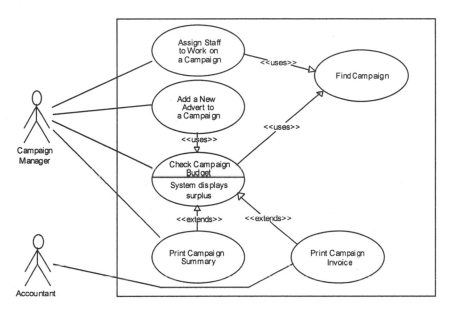

Use case description: **Check Campaign Budget**

Actor Action	System Response
1. The campaign manager enters the Client name.	2. Lists the titles of all campaigns for that client.
3. Selects the relevant campaign.	4. Displays the budget surplus for that campaign.
Alternative Courses Steps 1–3. The campaign manager knows the campaign name and enters it directly. After step 4, the campaign manager prints a campaign summary. After step 4, the campaign manager prints a campaign invoice.	

Use case description: **Add a New Advert to a Campaign**

Actor Action	System Response
1. Campaign manager checks campaign budget.	2. Displays campaign budget, total cost of adverts paid for and total estimated cost of adverts not paid for.
3. Campaign manager enters details of advert including estimated cost.	4. Creates new advert and adds to list of adverts for that campaign.
Alternative Courses Step 4. The advert will exceed the budget and a budget extension request is generated.	

Actor description: **Campaign Manager**

A Campaign manager is either a Director or an Account Manager. They are responsible for estimating the campaign cost and agreeing it with the client. They are responsible for assigning staff to the team and supervising their work, managing the progress of the campaign, conducting any further budget negotiations, and authorizing the final invoice.

Figure A2.1 *Use case diagrams and descriptions for selected Agate use cases, and actor description for* Campaign Manager.

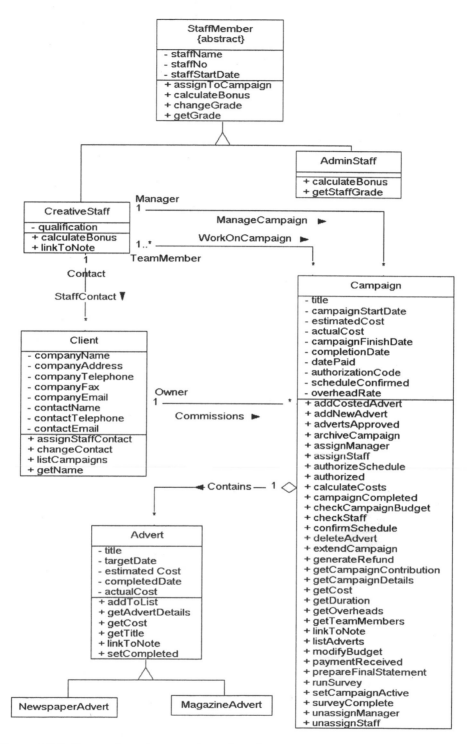

Figure A2.2 *Extract from Agate analysis class diagram.*

A2.3 Class Diagram

Part of the class diagram for Agate is shown in Figure A2.2. This shows the business classes involved in the use cases that have been included in Section A2.2. Some example data dictionary entries for the classes Advert and Campaign are shown in Figure A2.3.

Class description: **Advert**

Description: Each campaign consists of one or more adverts. An advert will be placed in a newspaper or magazine (possibly also TV, radio, cinema). The data held about adverts is concerned mainly with their cost and timing within the campaign. The actual design material is kept separately.

Synonyms: Advertisement.

Persistent: Yes.

Notes: Need to determine what different types of adverts are created by Agate and how different their attributes are. Investigate need for extended inheritance hierarchy.

Attribute description: **title**

Description: Name given to the advert by the client.

Format: String. Up to 50 characters.

Attribute description: **targetDate**

Description: Date on which the advert should appear or start to appear if it runs over a period of time.

Format: 4-digit year. Should conform to local date format, e.g. dd/mm/yyyy in UK.

Attribute description: **estimatedCost**

Description: Estimated cost of an advert before work begins on it. Used to budget for the advert and for the campaign.

Format: Money value in the local currency.

Attribute description: **completedDate**

Description: Date on which the advert completes its appearance or run.

Format: 4-digit year. Should conform to local date format, e.g. dd/mm/yyyy in UK.

Attribute description: **actualCost**

Description: Actual cost of the advert after all work on it is completed and external costs, e.g newspaper advertising charges have been accounted for.

Format: Money value in the local currency.

Class description: **Campaign**

Description: A campaign is undertaken for a client, and consists of one or more adverts. A campaign has an agreed budget that can be modified only with the consent of the client. Each campaign has a campaign manager, who sets its overhead rate and authorizes it to become active, and staff who are assigned to work on it. Once it is active, the campaign manager confirms the advert schedule. A campaign can be cancelled at any time prior to completion. On completion, a

Figure A2.3 *Selected attribute and operation specifications. (Continues on next page.)*

final statement is prepared for payment. A paid campaign can be archived.

Synonyms: None.

Persistence: Yes.

Notes: Can campaign budget be modified more than once? When budget is altered, do we need to record original budget together with changes? This would imply CampaignBudget class. There is also suggestion that payments for campaign should be recorded—do we need CampaignPayment class too?

Attribute description: **overheadRate**

Description: Each campaign has an overhead rate (may be unique). Rate is set by campaign manager before campaign becomes active. Used in calculating the actual cost.

Format: Numeric, two decimal places, max value 99.99.

Attribute specification: **scheduleConfirmed**

Description: Set for an active campaign to indicate that campaign manager has approved an advert schedule.

Format: Boolean.

Attribute specification: **estimatedCost**

Description: Records current agreed budget for campaign. May be changed only by agreement between campaign manager and client. Used to provide check on budget surplus. New adverts may only be added to schedule if total cost stays within budget.

Format: Money value in the local currency.

Attribute specification: **actualCost**

Description: Records current total committed cost (derived from advert costs and overhead rate). When campaign is completed this value is due for payment by client.

Format: Money value in the local currency.

Attribute specification: **authorizationCode**

Description: Code that is entered by campaign manager to indicate a campaign can become active. No costs can be incurred until this code is set.

Format: Maximum of 10 text characters. Access restriction required—this attribute may not be updated once it has been set.

Operation specification: **addCostedAdvert**

Operation intent: Add a new advert if within budget else raise budget extension request.

operation signature: Campaign::addCostedAdvert(anAdvert : Advert)

logic description (pre- and post-conditions):

```
pre:  anAdvert->notEmpty
      Advert.allInstances->forAll(a | a.title <>
         anAdvert.title) -- This advert has a unique title.
      not adverts->includes(anAdvert)
      anAdvert.title.oclType = String
```

Figure A2.3 *Selected attribute and operation specifications. (Continues on next page.)*

```
       anAdvert.targetDate.oclIsTypeOf(Date)
       anAdvert.estimatedCost.oclIsTypeOf(Money)
          -- Allow for local subclasses of Date and Money.
post: if ( estimatedCost > self.adverts@pre.getCost()->sum +
          anAdvert.estimatedCost )
     then ( result = (newAdvert = anAdvert) and
          adverts->includes(newAdvert) )
     else ( result = (aRequest.oclType = Request) )
     endif
```

Other operations called: Advert.Advert,   self.addNewAdvert,
                   Request.Request
Events transmitted to other objects: none
Attributes set: estimatedCost,   adverts
Response to exceptions: raises InvalidAdvertException
Non-functional requirements: Print Request within 24 hours.

Operation specification: checkCampaignBudget
  Operation intent: return campaign committed expenditure and budget.
  operation signature: Campaign::checkCampaignBudget()
  campaignBudget : Money, committedExpenditure : Money
  logic description (pre- and post-conditions):

```
     pre: self.id->exists
     post: campaignBudget = self.estimatedCost
          committedExpenditure = self.adverts.estimatedCost->sum
```

  Other operations called: Advert.getCost,   self.getOverheads
  Events transmitted to other objects: none
  Attributes set: none
  Response to exceptions: none defined
  Non-functional requirements: none defined

Figure A2.3 *Selected operation and attribute specifications. (Continued.)*

The operations here have been specified using pre- and post-conditions in OCL. They can also be specified in other ways, for example Structured English, activity diagrams or collaboration diagrams (see Figure A2.4).

A2.4 Interaction Diagrams

This section illustrates some of the ways that interaction diagrams are used to illustrate the behaviour of various parts of a model. Figure A2.4 shows a collaboration diagram for the use case Add a New Advert to a Campaign, while Figure A2.5 shows another for the operation checkCampaignBudget(). Figure A2.6 shows the sequence diagrams that correspond to these collaboration diagrams.

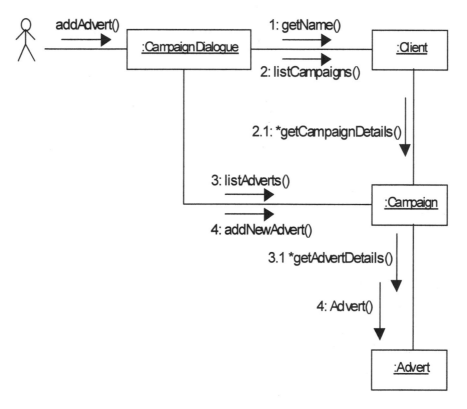

Figure A2.4 *Collaboration diagram for the use case* Add a New Advert to a Campaign *(without the branching shown in the sequence diagram).*

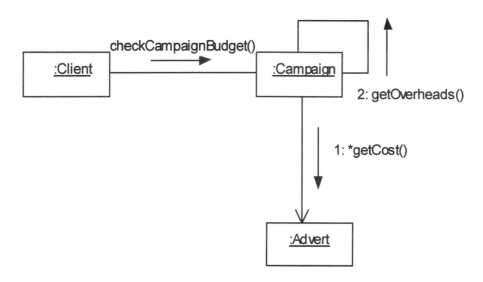

Figure A2.5 *Collaboration diagram for* Campaign.checkCampaignBudget().

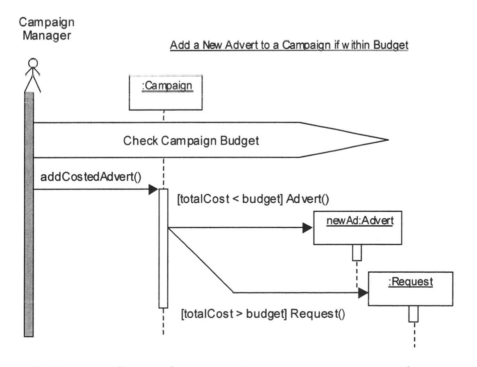

Figure A2.6 *Sequence diagrams for use cases* Check Campaign Budget *and* Add a New Advert to a Campaign if within Budget. *(Note that the annotated bar linking the second sequence diagram to the first is not part of UML.)*

A2.5 State and Activity Diagrams

Finally, in this section we show a statechart and activity diagram for the `Campaign` class, which is centrally involved in the interactions illustrated in earlier models. The activity diagram in Figure A2.7 analyses events that can affect a `Campaign` object according to their origin, while Figure A2.8 shows the complete statechart for `Campaign`.

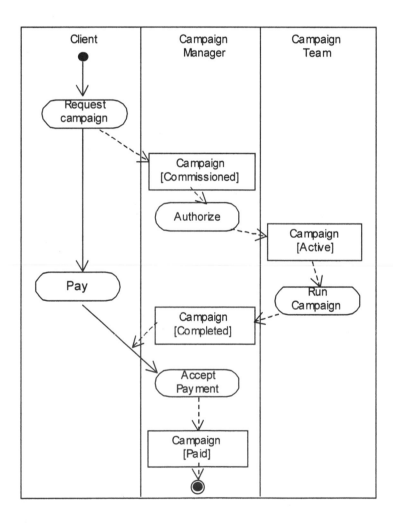

Figure A2.7 *Activity diagram for running a campaign.*

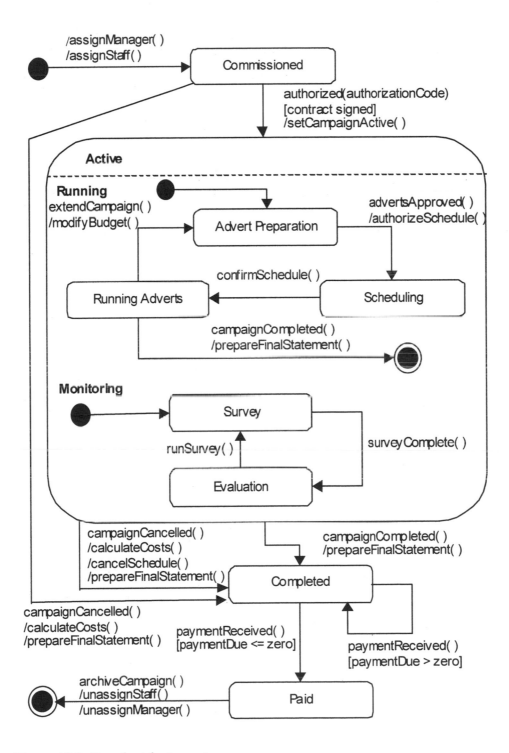

Figure A2.8 *Statechart for* Campaign

11

Moving into Design

Objectives
In this chapter you will learn:
- the difference between analysis and design;
- the difference between logical and physical design;
- the difference between system and detailed design;
- the characteristics of a good design;
- the need to make trade-offs in design.

11.1 Introduction

Two questions that may be asked by those who are new to systems analysis and design are 'What is the difference between analysis and design?' and 'Why treat analysis and design as separate activities?' In the development of information systems, as in the development of many kinds of system, the process of analysis is distinguished from the process of design (Section 11.2). Analysis is often said to be about the 'What?' of a system, and design is described as being about the 'How?'. Design can start before or after the decision has been made about the hardware and software to be used in implementing the system. Implementation-independent or logical design is distinguished from implementation-dependent or physical design (Section 11.3). Design also takes place at two main levels: system design, which addresses aspects that affect the overall system, and detailed design, which addresses the design of classes and the detailed working of the system (Section 11.4). In producing a design for a system, a designer will be working within a framework of general quality criteria and will also be trying to achieve measurable objectives for the design that are specific to the particular system (Section 11.5). Some of these objectives may conflict with one another, and constraints on the design may result in the need for trade-offs to be made.

11.2 How is Design Different from Analysis?

Design has been described by Rumbaugh (1997) as stating 'how the system will be constructed without actually building it'. The models that are produced during design show how the various parts of the system will work together; the models produced during analysis show what is in the system and how those parts are related to one another.

The word *analysis* comes from a Greek word meaning to break down into component parts. When we analyse an organization and its need for a new system, the analysis activity is characterized as asking *what* happens in the current system and *what* is required in the new system. It is a process of seeking to understand the organization and investigating its requirements. The result of this analysis activity is a specification of requirements for the proposed system.

Design is about producing a solution that meets the requirements that have been specified during analysis. The design activity is concerned with specifying *how* the new system will meet those requirements. There may be many possible design solutions, but the intention is to produce the best possible solution in the circumstances. Those circumstances may reflect constraints such as limits on how much can be spent on the new system or the need for the new system to work with an existing system. Jacobson et al. (1992) regard design as part of the construction process (together with implementation). The systems designer has his or her attention focused on the implementation of the new system, while the systems analyst is focused on the way the business is organized and a possible better organization; the foci of these two activities are very different.

A simple example of this can be seen in the Agate case study. During analysis, the fact that each `Campaign` has a `title` attribute has been identified and documented in the class diagram and the data dictionary. During design we shall determine how this will be entered into the system, displayed on screen and stored in some kind of database together with all the other attributes of `Campaign` and other classes.

Design can be seen either as a phase in the systems development life cycle or as an activity that takes place within the development of a system. In projects that follow the waterfall life cycle model, the analysis phase will be complete before the design phase begins. However, in projects that follow an iterative life cycle, design is not such a clear-cut phase, but is rather an activity that will be carried out on the evolving model of the system. Rumbaugh (1997) distinguishes between the idea of design as a phase in the waterfall life cycle and design as a stage that different parts of the model of the system will go through at different times.

In large-scale projects that follow a traditional systems development life cycle there are a number of advantages to making a clear break between analysis and design. These are concerned with:

- project management,
- staff skills and experience,
- client decisions, and
- choice of development environment.

Project management. The project manager will have an overall budget in terms of money and staff time within which the system must be developed. Some proportion of those resources will have been allocated to analysis and some to design. In order to manage and control the project effectively, the project manager will want to have a clear idea of how much time is spent on each of these activities. If the two activities are allowed to merge the management of the project becomes more difficult. If all the time is being spent on analysis, then the project will fall behind schedule, whereas if all the time is being spent on design, then it is likely that the requirements have not been properly understood.

Staff skills and experience. Analysis and design may be carried out by staff with different skills and experience. Staff with job titles such as business analyst and systems analyst will have the skills and expertise to carry out the analysis, while systems architects and systems designers will have an understanding of the technology available to deliver the solution and will carry out the design.

Client decisions. The clients will want to know what they are paying for. The end of analysis is often a decision point in a project. The clients will be provided with a specification of the system in terms of their requirements and they will have to agree to this or *sign it off* before work progresses to design. In some projects, the client may be presented with a number of alternative specifications that differ in scope (what parts of the system will be computerized). In this case, the client must choose which of these alternative systems to take forward into design.

Choice of development environment. In many projects the hardware and software that will be used to develop and deliver the finished system will not be known at the time of the analysis phase. There may be good reasons for delaying the choice of hardware and software until the requirements for the system have been determined by the analysis, especially in the rapidly changing world of information technology in which some new technology always seems to be due for release. Because the choice of hardware, development language and database will affect the design, it may be necessary to make a break between analysis and design so that decisions about the development environment can be made. This point is discussed in more detail in the next section.

Before object-oriented approaches were developed, structured analysis and design was the dominant approach to analysis and design. In structured approaches, a clear distinction between analysis and design was made in terms of the types of diagram that were used. During analysis data flow diagrams were used to model requirements, whereas structure charts or structure diagrams were used to model the design of the system and the programs in it. One of the arguments put forward for the use of object-oriented approaches is that the same model (the class diagram or object model) is used right through the life of the project. Analysis identifies classes, those classes are refined in design, and the eventual programs will be written in terms of classes. While this so-called *seamlessness* of object-oriented methods may seem like an argument for weakening the distinction between analysis and design, when we move into design different information is added to the class diagram, and other different diagrams are used to support the class diagram. Rumbaugh (1997) distinguishes between analysis and design in terms of the amount of detail that is included in the model. On a continuum, the analysis stage provides an abstract model of 'what to do' while the design stage documents 'exactly how to do it'. As the project moves from one end of this continuum to the other, additional detail is added to the model until a clear specification of 'how to do it' is provided. This additional detail is added in the form of diagrams such as collaboration diagrams, state diagrams and deployment diagrams that supplement the information in the class diagram. The class diagram is also enhanced during design.

11.3 Logical and Physical Design

At some point in the life of a systems development project a decision must be made about the hardware and software that are to be used to develop and deliver the system—the hardware and software platform. In some projects this is known right from the start. Many companies have an existing investment in hardware and software, and any new project must use existing system software (such as programming languages and database management systems) and will be expected to run on the same hardware. This is more often the case in large companies with mainframe computers. In such companies the choice of configuration has been limited in the past to the use of terminals connected to the mainframe. However, client–server architectures (see next section) and open system standards, that allow for different hardware and software to operate together, have meant that even for such companies, the choice of platform is more open. For many new projects the choice of platform is relatively unconstrained, and so at some point in the life of the project a decision must be made about the platform to be used.

Some aspects of the design of systems are dependent on the choice of platform. These will affect the system architecture, the design of objects and the interfaces with various components of the system. Examples include the following.

- The decision to create a distributed system with elements of the system running on different machines will require the use of some *middleware* such as is provided by CORBA® to allow objects to communicate with one another across the network. This will affect the design of objects.

- The decision to write programs in Java and to use a relational database that supports ODBC™ (Object Data Base Connectivity) will require the use of JDBC™ (Java Data Base Connectivity) and the creation of classes to map between the objects and the relational database.

- The choice of Java as a software development language will mean that the developer has the choice of using the standard Java AWT (Abstract Windowing Toolkit) or the Java Swing classes for designing the interface.

- Java does not support multiple inheritance; other object-oriented languages such as C++ do. If the system being developed appears to require multiple inheritance then in Java this will have to be implemented using Java's interface mechanism.

- If the system has to interface with special hardware, for example bar-code scanners, then it may be necessary to design the interface so that it can be written in C as a *native method* and encapsulated in a Java class, as Java cannot directly access low-level features of hardware.

Java has been used here as an example. The same kinds of issues will arise whatever platform is chosen.

It is also the case that there are many design decisions that can be made without knowledge of the hardware and software platform.

- The interaction between objects to provide the functionality of particular use cases can be designed using interaction diagrams or collaboration diagrams.

- The layout of data entry screens can be designed in terms of the fields that will be required to provide the data for the objects that are to be created or updated, and the order in which they will appear on the screen can be determined. However, the exact nature of a textbox and whether it is a Borland® C++ TEdit, a Delphi Edit, a Java TextField or something else can be left until later.

- The nature of commands and data to be sent to and received from special hardware or other systems can be determined without needing to design the exact format of messages.

Because of this, design is sometimes divided into two stages. The first is *implementation-independent* or *logical* design and the second is *implementation-dependent* or *physical* design. Logical design is concerned with those aspects of the system that can be designed without knowledge of the implementation platform; physical design deals with those aspects of the system that are dependent on the implementation platform that will be used.

In Syntropy, an object-oriented methodology (see Chapter 20 for discussion of methodologies), the development of a system is viewed from three perspectives: the essential model, the specification model and the implementation model. The essential model is equivalent to an analysis model, while the specification model is a logical design model and the implementation model is a physical design model.

In many projects, design begins after hardware and software decisions have been made. However, if this is not the case, then the project manager must ensure that the plan of work for the project takes account of this and that logical design activities are tackled first.

Review Questions

11.1   What are the advantages of separating the analysis and design phases of a project?

11.2   Users at Agate require a report of unpaid campaigns. Which of the following aspects of the report represents analysis, logical design and physical design?

The size of the paper and the position of each field in the report.

The fact that the user wants a report of completed campaigns that have not yet been paid for by the client.

The selection of the business objects and their attributes used by the report.

11.3   Which of the following sentences describing an element of the FoodCo system represents analysis, logical design and physical design?

The reason for stopping a run will be selected from one of the values displayed in a listbox (Java `Choice`*) in the Record Line Stop dialogue window.*

When a production line stops during a run, the reason for stopping will be recorded.

The reason for stopping a run will be entered into the system by selecting from a list of valid reasons.

11.4   What is meant by *seamlessness* in object-oriented systems development?

11.4 System Design and Detailed Design

Design of systems takes place at two levels: system design and detailed design. System design is concerned with the overall architecture of the system and the setting of standards, for example for the design of the human–computer interface; detailed design is concerned with designing individual components to fit this architecture and to conform to the standards. In an object-oriented system, the detailed design is mainly concerned with the design of objects. These two levels of design are described briefly in the next sections and in more detail in Chapters 12 and 13.

11.4.1 System design

During system design the designers make decisions that will affect the system as a whole. The most important aspect of this is the overall architecture of the system. Many modern systems use a client–server architecture in which the work of the system is divided between the clients (typically PCs on the users' desks) and a server (usually a Unix or Windows NT machine that provides services to a number of users). This raises questions about how processes and objects will be distributed on different machines, and it is the role of the systems designer or systems architect to decide on this. The design will have to be broken down into sub-systems and these sub-systems may be allocated to different processors. This introduces a requirement for communication between processors, and the systems designer will need to determine the mechanisms used to provide for this communication. Distributing systems over multiple processors also makes it possible for different sub-systems to be active simultaneously or concurrently. This concurrency needs to be designed into the system explicitly rather than left to chance.

Many organizations have existing standards for their systems. These may involve interface design issues such as screen layouts, report layouts or how on-line help is provided. Decisions about the standards to be applied across the whole system are part of system design, whereas the design of individual screens and documents (to comply with these standards) is part of detailed design.

When a new system is introduced into an organization, it will have an impact on people and their existing working practices. Job design is often included in system design and addresses concerns about how people's work will change, how their interest and motivation can be maintained, and what training they will require in order to carry out their new job. How people use particular use cases will be included in the detailed design of the human–computer interface.

11.4.2 What happens in traditional detailed design?

In the 1960s and 1970s, detailed design was seen as consisting of four main activities.

- designing inputs;
- designing outputs;
- designing processes; and
- designing files.

Designing inputs meant designing the layout of menus and data entry screens; designing outputs was concerned with the layout of enquiry screens, reports and printed documents; designing processes dealt with the choice of algorithms and ensuring that processes correctly reflected the decisions that the software needed to make; and designing files dealt with the structure of files and records, the file organization and the access methods used to update and retrieve data from the files.

The development of structured design methods made two major changes to the way in which design was carried out. Firstly, the work of Jackson (1975) provided a method for designers to design programs by using a technique to match the structure of the inputs and outputs with the structure of data to be read from or written to files.

The second major contribution came from Yourdon and Constantine (1979) who defined a series of criteria that could be used in breaking systems and programs down into modules to ensure that are easy to develop and maintain. These criteria concern two issues: *cohesion* and *coupling*. Criteria to maximize desirable types of cohesion have as their aim the production of modules—sections of program code in whatever language is used—that carry out a clearly defined process or a group of processes that are functionally related to one another. This means that all the elements of the module contribute to the performance of a single function. Poor cohesion is found when processes are grouped together in modules for other reasons. Examples of poor types of cohesion include:

- when processes are grouped together for no obvious reason (*coincidental cohesion*),
- because they handle logically similar processes such as inputs (*logical cohesion*),
- because they happen at the same time—for example when the system initializes— (*temporal cohesion*), and
- because the outputs of one process are used as inputs by the next (*sequential cohesion*).

By aiming to produce modules that are functionally cohesive, the designer should produce modules that are straightforward to develop, easy to maintain and have the maximum potential to be reused in different parts of the system. This will be assisted if coupling between modules is also reduced to the minimum.

Criteria to minimize the coupling between modules have as their aim the production of modules that are independent of one another and that can be amended without resulting in knock-on effects to other parts of the system. Good coupling is achieved if a module can perform its function using only the data that is passed to it by another module and using the minimum necessary amount of data. Poor coupling is found in the following circumstances:

- modules that rely on data in global variables or data in common blocks (used in languages such as COBOL and FORTRAN) that other modules may change,
- modules that are designed to need large amounts of data to be passed to them as parameters, and
- modules that are not cohesive because they perform several functions and therefore require control information as well as data to be passed as parameters so that a decision can be made within the module about which function is required.

Modules with low coupling and high cohesion are the aim of structured design and programming techniques. In Chapter 13 we shall explain how these issues still apply to the design of object-oriented systems. Objects can be viewed in terms of cohesion—each class in the class diagram should encapsulate data and functionality that clearly belongs together. It is no less possible to produce classes that reflect poor cohesion, for example temporal cohesion, as it is to produce modules with poor cohesion. Classes with poor cohesion will reflect the structure of the particular application and will not be reusable in other circumstances. The assignment of responsibilities to classes, that is determining which operations belong in which classes, contributes to the cohesiveness and reusability of classes. Cohesion and coupling also apply to the operations of classes. Larman (1998) discusses the assignment of responsibilities to objects in order to achieve high cohesion and low coupling, and these issues are discussed in more detail in Chapter 13.

11.4.3 What do we add in O-O detailed design?

Traditionally, detailed systems design has been about designing inputs, outputs, processes and file or database structures; these same aspects of the system also have to be designed in an object-oriented system. During the analysis phase of a project, concepts in the business will have been identified and elaborated in terms of classes, and use cases will have been identified and described. The classes that have been included in the class diagram will reflect the business requirements but they will not include classes to handle the interface with the user, the interface with other systems, the storage of data or the overall co-ordination of the other classes into programs. These classes will be added in design with greater or lesser degrees of detail depending on the hardware and software platform that is being used for the new system.

Different authors describe these additional aspects of the system in different ways. Coad and Yourdon (1991) call the business classes the problem domain component and develop three further components in the design phase:

- human interface component,
- data management component, and
- task management component.

Coad et al. (1997) propose a slightly different set of components:

- human interface component,
- data management component, and
- system interaction component.

In their designs, windows play the co-ordinating role played by the classes in the task management component. Larman (1998) proposes an architecture based on three layers:

- presentation layer (windows and reports),
- application logic layer, and
- storage layer.

However, his application logic layer includes both domain concepts—equivalent to the problem domain component—and services, which may include interfaces to the database and to other systems. Figure 11.1 shows these layers as UML packages.

What should be apparent, whatever the terminology, is that it is necessary to decide on the architecture that will be used and that this will involve the designers in designing classes to handle these aspects of the system. The amount of work involved may vary from system to system and component to component, depending on the hardware and software platform to be used. In particular, the design of these components offers opportunities to reuse existing classes.

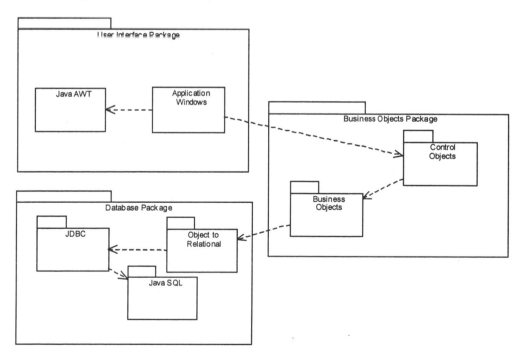

Figure 11.1 *UML packages representing layers in the three-tier architecture.*

11.4.4 Aspects requiring attention in O-O detailed design

Certain aspects of the detailed design require special attention in the development of object-oriented systems. These include reuse and assignment of responsibilities to classes.

One of the arguments for the use of object-oriented languages is that they promote reuse through encapsulation of functionality and data together in classes and through the use of inheritance. This is not just a programming issue, but one that also affects analysis and design. There is a growing recognition of the need to reuse analysis results in object-oriented systems development. Design reuse already takes place at two levels: firstly through the use of design patterns, which are discussed in detail in Chapter 14; and secondly by recognizing during design that business classes that have been identified during analysis may be provided by reusing classes that have already been designed within the organization, or even bought in from outside vendors. There is a move in the software industry towards the use of *components* that provide this kind of functionality and that can be bought from vendors (Allen and Frost, 1998). CASE tool suppliers such as SELECT Software Tools have added component management software to their range of products, and IBM, through its San Francisco™ project, is making general-purpose business classes available as part of component packages. This issue is addressed in more detail in Chapter 19.

The assignment of responsibilities to classes is an issue that is related to reuse. Larman (1998) highlights this activity as the main task in design. In an object-oriented system, it is important to assign responsibility for operations to the right classes, and there is often a choice. In the FoodCo system, there will be a need to produce invoices for customers that include the calculation of Value Added Tax (VAT). (Value Added Tax is a tax used throughout Europe that is applied at each stage of the supply chain and not just as a purchase tax paid by the final end-user or consumer.) The calculation of VAT could be carried out by one of a number of classes in the model (Figure 11.2).

- `Invoice`—which organizes the total information for the whole sale.
- `InvoiceLine`—which contains the detail of each item sold and to which the tax applies.
- `Product`—to which different VAT rates may apply.
- `TaxRate`—which carries the details of the percentage that applies for each valid rate.

If the designer makes the wrong decision, the resulting class will be less reusable and may constrain the design of other classes. If the responsibility for tax calculation is allocated to `Invoice` or `InvoiceLine`, then this has implications for `CreditNote` and

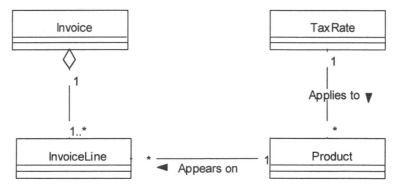

Figure 11.2 *Partial class diagram for FoodCo.*

`CreditNoteLine`, which may also need to calculate tax. If it is assigned to `Product`, then it cannot be reused in the Agate project where VAT applies to services as well as products. Clearly it needs to be assigned to `TaxRate` in order to maximize the reuse that can be made of the classes in this design.

Review Questions

11.5   What are the differences between system design and detailed design?
11.6   Explain the difference between cohesion and coupling.
11.7   What aspects of the system are added to the class diagram(s) in object-oriented detailed design?

11.5 Qualities and Objectives of Analysis and Design

At the end of the previous section, reusability was highlighted as something that designers of object-oriented systems are trying to achieve in their design. Reusability is not the only objective of design. There are a number of other criteria for a good design. Perhaps the most obvious measure of design quality is whether the finished application is of high quality. This assumes that the analysis that preceded the design work was itself of high quality. However, this is a rather vague and circular way of assessing quality: in this section we look briefly at aspects of quality in analysis and in more detail at quality in design. Design quality cannot be measured only against general criteria, and in Section 11.6 we consider more specific measures that can be applied to particular projects.

11.5.1 What makes for good analysis?

The cost of fixing faults in a system increases as the system progresses through the systems development life cycle. If an error occurs in the analysis of a system, it is cheaper to fix it during the analysis phase than it is later when that error may have propagated through numerous aspects of the design and implementation. It is most expensive to fix it after the system has been deployed and the error may be reflected in many different parts of the system. The quality of the design is, therefore, dependent to a large extent on the quality of the analysis.

Some methodologies have explicit quality criteria that can be applied to the products of every stage of the life cycle, but these quality criteria typically check syntactic aspects of the products, that is whether the notation is correct in diagrams, rather than semantic aspects, that is whether the diagrams correctly represent the organization's requirements. To provide a sound foundation for design, analysis should meet the following four criteria:

- correct scope,
- completeness,
- correct content, and
- consistency.

These are described in more detail below.

Correct scope. The scope of a system determines what is included in that system and what is excluded. It is important, firstly that the required scope of the system is clearly understood, documented and agreed with the clients, and secondly that everything that is in the analysis models *does* fall within the scope of the system. In the case of the Agate system, it is not a requirement to replace the existing accounting system that is used to invoice clients. It is, however, a requirement that the new system should interface with the accounting system to provide for the transfer of data relating to financial aspects of advertising campaigns. The scope of the system therefore excludes use cases for accounting but should include use cases both to handle the entry of data that will be transferred to the accounting system and to handle the transfer itself. Coad et al. (1997) include a *not this time* component with their other four components (problem domain, human interface, data management and system interaction). The not this time component is used to document classes and business services that emerge during the analysis but are not part of the requirements this time. This is a useful way of forcing consideration of the scope of the system.

Clearly determining the scope of a system before design starts is a strong argument for making a distinction and a break between the analysis and design phases of a project. If the scope of the Agate system is agreed before design begins, then the project manager can plan more accurately for the work that will have to take place during the design phase. If the specification is allowed to evolve iteratively, then the management of Agate could introduce new requirements well into the life of the project that could have a severe impact on the ability to complete the project on time. An example could be the requirement to store not just data about adverts but digital multimedia versions of the components of the adverts themselves so that mock-ups of the adverts can be edited together and shown to clients on screen. An additional requirement of this nature would transform the project from a relatively straightforward business information system to a complex multimedia system.

Completeness. Just as there is a requirement that everything that is in the analysis models is within the scope of the system, so everything that is within the scope of the system should be documented in the analysis models. Everything that is known about the system from the requirements capture should be documented and included in appropriate diagrams. Often the completeness of the analysis is dependent on the skills and experience of the analyst. Knowing what questions to ask in order to elicit requirements comes with time and experience. However, analysis patterns and strategies, as proposed by Coad et al. (1997) and Fowler (1997), can help the less experienced analyst to identify likely issues. (The use of patterns, which draw on past experience, can be a good way of ensuring that the analysis is effective.)

Non-functional requirements should be documented even though they may not affect the analysis models directly. Rumbaugh (1997) suggests that some of the requirements found during analysis are not analysis requirements but design requirements. These should be documented, but the development team may only have to consider them once the design phase has begun. An example in the Agate system is the requirement that the system should be usable in different offices around the world and should handle multiple currencies. This would be noted during analysis; during design, it will mean that the system must be designed to support localization (adaptation to local needs) and to display different national currency symbols (perhaps using the Unicode™ standard).

Correct content. The analysis documentation should be correct and accurate in what it describes. This applies to textual information, diagrams and also to quantitative features of the non-functional requirements. Examples include correct descriptions of attributes and any operations that are known at this stage, correct representation of associations between classes, particularly the multiplicity of associations, and accurate information about volumes of data. Accuracy should not be confused with precision. FoodCo owns 1,500 acres of land (to the nearest 100 acres). To state that the company owns 1,700 is inaccurate. To state that it owns 1,523 is more precise. To state that it owns 1,253 is still inaccurate, although the precision gives a spurious impression of accuracy.

Consistency. Where the analysis documentation includes different models that refer to the same things (use cases, classes, attributes or operations) the same name should be used consistently for the same thing. Errors of consistency can result in errors being made by designers, for example, creating two attributes with different names that are used in different parts of the system but should be the same attribute. If the designers spot the inconsistency, they may try to resolve it themselves, but may get it wrong because the information they have about the system is all dependent on what they have received in the specification of requirements from the analysts.

Errors of scope or completeness will typically be reflected in the finished product not doing what the users require; the product will either include features that are not required or lack features that are. Errors of correctness and consistency will typically be reflected in the finished product performing incorrectly. Errors of completeness and consistency will most often result in difficulties for the designers; in the face of incomplete or inconsistent specifications, they will have to try to decide what is required or refer back to the analysts.

One general way of ensuring that the analysis models reflect the requirements is to use walkthroughs. Walkthroughs are described by Yourdon (1985) (and also in an appendix to Yourdon, 1989). They provide a structured review with other analysts and interested parties of the products of the analysis phase.

11.5.2 What makes for good design?

The quality of the design will clearly be reflected in the quality of the finished system that is delivered to the clients. Moreover, in the same way as the quality of analysis affects the work of designers, the quality of the design has an impact on the work of the programmers who will write the programme code in order to implement the system based on the design. Some of the criteria given below for a good design will bring benefits to the developers, while some will provide benefits for the eventual users of the system.

11.5.3 Objectives and constraints

The designers of a system seek to achieve many objectives that have been identified as the characteristics of a good design since the early days of information systems development. Yourdon and Constantine (1979) cite efficiency, flexibility, generality, maintainability and reliability; DeMarco (1979) proposes efficiency, maintainability and buildability; and Page-Jones (1988) suggests that a good design is efficient, flexible, maintainable, manageable, satisfying and productive. These latter two points highlight issues concerned with human–computer interaction and remind us of the need for the design to produce a usable system. Other characteristics of a good design are that it should be functional, portable, secure and economical; in the context of object-oriented systems, reusability is a priority objective.

Functional. When we use a computer system, we expect it to perform correctly and completely those functions that it is claimed to perform; when an information system is developed for an organization, the staff of that organization will expect it to meet their documented requirements fully and according to specification. So, for example, the staff of Agate will expect their system to provide them with the functionality required to document advertising campaigns, record notes about campaigns and store information about the advertisements to be used in those campaigns. If it does not perform these functions, it is not fully functional. Referring back to Rumbaugh's definition of design as 'how the system will be constructed without actually building it', a functional design should show how every element of the required system will work.

Efficient. It is not enough that a system performs the required functionality; it should also do so efficiently, in terms both of time and resources. Those resources can include disk storage, processor time and network capacity. This is why design is not just about producing any solution, but about producing the best solution. This objective may apply to Agate's requirement to store textual notes about ideas for campaigns and advertisements. A sample two lines of text containing twenty words takes up 97 bytes in text format, but stored in the format of a well-known word-processor takes up 12,800 bytes (this is without the use of any special fonts or styles). A poor design might use object linking and embedding (OLE) to handle the word-processing of the notes but would pay a severe penalty in increased storage requirements!

Economical. Linked to efficiency is the idea that a design should be economical. This applies not only to the fixed costs of the hardware and software that will be required to run it, but also to the running costs of the system. The cost of memory and disk storage is very low compared to twenty years ago, and most small businesses using Microsoft® Windows probably now require more disk space for their programs than they do for their data. However, the growth of multimedia systems for business purposes may make it once more important to calculate the storage requirements of a system carefully.

Reliable. The system must be reliable in two ways: firstly, it should not be prone to either hardware or software failure; secondly it should reliably maintain the integrity of the data in the system. Hardware reliability can be paid for: some manufacturers provide systems with redundant components that run in parallel or that step in when an equivalent component fails; RAID (redundant arrays of inexpensive disks) technology can provide users with disk storage that is capable of recovering from failure of one drive in an array. The designers must design software reliability into the system. In physical design, detailed knowledge of the development environment is likely to help ensure reliability.

Reliability depends to some extent on the ability of the system to be tested thoroughly. A well analysed and designed system will specify the valid and invalid combinations of data that can be handled. It will also show clearly the structure of the system and which elements of the system are dependent on others so that testing can work up through classes, groups of classes, sub-systems and eventually the whole system.

Secure. Systems should be designed to be secure against malicious attack by outsiders and against unauthorized use by insiders. System design should include considerations of how people are authorized to use the system and policies on passwords. It should also cover protection of the system from outsiders, including firewalls in either hardware or software to protect the system from access via public networks, such as the Internet. In European countries that are members of the European Union, there are data protection laws that are designed to protect the interests of individuals about whom data is held in information systems, such as the Data Protection Acts (1984 and 1998) and the Computer

Misuse Act (1990) in the UK. Where such legislation exists, the designer should ensure that the design of the system will comply with its requirements.

Flexible. Some authors treat flexibility as the ability of the system to adapt to changing business requirements as time passes. Yourdon and Constantine call this feature *modifiability*. By flexibility they mean the ability to configure the system to handle different circumstances based on control values that are not compiled into the system but are available for the user to set at run-time. In the Agate system, this could be reflected in the choice of ODBC as the means to access the database. This provides a standard mechanism for accessing databases, and changing the ODBC driver used would allow the system to access a local or a remote version of the database, or for the system to be migrated to a different database engine at a later date. Another possibility would be to ensure that all the prompts and error messages used by the system are held in an external data file that can be loaded up when the program runs or in response to a menu option. This allows for the creation of multiple files of messages and would enable users to set the language that they wish to use. Java's use of Unicode, which provides a character set that includes ideographic characters (for example, Chinese, Japanese and Korean) as well as all the world's alphabets, would enable a system to be developed for Agate that could be localized for each location in which it is used.

General. Generality describes the extent to which a system is general-purpose. It is more applicable to utility programs than to large information systems. However, it includes the issue of *portability*, which applies to the Agate system that is to be developed in Java so that it can run on different hardware (PCs and Apple Macs). The system may also exhibit generality from the point of view of the developers who may wish to use the same system for other clients in the advertising industry. Reuse is discussed below.

Buildable. From the perspective of the programmer who has to write the program code to build the system, it is important that the design is clear and not unnecessarily complex. In particular the physical design should relate closely to the features that are available in the development language. Not all object-oriented languages offer the same features, for example in the visibility of attributes and operations (public, private, protected, friend etc.), in the ability to handle multiple inheritance, or in the availability of utility classes such as collections or linked lists in the base language. Designs that rely on features such as these will force the programmer to work around them if a different language is being used from the one that the designer had in mind.

Manageable. A good design should allow the project manager to estimate the amount of work involved in implementing the various sub-systems. It should also provide for sub-systems that are relatively self-contained and can be marked off as completed and passed on for testing without fear that changes to other parts of the system still in development will have unforeseen consequences on them.

Maintainable. Maintenance is cited as taking up as much as 60% of the data-processing budget of organizations. Maintenance activities include fixing bugs, modifying reports and screen layouts, enhancing programs to deal with new business requirements, migrating systems to new hardware and fixing the new bugs that are introduced by all of the above. A well-designed and documented system is easier to maintain than one that is poorly designed and documented. If maintenance is easy then it is less costly. It is particularly important that there is a close match between the developed program code and the design. This makes it easier for the maintenance programmer to understand the intention of the designer and to ensure that it is not subverted by the introduction of new code.

Usable. Usability includes a range of aspects including the idea, mentioned above, that a system should be both satisfying and productive. It may seem odd to suggest that people should enjoy using their computer systems and find it a satisfying experience. However, if you think about the times that you have used a computer system and have found it a source of dissatisfaction, then you can perhaps imagine a satisfying system as one with an absence of dissatisfying features. Many of the features that contribute to user satisfaction are characteristic of good human–computer interface (HCI) design. For example, the concept of *affordance* (meaning that objects on the interface suggest their function) can reduce the number of errors made by users. Reducing error rates and ensuring that if users do make an error it is clear both where they went wrong and how to recover from the error can contribute to the satisfaction of users. Productivity can be enhanced by ensuring that the tasks that users wish to carry out using the system are straightforward to carry out and do not introduce an overhead of keystrokes or mouse-clicks to achieve. If usability requirements have been captured (see Section 5.2.2), then the design should take these into account. Usability is considered in more detail in Chapter 15.

Reusable. Reusability is the Holy Grail of object-oriented development. Many of the features of object-oriented systems are geared to improve the possibility of reuse. Reuse affects the designer in three ways. Firstly, he or she will consider how economies can be made by designing reuse into the system through the use of inheritance; secondly, he or she will look for opportunities to use design patterns, which provide templates for the design of reusable elements; and thirdly, he or she will seek to reuse existing classes either directly or by sub-classing them. Design patterns are described in detail in Chapter 14. Existing classes could be classes that have been developed for other projects, classes in class libraries that are associated with the development language (such as the Java AWT) or classes that are bought in from outside vendors. To date, object-oriented development has not achieved the levels of reuse that were expected. In order to reuse a software class, a designer must be aware of the existence of the class, and be able to determine both that its interface matches the interface for the class that he or she requires and that the methods of the class match those required. It is arguable that in order to determine whether an available class matches requirements, the required class must already have been designed. The economies from reuse thus appear during the construction of the software and require a change to a culture of project management that supports reuse; this means that project managers must be able to recognize the effort that goes into **not** writing and testing lines of code (because a class is being reused). The development of strategies to parcel up classes as components and the provision of component management software are an attempt to develop support for reuse (Allen and Frost, 1998). Chapter 19 covers some of these issues.

There is clearly some overlap between the categories that have been listed here. Aspects of maintainability overlap with flexibility, generality with reuse, efficiency with economy. What is often the case however is that some design objectives will conflict with one another. This more often happens at the level of specific objectives rather than general ones such as those described above. However, it should be possible to see that functionality, reliability and security could all conflict with economy. Many of the conflicts result from constraints that are imposed on the system by the users' non-functional requirements.

Constraints arise from the context of the project as well as from the users' requirements. The clients' budget for the project, the timescale within which they expect the system to be delivered, the skills of staff working on the project, the need to integrate the new system with existing hardware or systems, and standards set as part of the overall systems design process can all constrain what can be achieved. Resolving conflicts between requirements and constraints results in the need for compromises or trade-offs in design. A couple of examples should illustrate how these can occur.

- If the users of Agate's new system require the ability to change fonts in the notes that they write about campaigns and adverts, then they will want to be able to edit notes with the same kind of functionality that would be found in a word-processor. As pointed out above, this will seriously impact the storage requirements for notes. It will also have an effect on network traffic, as larger volumes of data will need to be transferred across the network when users browse through the notes. The designers will have to consider the impact of this requirement. It may be that the users will have to accept reduced functionality or the management of Agate will have to recognize that their system will have higher costs for storage than first envisaged. Compromise solutions may involve only transferring the text of a note (without the overhead of all the formatting information) when users are browsing a note and transferring the full file only when it needs to be viewed or edited. However, this will increase the processing load on the server. Another compromise solution might be to use a different file format such as RTF (rich text format) rather than the word-processor format. For the short text file discussed above this reduces the byte count to 1,770 while retaining formatting information.

- Agate would like the system to be configurable so that prompts, help and error messages are displayed in the language of the user. This means that each prompt and error message must be read into the programs from data files or the database. While this is good software design practice and makes the system more flexible, it will increase the workload of the designers when they design elements of the interface. Without this requirement, it is enough for each designer to specify that messages such as 'Campaign' or 'Not on file' appear on screen; there is a minimal need for liaison between designers. With this requirement, the designers will need to draw up a list of prompts, labels and error messages that can be referred to by number so that the same message is used consistently wherever it is applicable. This means that the programmers will not hard code messages into the system, but will refer to them as elements in an array of messages, for example. While this increases the flexibility and to some extent the maintainability of the system, it is likely to increase the cost of the design phase.

It is important that these design decisions are clearly documented and the reasoning behind compromises and trade-offs is recorded.

11.6 Measurable Objectives in Design

In the previous section, we discussed some of the general objectives of the designers in a systems development project. Some objectives are specific to a particular project, and it is important to be able to assess whether these objectives have been achieved. One way of doing this is to ensure that these objectives are expressed in measurable terms so that they can be tested by simulation during the design phase, in prototypes that are built for this purpose or in the final system.

Measurable objectives often represent the requirements that we referred to as non-functional requirements in Chapter 5. They also reflect the fact that information systems are not built for their own sake, but are developed to meet the business needs of some organization. The system should contribute to the strategic aims of the business, and so should help to achieve aims such as:

- provide better response to customers, or
- increase market share.

However, such aims are vague and difficult to assess. If they are expressed in measurable terms, then it is possible to evaluate whether they can be achieved by the design, or whether they have been achieved by the finished system. Ideally they should be phrased in a way that shows how these objectives are attributable to the system. If a company expects to increase its market share as a result of introducing a new computer system but does not, it should be possible to tell whether this is a failure of the new system or the outcome of other factors outside the control of the system developers such as economic recession. The system may contribute to business objectives such as those above by providing better information or more efficient procedures, but for the objectives to be measurable they need to be phrased as operational objectives that can be quantified, such as:

- to reduce invoice errors by one-third within a year, or
- to process 50% more orders during peak periods.

These set clear targets for the designers and a way of checking whether these objectives can be achieved (within the constraints on the system) and whether they have been achieved once the system is up and running.

11.7 Planning for Design

The transition from the analysis phase to the design phase also gives the project manager the opportunity to plan for the activities that must be undertaken during design.

- If the hardware and software platform has not been decided upon, then design will begin as logical design, but the project manager must plan for the time when the platform is known and physical design can begin.
- The system architecture must be agreed and system standards must be set that will affect the design of individual sub-systems.
- If the designers are not familiar with all aspects of the platform, time must be allowed for training, or additional staff must be brought in, perhaps contractors with expertise in particular aspects of the hardware or software.
- Design objectives must be set and procedures put in place for testing those that can be tested by simulation during the design process.
- Procedures must also be put in place for resolving conflicts between constraints and requirements and for documenting and agreeing any trade-offs that are made.
- The amount of time to be spent on design of different aspects of the system (described in Chapters 12, 13, 15, 16 and 17) must be agreed.

All this provides a further case for a clear separation between analysis and design.

11.8 Summary

While analysis looks to the business in order to establish requirements, design looks to the technology that will be used to implement those requirements. An effective design will meet general objectives that will make the system easier to build and maintain, and more usable and functional for the end-users. The design of a system should also meet specific objectives relating to the business needs of the users, and these specific objectives should be phrased in quantifiable, operational terms that allow them to be tested. This process of design takes place in the context of constraints that are imposed by the users, their budget and existing systems, the available technology and the skills and knowledge of the design and development team.

In Chapters 12 and 13 we describe system design and detailed design of the classes in the required system. Chapter 14 explains how patterns can be used to assist the design process. Chapters 15 and 16 look specifically at the design of the human–computer interface, and Chapter 17 discusses the design of data storage.

Review Questions

11.8   List four quality criteria for good analysis.

11.9   List twelve quality criteria for good design.

11.10   Re-read the description of the FoodCo case study in Case Study B1. Identify any constraints that you think might be imposed on the design of the new system.

11.11   Based on the same information try to identify possible measurable objectives for the new FoodCo system.

11.12   Agate wants the new system to provide access to the same data from every office around the world. Maintaining a network that is constantly connected between all the offices is considered too expensive, and using a network that dials up remote offices as required would provide response times that are too slow. What kind of compromise solution can you come up with to this problem?

Case Study Work, Exercises and Projects

11.A   FoodCo require a data entry screen for entering details of staff holidays. Without knowing what software or hardware is going to be used to develop this data entry screen, list as many features of the design as you can that are not dependent on the implementation platform.

11.B   Design applies to a wide range of artefacts, for example cars, buildings, books and packaging. Choose some artefact that you use and try to identify what makes for a good design in this context. Are there aspects that do not apply to systems design? Are there aspects of systems design that should perhaps apply to the design of artefacts that you use?

11.C   Find out what laws (if any) exist in your country to protect computer systems against malicious attack from hackers. What implications does the law have for the design of systems?

11.D   One aspect of system design is concerned with the setting of standards across the system. Choose a system that you use regularly as part of your work or study and try to identify standards that are applied to the design of the interface. (This could be a software package you use, the library system you use to access your library catalogue, or even Windows.)

11.E   In Section 11.5.2 we pointed out that some criteria for good quality in design will bring benefits to the designers, while others will bring benefits to the eventual users of the system. Try to decide which of the characteristics discussed in Section 11.5.3 bring benefits to the designers as well as the end users.

Further Reading

For those with an interest in the historical development of systems design, the classics of structured design are Jackson (1975) and Yourdon and Constantine (1979). DeMarco (1979) deals with structured analysis. Two more recent books that are likely to be more easily available are Yourdon (1989) and Page-Jones (1988). Most standard textbooks on systems analysis and design take either a structured approach or a toolkit approach, drawing on a range of different techniques.

If you are interested in an approach to the analysis and design of requirements that is completely different from object-oriented approaches, SSADM (Structured Systems Analysis and Design Method) makes a very clear distinction between requirements analysis, logical design and physical design. A separate stage in SSADM is used to carry out the choice of development environment (Technical System Options). Skidmore, Mills and Farmer (1994), Goodland and Slater (1995) or any other book on SSADM Version 4 explains the way in which these stages are handled.

More detail is provided in Chapters 12 to 17 on specific aspects of the design task.

12

System Design

Objectives
In this chapter you will learn:
- the major concerns of system design;
- the main aspects of system architecture, in particular what is meant by subdividing a system into layers and partitions;
- how to apply the MVC architecture;
- which architectures are most suitable for distributed systems;
- how design standards are specified.

12.1 Introduction

In Chapter 11 we saw how the move from analysis to design requires the consideration of many issues that are not directly related to the functional requirements of the information system. These are the non functional requirements, and they reflect the way that the system is to be constructed and used, rather than what it does. Design is concerned with establishing how to deliver the functionality that was specified in analysis while, at the same time, meeting non-functional requirements that may sometimes conflict with each other.

System design is focused on making high-level decisions concerning the overall structure of the system. The main activities in system design are introduced in Section 12.2, and possible software architectures are discussed in Section 12.3. The design of those parts of the system that have to operate concurrently is addressed in Section 12.4. Other aspects of system design include the allocation of different parts of a system to run on different computers (Section 12.5) and the selection of suitable data management strategies (Section 12.6). System design also involves establishing protocols and standards for the design activities and the products that are produced. For example, a standard naming format should be used throughout the project, and program code must be written in an agreed style (Section 12.7). Finally, we consider the importance of prioritizing design trade-offs (Section 12.8), and how implementation issues impact on design (Section 12.9). The technology available for the construction of the application has a significant impact upon the design process as a whole and upon the decisions that are made during system design. It is a truism, but none the less worth stating, that the design activity requires detailed knowledge of the technologies that are available for use. The decisions that are made during system design inform the rest of the design process.

12.2 The Major Elements of System Design

The system design activity specifies the context within which detailed design will occur. A major part of system design is defining the *system architecture*. The meaning and scope of the term architecture for computerized information systems is much debated but it is generally accepted that it is an important feature of the delivered system. Evidence of this shift can be found in the employment market in the increasing number of jobs advertised that have the title of System Architect. The architecture of a system is concerned with its overall structure, the relationships among its major components and their interactions. If the system being considered contains human, software and hardware elements then its architecture includes how these elements are structured and how they interact. On the other hand, if the system being considered comprises software and hardware, then its architecture only concerns these elements. It is important to consider the structure of the software elements of the system and this is termed the *software architecture*. The hardware architecture of a system is discussed in this book in terms of the computers and peripherals required for the system and how software is allocated to them.

The detailed software architecture of a computerized information system develops as the design process continues into object design but it is important to identify an overall system architecture within which the detail can be refined. High-level architectural decisions that are made during system design determine how successfully the system will meet its non-functional objectives (e.g. performance, extensibility) and thus its long-term utility for the client. Reuse is one of the much-vaunted benefits of object-orientation and poor software architecture usually reduces both the reusability of the components produced and the opportunity to reuse existing components.

During system design the following activities are undertaken.

- Sub-systems and major components are identified.
- Any inherent concurrency is identified.
- Sub-systems are allocated to processors.
- A data management strategy is selected.
- A strategy and standards for human–computer interaction are chosen.
- Code development standards are specified.
- The control aspects of the application are planned.
- Test plans are produced.
- Priorities are set for design trade-offs.
- Implementation requirements are identified (for example, data conversion).

A similar list of activities is proposed by Rumbaugh et al. (1991), and equivalent activities for real-time structured design are cited by Goldsmith (1993). The product of the system design process is not a detailed specification. It is rather a specification of the design context that can guide the developers as they make detailed design decisions regarding the design specification from which construction will proceed.

12.3 Software Architecture

Software architecture, like system architecture has no generally agreed definition and can be interpreted differently depending upon the context. For Booch (1994), the architecture of a system is its class structure. He suggests that part of the architecture is the way in which classes are grouped together. Rumbaugh et al. (1991) use the term system architecture to describe the overall organization of a system into sub-systems. Buschmann et al. (1996) give the following definition of a software architecture.

> A *software architecture* is a description of the subsystems and components of a software system and the relationships between them. Subsystems and components are typically specified in different views to show the relevant functional and non-functional properties of a software system. The software architecture of a system is an artefact. It is the result of the software design activity.

A contrasting view is offered by Soni et al. (1995), who identify the four different aspects of software architecture shown in Figure 12.1.

Type of architecture	Examples of elements	Examples of relationships
Conceptual	Components	Connectors
Module	Sub-systems, modules	Exports, imports
Code	Files, directories, libraries	Includes, contains
Execution	Tasks, threads, object interactions	Uses, calls

Adapted from Weir and Daniels (1998)

Figure 12.1 *Four aspects of software architecture according to Soni et al.*

In terms of object-oriented development, the conceptual architecture is concerned with the structure of the static class model and the connections between the components of the model. The module architecture describes the way the system is divided into sub-systems or modules and how they communicate by exporting and importing data. The code architecture defines how the program code is organized into files and directories and grouped into libraries. The execution architecture focuses on the dynamic aspects of the system and the communication between components as tasks and operations execute.

There are alternative ways of delineating different aspects of architecture. For example, a logical architecture might comprise the class model, while a physical architecture is concerned with mapping the software onto the hardware components. However, in all cases these different aspects combine to define a software architecture for the system. There are various architectural styles, each of which has characteristics that make it more or less suitable for certain types of application. We will consider some of the major alternatives.

12.3.1 Sub-systems

A sub-system typically groups together elements of the system that share some common properties. An object-oriented sub-system encapsulates a coherent set of responsibilities in order to ensure that it has integrity and can be maintained. For example, the elements of one sub-system might all deal with the human–computer interface, the elements of another might all deal with data management and the elements of a third may all focus on a particular functional requirement.

The sub-division of an information system into sub-systems has the following advantages.

- It produces smaller units of development.
- It helps to maximize reuse at the component level.

- It helps the developers to cope with complexity.
- It improves maintainability.
- It aids portability.

Each sub-system should have a clearly specified boundary and fully defined interfaces with other sub-systems. A specification for the interface of a sub-system defines the precise nature of the sub-system's interaction with the rest of the system but does not describe its internal structure (this is a high level use of contracts, which are described in Chapter 9). A sub-system can be designed and constructed independently of other sub-systems, simplifying the development process. Sub-systems may correspond to increments of development that can be delivered individually as part of an incremental life cycle (if the developers are using the spiral life cycle model discussed in Chapter 3).

Dividing a system into sub-systems is an effective strategy for handling complexity. Sometimes it is only feasible to model a large complex system piece by piece, with the sub-division forced on the developers by the nature of the application. Splitting a system into sub-systems can also aid reuse as each sub-system may correspond to a component that is suitable for reuse in other applications. A judicious choice of sub-systems during design can reduce the impact on the overall system of a change to its requirements. For example, consider an information system that contains a presentation sub-system which deals with the human–computer interface (HCI). A change to the data display format need not affect other sub-systems. Of course there may still be some changes to the requirements that affect more than one sub-system. The aim is to localize the consequences of change, so that a change in one sub-system does not trigger changes in other sub-systems (sometimes referred to as the ripple effect). Moving an application from one implementation platform to another can be much easier if the software architecture is appropriate. An example of this would be the conversion of a Windows 98 application so that it could run in a Unix environment. This would require changes to the software that implements the human–computer interface. If this were dealt with by specialized sub-systems then the overall software change is localized to these sub-systems. As a result, the system as a whole is easier to port to a different operating environment.

Each sub-system provides services for other sub-systems, and there are two different styles of communication that make this possible. These are known as *client–server* and *peer-to-peer* communication (Figure 12.2).

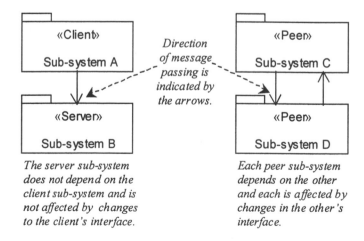

The server sub-system does not depend on the client sub-system and is not affected by changes to the client's interface.

Each peer sub-system depends on the other and each is affected by changes in the other's interface.

Figure 12.2 *Styles of communication between sub-systems.*

Client–server communication requires the client to know the interface of the server sub-system, but the communication is only in one direction. The client sub-system requests services from the server sub-system and not vice versa. Peer-to-peer communication requires each sub-system to know the interface of the other, thus coupling them more tightly. The communication is two way since either peer sub-system may request services from the other.

In general client–server communication is simpler to implement and to maintain, as the sub-systems are less tightly coupled than they are when peer-to-peer communication is used. In Figure 12.2 the sub-systems are represented using packages that have been stereotyped[1] to indicate their role. Component and deployment diagrams can also be used to model the implementation of sub-systems (see Chapter 18).

12.3.2 Layering and partitioning

There are two general approaches to the division of a software system into sub-systems. These are known as *layering*—so called because the different sub-systems usually represent different levels of abstraction—and *partitioning*, which usually means that each sub-system focuses on a different aspect of the functionality of the system as a whole. In practice both approaches are often used together on one system, so that some of its sub-systems are divided by layering, while others are divided by partitioning.

Layered Sub-systems

Layered architectures are among the most frequently used high-level structures for a system. A schematic of the general structure is shown in Figure 12.3.

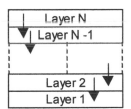

Closed architecture—
messages may only be
sent to the adjacent
lower layer.

Open architecture—
messages can be sent
to any lower layer.

Figure 12.3 *Schematic of a layered architecture.*

Each layer corresponds to one or more sub-systems, which may be differentiated from each other by differing levels of abstraction or by a different focus of their functionality. It works like this: the top layer uses services provided by the layer immediately below it. This in turn may require the services of the next layer down. Layered architectures can be either open or closed, and each has its particular advantages. In a closed layered architecture a certain layer (say layer N) can only use the services of the layer immediately below it (layer

---

[1] A stereotype is a specialized UML modelling element. The stereotype name is contained within matched guillemets «...». Another example of a package stereotype might be an interface package.

N - 1). In an open layered architecture layer N may directly use the services of any of the layers that lie below it.

A closed architecture minimizes dependencies between the layers and reduces the impact of a change to the interface of any one layer. An open layered architecture produces more compact code since the services of all lower level layers can be accessed directly by any layer above them without the need for extra program code to pass messages through each intervening layer. However this breaks the encapsulation of the layers, increases the dependencies between layers and increases the difficulty caused when a layer needs to be changed.

Networking protocols provide some of the best known examples of layered architectures. A network protocol defines how computer programs executing on different computers communicate with each other. Protocols can be defined at various levels of abstraction and each level can be mapped onto a layer. The OSI (Open Systems Interconnection) 7 Layer Model was defined by the International Standardization Organization (ISO) as a standard architectural model for network protocols (Tanenbaum, 1992). The structure provides flexibility for change since a layer may be changed internally without affecting other layers, and it enables the reuse of layer components. The OSI 7 Layer Model is illustrated in Figure 12.4.

Figure 12.4 *OSI 7 Layer Model (adapted from Buschmann et al., 1996).*

Buschmann et al. (1996) suggest that a series of issues need to be addressed when applying a layered architecture in an application. These include:
- maintaining the stability of the interfaces of each layer;
- the construction of other systems using some of the lower layers;
- variations in the appropriate level of granularity for sub-systems[2];

---

[2] The context determines an appropriate size for the sub-systems.

- the further sub-division of complex layers;
- performance reductions due to a closed layered architecture.

The OSI model has 7 layers only because it covers every aspect of the communication between two applications, ranging from application-oriented processes to drivers and protocols that directly control network hardware devices. Many layered architectures are much simpler than this. Figure 12.5 shows a simple example of a three layer architecture.

Figure 12.5 *Simple layered architecture.*

The lowest layer of the architecture in Figure 12.5 consists of data management library classes. The layer immediately above this, the data formatting layer, uses services that are provided by the data management library classes in order to get data from a database management system. This data is formatted before it is passed upwards to the application layer. Supposing this system were to be modified to allow it to use a different database management system, the layered architecture limits major changes to the data management library class layer with some possible changes to the data formatting layer.

The following steps are adapted from Buschmann et al. (1996), and provide an outline process for the development of a layered architecture for an application. Note that this does not suggest that the specification of a system's architecture is a rule-based procedure. The steps offer guidelines on the issues that need to be addressed during the development of a layered architecture.

1. Define the criteria by which the application will be grouped into layers. A commonly used criterion is level of abstraction from the hardware. The lowest layer provides primitive services for direct access to the hardware while the layers above provide more complex services that are based upon these primitives. Higher layers in the architecture carry out tasks that are more complex and correspond to concepts that occur in the application domain.
2. Determine the number of layers. Too many layers will introduce unnecessary overheads while too few will result in a poor structure.
3. Name the layers and assign functionality to them. The top layer should be concerned with the main system functions as perceived by the user. The layers below should provide services and infrastructure that enable the delivery of the functional requirements.
4. Specify the services for each layer. In general it is better in the lower layers to have a small number of low-level services that are used by a larger number of services in higher layers.
5. Refine the layering by iterating through steps 1 to 4.
6. Specify interfaces for each layer.
7. Specify the structure of each layer. This may involve partitioning within the layer.

8. Specify the communication between adjacent layers (this assumes that a closed layer architecture is intended).
9. Reduce the coupling between adjacent layers. This effectively means that each layer should be strongly encapsulated. Where a client-server communication protocol will be used, each layer should have knowledge only of the layer immediately below it.

One of the simplest application architectures has only two layers—the application layer and a database layer. Tight coupling between the user interface and the data representation would make it more difficult to modify either independently so a middle layer is often introduced in order to separate the conceptual structure of the problem domain. This gives the architecture shown in Figure 12.6, which is commonly used for business-oriented information systems. Note that it is essentially the same as the three-tier architecture suggested in Figure 11.1 (user interface, business objects and database tiers).

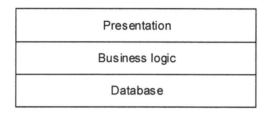

Figure 12.6 *Three layer architecture.*

A common four layer architecture separates the business logic layer into application logic and domain layers, and this is illustrated in Figure 12.7.

| Presentation |
| Application logic |
| Domain |
| Database |

Figure 12.7 *Four layer architecture.*

Such an approach may be justified because several applications share (or are likely to share) one domain layer, or because the complexity of the business objects forces a separation into two layers. It can also be used when the objects are physically distributed (see Chapter 18). However, it must be emphasized that there is no perfect solution to this kind of design problem. There are only solutions that have different characteristics (perhaps different levels of efficiency or maintainability). A good design solution is one that balances competing requirements effectively.

Partitioning

As suggested earlier, some layers within a layered architecture may have to be decomposed because of their intrinsic complexity. Figure 12.8 shows a four layer architecture for part of Agate's campaign management system that also has some partitioning in the upper layers. In this example the application layer corresponds to the analysis class model for a single application, and is partitioned into a series of sub-systems. These sub-systems are loosely coupled and each should deliver a single service or coherent group of services. The Campaign Database layer provides access to a database that contains all the details of the campaigns, their adverts and the campaign teams. The Campaign Domain layer uses the lower layer to retrieve and store data in the database and provides common domain functionality for the layers above. For example, the Advert sub-system might support individual advert costing while the Campaign Costs sub-system uses some of the same common domain functionality when costing a complete campaign. Each application sub-system has its own presentation layer to cater for the differing interface needs of different user roles[3].

A system may be split into sub-systems during analysis because of the system's size and complexity. However, the analysis sub-systems should be reviewed during design for coherence and compatibility with the overall system architecture.

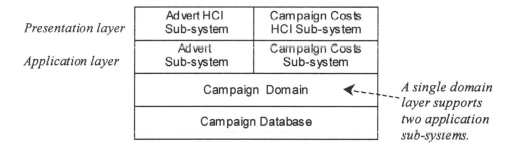

Figure 12.8 *Four layer architecture applied to part of the Agate campaign management system.*

The sub-systems that result from partitioning should have clearly defined boundaries and well specified interfaces, thus providing high levels of encapsulation so that the implementation of an individual sub-system may be varied without causing dependent changes in the other sub-systems. The process of identifying sub-systems within a particular layer can be detailed in much the same way as for sub-system layers.

Review Questions

12.1 Why is an open layered architecture more difficult to maintain?

12.2 What are the disadvantages of the closed layer architecture?

---

[3] This example is for illustrative purposes only. Our analysis class model for Agate is too small to justify this kind of partitioning in practice.

12.3   What advantages would there be if the Advert HCI sub-system in Figure 12.8 were designed to have direct access to the Campaign Database layer?

12.3.3 Model-View-Controller

Many interactive systems use the Model-View-Controller (MVC) architecture. This structure was first used with Smalltalk but has since become widely used with many other object-oriented development environments. The MVC architecture separates an application into three major types of component: models that comprise the main functionality, views that present the user interface, and controllers that manage the updates to views. This structure is capable of supporting user requirements that are presented through differing interface styles, and it aids maintainability and portability.

It is common for the view of an information system that is required for each user to differ according to their role. This means that the data and functionality available to any user should be tailored to his or her needs. The needs of different types of user can also change at varying rates. For both these reasons it makes sense to give each user access to only the relevant part of the functionality of the system as a whole. For example, in the Agate case study many users need access to information about campaigns, but their perspectives vary. The campaign manager needs to know about the current progress of a campaign. She is concerned with the current state of each advertisement and how this impacts on the campaign as a whole—is it prepared and ready to run, or is it still in the preparation stage? If an advert is behind schedule does this affect other aspects of the campaign? The creative artist also needs access to adverts but he is likely to need access to the contents of the advert (its components and any notes that have been attached to it) as well as some scheduling information. A director may wish to know about the state of all live campaigns and their projected income over the next six months. This gives at least three different perspectives on campaigns and adverts, each of which might use different styles of display. The director may require charts and graphs that summarize the current position at quite a high level. The campaign manager may require lower level summaries that are both textual and graphical in form. The graphic designer may require detailed textual displays of notes with a capability to display graphical images of an advert's content. Ideally, if any information about a campaign or an advert is updated in one view then the changes should also be immediately reflected in all other views. Figure 12.9 shows a possible architecture, but some problems remain.

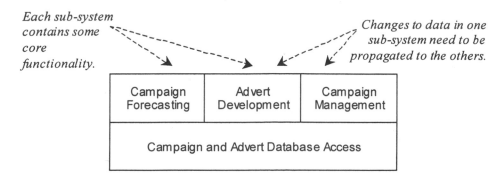

Figure 12.9 *Multiple interfaces for the same core functionality.*

The design of such varied and flexible user interfaces that still incorporate the same core functionality is likely to be expensive because elements of functionality may have to be duplicated for different interfaces. This makes the software more complex and thus also more error prone. There is an impact on maintainability too, since any change to core functionality will necessitate changes to each interface sub-system.

We repeat below some of the difficulties that need to be resolved for this type of application.
- The same information should be capable of presentation in different formats in different windows.
- Changes made within one view should be reflected immediately in the other views.
- Changes in the user interface should be easy to make.
- Core functionality should be independent of the interface to enable multiple interface styles to co-exist.

While the four layer architecture in Figure 12.8 resolves some of these problems it does not handle the need to ensure that all view components are kept up to date. The MVC architecture solves this through its separation of core functionality (model) from the interface and through its incorporation of a mechanism for propagating updates to other views. The interface itself is split into two elements: the output presentation (view) and the input controller (controller).

Figure 12.10 shows the basic structure of the MVC architecture.

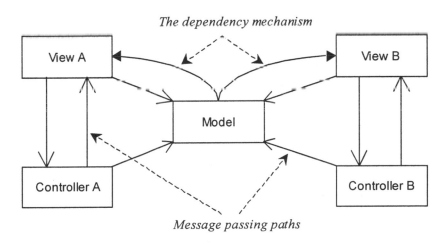

Figure 12.10 *General structure of Mode-View-Controller (adapted from Hopkins and Horan, 1995).*

The responsibilities of the components of an MVC architecture are listed below.
Model. The model provides the central functionality of the application and is aware of each of its dependent view and controller components.
View. Each view corresponds to a particular style and format of presentation of information to the user. The view retrieves data from the model and updates its presentations when data has been changed in one of the other views. The view creates its associated controller.

Controller. The controller accepts user input in the form of events that trigger the execution of operations within the model. These may cause changes to the information and in turn trigger updates in all the views ensuring that they are all up to date.

Propagation Mechanism. This enables the model to inform each view that the model data has changed and as a result the view must update itself. It is also often called the dependency mechanism.

Figure 12.11 represents the capabilities offered by the different MVC components as they might be applied to part of the campaign management system at Agate.

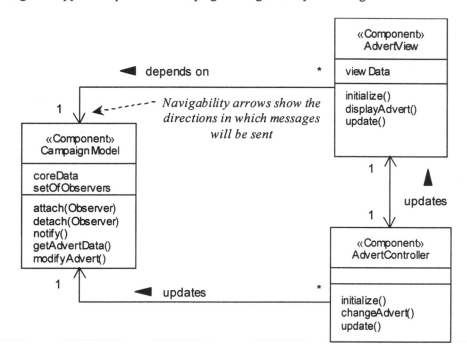

Figure 12.11 *Responsibilities of MVC components, as applied to Agate.*

The operation `update()` in the `AdvertView` and `AdvertController` components triggers these components to request data from the `CampaignModel` component. This model component has no knowledge of the way that each view and controller component will use its services. It need only know that all view and controller components must be informed whenever there is a change of state (a modification either of object attributes or of their links).

The `attach()` and `detach()` services in the `CampaignModel` component enable views and controllers to be added to the `setOfObservers`. This contains a list of all components that must be informed of any change to the model core data. In practice there would be separate views, each with its own controller, to support the requirements of the campaign manager and the director.

The interaction sequence diagram in Figure 12.12 illustrates the communication that is involved in the operation of an MVC architecture. An AdvertController component receives the interface event `changeAdvert()`. In response to this event the controller invokes the `modifyAdvert()` operation in the `CampaignModel` object. The execution of this operation causes a change to the model.

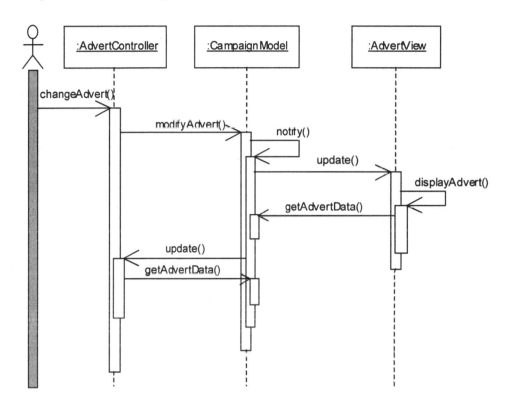

Figure 12.12 *MVC component interaction (adapted from Buschmann et al., 1996).*

For example, the target completion date for an advertisement is altered. This change of state must now be propagated to all controllers and views that are currently registered with the model as active. To do this the `modifyAdvert()` operation invokes the `notify()` operation in the model that sends an `update()` message to the view. The view responds to the `update()` message by executing the `displayAdvert()` operation which requests the appropriate data from the model via the `getAdvertData()` operation. The model also sends an `update()` message to the `AdvertController` that then requests the data it needs from the model.

One of the most important aspects of the MVC architecture is that each model knows only which views and controllers are registered with it, but not what they do. The `notify()` operation causes an update message to all the views and controllers (for clarity, only one view and one controller are shown in the diagram, but interaction with the others would be similar). The `update()` message from the model is in effect saying to the views and controllers 'I have been updated and you must now ensure that your data is consistent'. Thus the model, which should be the most stable part of the application, is unaffected by changes in the presentation requirements of any view or controller. The change propagation mechanism can be structured so that further views and controllers can be added without causing a change to the model. Each of these may support different interface requirements but require the same model functionality. However, since views and controllers need to know how to access the model in order to get the information they require, some changes in the model will inevitably still cause changes in other components.

Other kinds of communication may take place between the MVC components during the operation of the application. The controller may receive events from the interface that require a change in the way that some data is presented to the user but do not cause a change of state. The controller's response to such an event would be to send an appropriate message to the view. There would be no need for any communication with the model.

12.3.4 Architectures for distributed systems

Distributed information systems are becoming more common as communications technology improves and becomes more reliable. An information system may be distributed over computers at the same location or different locations. Since Agate has offices around the world, it may need information systems that use data that is distributed between different locations. If Agate grows, it may also open new offices and require new features from its information systems. An architecture that is suitable for distributed information systems needs also to be flexible so that it can cope with change. A distributed information system may be supported by software products such as distributed database management systems or object request brokers (these are discussed in Chapter 18).

A general *broker* architecture for distributed systems is described by Buschmann et al. (1996). A simplified version of the broker architecture is shown in Figure 12.13.

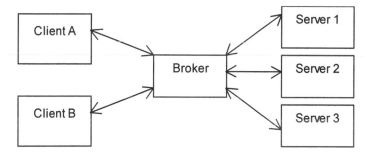

Figure 12.13 *Simplified broker architecture.*

A broker component increases the flexibility of the system by decoupling the client and server components. Each client sends its requests to the broker rather than communicating directly with the server component. The broker then forwards the service request to an appropriate server. A broker may offer the services of many servers and part of its task is to identify the relevant server to which a service request should be forwarded. The advantage offered by a broker architecture is that a client need not know where the server is located, and it may therefore be stored on either a local or a remote computer. Only the broker needs to know the location of the servers that it handles.

Figure 12.14 shows a sequence diagram for client-server communication using the broker architecture. In this example the server sub-system is on a local computer. In addition to the broker itself, two additional *proxy* components have been introduced to insulate the client and server from direct access with the broker. On the client side a ClientSideProxy receives the initial request from the client and packs the data in a format suitable for transmission. The request is then forwarded to the Broker which finds an appropriate server and invokes the required service via the ServerSideProxy.

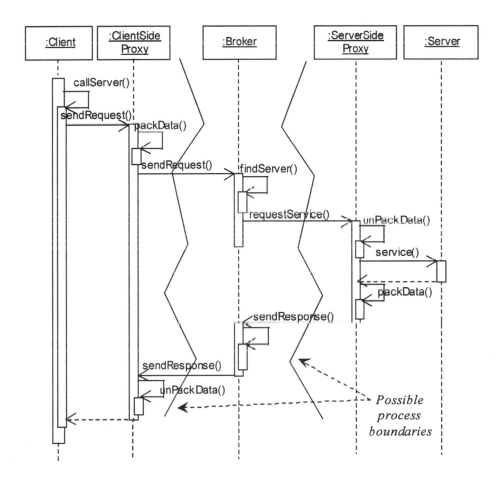

Figure 12.14 *Broker architecture for local server (adapted from Buschmann et al., 1996).*

The ServerSideProxy then unpacks the data and issues the service request sending the service() message to the Server object. The service() operation then executes and on completion control returns to the ServerSideProxy (this is shown explicitly by the dashed message return arrow). The response is then sent to the Broker which forwards it to the originating ClientSideServer. Note that these are both new messages and not returns. The reason for this is that a broker does not wait for each response before handling another request. Once its sendRequest activation has been completed the broker will in

all probability deal with many other requests, and thus requires a new message from the
`ServerSideProxy` object that causes it to enter a new activation. Unlike the broker, the
`ClientSideProxy` has remained active; this then unpacks the message and the response
becomes available to the `Client` as control returns (again shown explicitly for clarity).
Note that the diagram is drawn with uncommitted messages throughout, as described in
Chapter 8. This means only that decisions on concurrency have been deferred.

Figure 12.15 shows a schematic broker architecture that uses *bridge* components to
communicate between two remote processors. Each bridge converts service requests into a
network specific protocol so that the message can be transmitted.

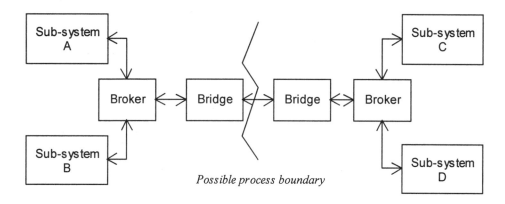

Figure 12.15 *Broker architecture using bridge components.*

12.3.5 Organization structures for architecture and development

Dividing a system into sub-systems has benefits for project management. Each sub-
system can be allocated to a single development team that can operate independently of
other teams, provided that they adhere to the interface requirements for their sub-system.
Where a sub-system must be split between two development teams there is a heavy
communications overhead that is incurred in ensuring that the different parts of the sub-
system are constructed to consistent standards. In such cases the structure of either the
organization or of the software tends to change so that they become more closely aligned
with each other; this helps to minimize the communications overhead and is sometimes
known as Conway's Law (Coplien, 1995). Where a sub-system that is being developed by
more than one team is cohesive, and the way it is split between teams has no apparent
functional basis, in practice the teams may coalesce and operate as one. Teams that are
working on the same sub-system are sometimes inhibited from merging—say, because they
are located on different continents. The sub-system should then be treated as if it were two
separate sub-systems. An interface between these two new sub-systems can be defined and
the teams can then operate autonomously. Where the allocation of one sub-system to two
teams is such that one team deals with one set of requirements, and another deals with a
different set of requirements, the sub-system can also be treated as if it were actually two
sub-systems, with a defined interface between them.

Review Questions

12.4 What are the main differences between the MVC architecture and the layered architecture?

12.5 Explain how the update propagation mechanism works in the MVC architecture.

12.6 In what sense does a broker decouple two sub-systems that need to communicate with each other? How does this work?

12.4 Concurrency

In most systems there are many objects that do not need to operate concurrently but some may need to be active simultaneously. Object-oriented modelling captures any inherent concurrency in the application principally through the development of interaction diagrams and statecharts. The examination of use cases also helps with the identification of concurrency. There are several ways of using these models to identify circumstances where concurrent processing may be necessary. Firstly, a use case may indicate a requirement that the system should be able to respond simultaneously to different events, each of which triggers a different thread of control. Secondly, if a statechart reveals that a class has complex nested states which themselves have concurrent substates, then the design must be able to handle this concurrency. The statechart for the class Campaign has nested concurrent states within the Active state (see Figure 10.13). In this particular example, the concurrent activity that occurs in the real world need not necessarily be represented as concurrent processing in the computerized information system.

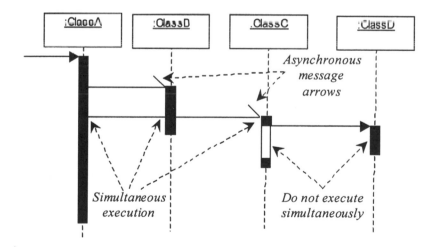

Figure 12.16 Concurrent activity in an interaction diagram.

In cases where an object is required to exhibit concurrent behaviour it is sometimes necessary to split the object into separate objects in order to avoid the need for concurrent activity within any one object. Thirdly, interaction diagrams may reveal that a single thread of control requires that operations in two different objects should execute simultaneously, perhaps because of asynchronous invocation. This essentially means that one thread of control is split into two or more active threads. An example of this is shown in Figure 12.16.

Different objects that are not active at the same time can be implemented on the same logical processor (and thus also on the same physical processor—this distinction is explained below). Objects that must operate concurrently must be implemented on different logical processors (though perhaps still on the same physical processor).

The distinction between logical and physical concurrency is as follows. There are a number of ways of simulating the existence of multiple processors using only a single physical processor. For example, some operating systems (Unix and Windows NT) allow more one than one task to appear to execute at the same time, and are thus called multi-tasking operating systems. In fact, only one task really takes place at any one time, but the operating system shares the processor between different tasks so quickly that the tasks appear to execute simultaneously. Where there are no tight time constraints a multi-tasking operating system can provide a satisfactory implementation of concurrency. But it is important to ensure that the hardware configuration of the computer can cope with the demands of multi-tasking.

When there are tight time constraints a scheduler sub-system can be introduced that ensures that each thread of control operates within the constraints on its response time. Figure 12.17 illustrates a possible relationship between a scheduler and the other parts of a system. Events that are detected by the I/O (input/output) sub-systems generate interrupts in the scheduler. The scheduler then invokes the appropriate thread of control. Further interrupts may invoke other threads of control and the scheduler allocates a share of physical processor time to each thread.

Figure 12.17 *Scheduler handling concurrency.*

Another way of implementing concurrency is to use a multi-threaded programming language (such as Java). These permit the direct implementation of concurrency within a single processor task. Finally, a multi-processor environment allows each concurrent task to be implemented on a separate processor.

Most concurrent activity in a business information system can be supported by a multi-user environment. These are designed to allow many users to perform tasks simultaneously. Multi-user concurrent access to data is normally handled by a separate database management system (DBMS)—these are introduced briefly in Section 12.6 and are discussed in more detail in Chapter 17.

12.5 Processor Allocation

In the case of a simple, single-user system it is always almost appropriate for the complete system to operate on a single computer. The software for a multi-user information system (all or part of it) may be installed on many computers that all use a common file-server. More complex applications sometimes require the use of more than one type of computer, where each provides a specialized kind of processing capability for a specific sub-system. An information system may also be partitioned over several processors either because sub-systems must operate concurrently, or because some parts of the application need to operate in different locations (in other words, it is a distributed system). Information systems that use the Internet or company intranets for their communications are being built more frequently. Such distributed information systems operate on diverse computers and operating systems.

The allocation of a system to multiple processors on different platforms involves the following steps.
- The application should be divided into sub-systems.
- Processing requirements for each sub-system should be estimated.
- Access criteria and location requirements should be determined.
- Concurrency requirements for the sub-systems should be identified.
- Each sub-system should be allocated to an operating platform—either general purpose (PC or workstation) or specialized (embedded micro-controller or specialist server).
- Communication requirements between sub-systems should be determined.
- The communications infrastructure should be specified.

The estimation of processing requirements requires careful consideration of such factors as event response times, the data throughput that is needed, the nature of the I/O that is required and any special algorithmic requirements. Access and location factors include the difficulties that may arise when a computer will be installed in a harsh operating environment such as a factory shop floor.

12.6 Data Management Issues

Suitable data management approaches for an information system can vary from simple file storage and retrieval to sophisticated database management systems of various types. In some applications where data has to be accessed very rapidly, the data may be kept in main memory while the system executes. However, most data management is concerned with storing data, often large volumes, so that it may be accessed at a later stage either by the same system or by another. A file-based approach to data management is suitable for simple data management, but the data storage and retrieval must be coded explicitly within the application. If the data is to be accessed by multiple users simultaneously, the management of the data storage and retrieval directly in the application becomes complex. However, fast access to data may justify the use of a file-based approach and its resulting code complexity.

Database management systems (DBMS) provide various facilities that are useful in many applications. A DBMS typically offers support for:

- different views of the data by different users,
- control of multi-user access,
- distribution of the data over different platforms,
- security,
- enforcement of integrity constraints,
- access to data by various applications,
- data recovery,
- portability across platforms,
- data access via query languages,
- query optimization.

These capabilities make a DBMS the obvious choice for many applications. However, a DBMS does have a significant performance overhead and the standard data access mechanisms may be inappropriate for specialized systems. Once a decision has been made to use a DBMS the most appropriate type must be selected. A relational DBMS is likely to be appropriate if there are large volumes of data with varying (perhaps ad hoc) access requirements. An object-oriented DBMS is more likely to be suitable if specific transactions require fast access or if there is a need to store complex data structures and there is not a need to support a wide range of transaction types. A third type of DBMS is emerging—the object-relational DBMS—that is similar to an object-oriented DBMS in its support for complex data structures, but that also provides effective querying facilities. In some systems there may be different data management requirements for different sub-systems and it may be best then to use a mix of DBMS types. These issues and the detailed design consequences of the choice of DBMS are explored further in Chapter 17.

The DBMS used by the application will typically be accessed from programs using specialized class libraries to provide the I/O functionality. It is rarely the case today that a developer must construct low-level primitives in order to access a DBMS since suitable commercial class libraries are widely available.

12.7 Development Standards

All information systems development projects should operate with clearly defined guidelines within which all members of the development team work. Many organizations will have corporate style guides that govern the production of software development artefacts, including the delivered system. In some organizations these corporate guides may be adapted for particular development projects. The development standards for analysis have been discussed in Chapter 5. From the design perspective it is important to specify guidelines for the development of I/O sub-systems and their interfaces and standards for the development of code.

12.7.1 HCI guidelines

Standards for the human–computer interface are an important aspect of the design activity, since it is with the interface that users actually interact. Some characteristics of good dialogues and the subject of style guides for HCI are discussed in Chapter 15.

12.7.2 Input/output device guidelines

Where an application interacts with mechanical or electronic devices such as temperature and pressure sensors or actuators that control heaters or motors, it is equally important to develop guidelines. The objective is to use a standard form of interface with the devices so that hardware may be changed or updated without occasioning any changes to the core system functionality. Sensor and controller devices usually have fully specified communications protocols and standardization is probably best achieved by encapsulating all direct access with each device in a single I/O device object. An I/O device class can be subclassed so that for each particular device involved with the application there is a class that deals with its communications protocol. Since all the I/O device classes would then be subclasses of one inheritance hierarchy they can easily be constructed to provide consistent interfaces to the rest of the application. One way of achieving this is by the use of polymorphism, as shown in Figure 12.18. In this diagram, each subclass overrides the abstract operation `handleEvent()`, so that each can respond to a `handleEvent()` message in its own appropriate way, but other objects in the system do not necessarily need to know which particular type of object will receive the message.

Figure 12.18 *I/O device hierarchy.*

12.7.3 Construction guidelines

Construction guidelines might not appear relevant at this stage in a systems development project. However, they are pertinent to system design because there is a growing tendency for developers to use CASE environments that have code generation capabilities. It is also the case that when a rapid development approach is followed, the demarcation between design and construction activities can become blurred.

Construction guidelines will normally include advice on the naming of classes, of operations and of attributes, and where this is the case these guidelines are also applicable during the analysis activity. Wherever possible, consistent naming conventions should be enforced throughout the project since this makes it easier to trace an analysis class directly through to its implementation. Other guidelines for construction might relate to the use of particular software features (for example, using only standard language constructs in order to aid portability) and the layout of the code. These issues are addressed in more detail in Chapter 18.

12.8 Prioritizing Design Trade-offs

Design frequently involves choosing the most appropriate compromise. The designer is often faced with design objectives that are mutually incompatible and he or she must then decide which objective is the more important. The requirements specification may indicate the relative priorities of the different objectives but, if it does not, then it is useful to prepare general guidelines. These guidelines must be agreed with clients since they determine the nature of the system and what functionality will be delivered. Guidelines for design trade-offs ensure consistency between the decisions that are made at different stages of development. They also ensure consistency between different sub-systems. However, no guidelines can legislate for every case. Design experience and further discussions with the client will remain necessary to resolve those situations that cannot be anticipated—at least some of these occur on almost every project.

12.9 Design for Implementation

The introduction of an information system within an organization frequently requires some form of data conversion. System design must consider how best to deal with this and other system initialization issues. Special programs may be designed and written that convert data from one format to another and procedures may have to be specified for the manual entry of data that configures the final system. A new system may need a batch input capability so that existing data can be loaded, even where the day-to-day use of the system is intended to be fully interactive with on-line update of all data.

The design and construction of the data conversion or initialization sub-systems may represent a significant part of development and can be complex. Logistical considerations such as the overall volumes of data and the time available for conversion and initialization will have a significant impact upon the approach that is taken. The data of an enterprise represents one of its most valuable resources and every care must be taken not to corrupt or lose it.

12.10 Summary

System design focuses on determining a suitable architectural structure for the system and defines the context within which the remaining design activity is performed. We have identified the following activities as germane to system design.

- Identify sub-systems and major components. Splitting a system into sub-systems provides many benefits, both technical and managerial. Three major software architectures were discussed, layering and partitioning, the MVC and the broker architecture.
- Identify any inherent concurrency. UML provides models of the system that help to identify concurrent behaviour. Different design strategies have to be adopted depending upon the implementation environment and the response requirements for the system.
- Allocate sub-systems to processors. Computerized information systems commonly operate on multiple computers. In some cases it is important to partition a system across a series of computers (specialized by capability or location) to ensure that user requirements are addressed.
- Select data management strategies. The alternatives range from a simple file storage and retrieval mechanism to relational, object-oriented or object-relational DBMS.
- Specify development standards. Style guides are applicable to almost all systems development products. They ensure a uniformity of presentation and help ensure quality.
- Set priorities for design trade-offs. A clearly defined set of priorities is advisable to ensure that appropriate design decisions are made.
- Identify implementation requirements (e.g. data conversion). The requirements for the installation of the system are often ignored during system design.

System design determines the degree to which a system satisfies its non functional objectives and consequently the degree to which it can effectively support the user requirements both now and in the future.

Review Questions

12.7   Why is it sometimes necessary to design information systems that have explicitly concurrent behaviour?

12.8   What ways are there of simulating concurrency in the execution of a system?

12.9   How should you go about allocating system tasks to processors?

12.10 What facilities are typically offered by a DBMS?

Case Study Work, Exercises and Projects

12.A   Develop a series of steps for the identification of sub-systems within a layer in a layered architecture. Use the process for the identification of layers described above as a starting point. Highlight any significant differences that you feel exist between the two processes.

12.B   Suggest a suitable layered architecture with any necessary partitioning for the FoodCo case study by following the procedures defined above.

12.C   Identify a suitable hardware architecture or architectures for the FoodCo case study and justify your decisions.

12.D   Identify a suitable data management approach for the FoodCo case study and, if appropriate, specify the type of DBMS to be used.

Further Reading

Buschmann et al. (1996) provides further details of the architectures discussed in this chapter and describe other interesting alternatives. Both Rumbaugh et al. (1991) and Booch (1994) offer useful advice on system design issues. Goldsmith (1993) provides a non-object-oriented perspective on system design and it is interesting to consider the many issues that are the same in both approaches. Sommerville (1992) is an excellent text that covers a broad range of software engineering issues. Shlaer and Mellor (1988 and 1992) describe an alternative approach to object-oriented system design. Rather than the elaboration of the analysis models that we describe they suggest an approach called Recursive Design to produce the design model.

13

Object Design

Objectives
In this chapter you will learn:
- how to apply criteria for good design;
- how to design associations;
- the impact of integrity constraints on design;
- how to design operations;
- the role of normalization in object design.

13.1 Introduction

Object design is concerned with the detailed design of the objects and their interactions. It is completed within the overall architecture defined during system design and according to agreed design guidelines and protocols. Object design is particularly concerned with the specification of the attribute types, how operations function and how objects are linked to other objects. There are many sources of information that guide the object design process (Figure 13.1). During the design process the analysis models undergo some degree of transformation. There is a commonly accepted view that changes made to analysis artefacts to produce the design model should be kept to a minimum, as the analysis model is a coherent and consistent description of the requirements (Ward and Mellor, 1985, Goldsmith, 1993). Object design produces a detailed specification of the classes and uses the UML notation described in Section 13.2 for attributes and operation signatures. An important aspect of every class is what attributes (if any) and operations are generally accessible, and this is considered in Section 13.3. In Section 13.4 we consider a series of criteria, including coupling and cohesion, that help to produce good object-oriented designs.

Much design activity focuses on adding detail to the analysis specification and makes no change to the structures identified during analysis. However, there may be a need to modify the analysis structures for various reasons. The associations modelled in analysis have to be designed to minimize coupling. A series of common designs for associations are discussed in Section 13.5. Analysis will have identified a series of rules of integrity constraints that have to be adhered to by the information system if it is to maintain its consistency and integrity. Various strategies for the design of integrity constraints are examined in Section 13.6. The functionality of the information system is defined in the operation specifications. These have now to be translated into suitable designs (Section 13.7). If the information system is going to utilize relational database technology it may be appropriate to normalize the data structures (Section 13.8). This technique may also be useful for decomposing complex classes. Concurrent activities may be placed in separate sub-systems and may require an analysis class to be split into two design classes. As new classes are added to the design model to support data management and the operation of the interface, analysis classes have to be modified to build associations and communication pathways to these new

Figure 13.1 *Information sources for object design.*

sub-systems. As we shall see in Chapter 14 the use of design patterns offers structures and modes of interaction that support non-functional requirements.

13.2 Class Specification

13.2.1 Attributes and operation signatures

During analysis we have not considered in detail the data types of the attributes, although, on occasions, it may be useful to record data type information in analysis. For example, an attribute temperature might be a floating-point data type if it holds the temperature in Centigrade or it might an enumerated data type if it holds one the values, 'hot' or 'cold'. The attribute has a different meaning and would be manipulated differently for each of these data types and it is important to determine during analysis, which meaning is appropriate.

Common primitive data types include Boolean (true or false), Character (any alphanumeric or special character), Integer (whole numbers) and Floating-Point (decimal numbers).[1] In most object-oriented languages more complex data types, such as Money, String, Date, or Name can be constructed from the primitive data types. An attribute's data type is declared in UML using the following syntax:

```
name ':' type-expression '=' initial-value '{'property-string'}'
```

The name is the attribute name, the type-expression is its data type, the initial-value is the value the attribute is set to when the object is first created and the property-string describes a property of the attribute, such as constant or fixed. The characters in single quotes are literals. The attribute name is the only feature of its declaration that is compulsory.

Figure 13.2 shows the class BankAccount with attribute data types declared. The attribute balance in a BankAccount class might be declared with an initial value of zero using the syntax:

```
balance:Money = 0.00
```

The attribute accountName might be declared with the property string indicating that it must have a value and may not be null using the syntax:

```
accountName:String {not null}
```

Attribute declarations can also include arrays. For example, an Employee class might include an attribute to hold a list of qualifications that would be declared using the syntax:

```
qualification[0..10]:String
```

This declaration states that the attribute qualification may hold zero or up to 10 qualifications.

Each operation also has to be specified in terms of the parameters that it passes and returns. The syntax used for an operation is:

```
operation name '('parameter-list ')' ':' return-type-expression
```

An operation's *signature* is determined by the operation's name, the number and type of its parameters and the type of the return value if any. The BankAccount class might have a

---

[1] A list of Java primitive data types can be found in Deitel and Deitel (1997).

`credit()` operation that passes the amount being credited to the receiving object and has a Boolean return value. The operation would be defined using the syntax:

```
credit(amount:Money):Boolean
```

A `credit()` message sent to a `BankAccount` object would have the format:

```
creditOK = accObject.credit(500.00)
```

`creditOK` holds the Boolean return value that is available to the sending object when the `credit()` operation has completed executing. This Boolean value may be tested to determine whether the `credit()` operation performed successfully.

13.2.2 Object visibility

The concept of encapsulation was discussed in Chapter 4 and is one of the fundamental principles of object-orientation. During analysis various assumptions have been made regarding the encapsulation boundary for an object and the way that objects interact with each other. For example, it is assumed that the attributes (or more precisely the values of the attributes) of an object cannot be accessed directly by other objects but only via 'get' and 'set' operations (primary operations) that are assumed to be available for each attribute. Moving to design involves making decisions regarding which operations (and possibly attributes) are publicly accessible. In other words we must define the encapsulation boundary.

Figure 13.2 shows the class `BankAccount` with the types of the attributes specified and the operation parameters defined. The class has the attribute `balance`, which we might assume during analysis, can be accessed directly by the simple primary operations `getBalance()` and `setBalance()`. (Note that the primary operations are not normally listed in analysis class diagrams.) However, the balance should be updated through the operations `credit()` and `debit()` that contain special processing to check whether these transactions should be permitted and to ensure that the transactions are logged in an audit trail[2]. In these circumstances, it is important that changes to the value of the `balance` attribute can only occur through the `debit()` and `credit()` operations. The operation `setBalance()` should not be publicly available for use by other classes.

Meyer (1997) introduces the term 'secret' to describe those features that are not available in the public interface. Programming languages designate the non-public parts of a class, which may include attributes and operations, in various ways. The three commonly accepted terms[3] used to describe *visibility* are listed in Figure 13.3.

To enforce encapsulation the attributes of a class are normally designated private (Figure 13.4). The operation `setBalance()` is also designated private to ensure that objects from other classes cannot access it directly and make changes that are not recorded in the audit trail. Private operations can, of course, be invoked from operations in the same class.

---

[2] In general, an audit trail records the details of transactions so that they can be checked to ensure that, for example, fraud has not occurred.

[3] The precise meaning of the non-public categories of visibility depends on the programming language being used. When a designer is determining the visibility of parts of a model, he or she must be aware of the visibility (or scoping) offered by the implementation environment.

Commonly, complex operations are simplified by factoring out procedures into private operations.

```
┌─────────────────────────────────────────────┐
│                BankAccount                    │
├─────────────────────────────────────────────┤
│ accountNumber:Integer                         │
│ accountName:String {not null}                 │
│ balance:Money = 0                             │
│ overdraftLimit:Money                          │
├─────────────────────────────────────────────┤
│ open(accountName:String):Boolean              │
│ close():Boolean                               │
│ credit(amount:Money):Boolean                  │
│ debit(amount:Money):Boolean                   │
│ getBalance():Money                            │
│ setBalance(newBalance:Money)                  │
│ getAccountName():String                       │
│ setAccountName(newName:String)                │
└─────────────────────────────────────────────┘
```

Figure 13.2 `BankAccount` *class.*

Visibility symbol	Visibility	Meaning
+	Public	The feature (an operation or an attribute) is directly accessible by any class.
-	Private	The feature may only be used by the class that includes it.
#	Protected	The feature may be used either by the class that includes it or by a subclass of that class.

Figure 13.3 *Visibility.*

In Figure 13.4 the operation `getBalance()` is assigned protected visibility so that subclasses of `BankAccount` can examine the value of the `balance` attribute. For example, the `debit()` operation might be redefined polymorphically in a `JuniorBankAccount` subclass. The redefined operation would use `getBalance()` to access the balance and check that a debit would not result in a negative balance.

The attribute `nextAccountNumber` in Figure 13.4 is an example of a class-scope attribute (indicated by underlining). A class-scope attribute occurs only once and is attached to the class not to any individual object. In this example `nextAccountNumber` holds the account number for the next new `BankAccount` object created. When a new `BankAccount` is created `nextAccountNumber` is incremented by one. The attribute `accountNumber` is an example of an instance-scope attribute (hence no underlining). Each `BankAccount` object has an instance-scope `accountNumber` attribute which holds its unique account number.

```
                    ┌─────────────────────────────────────────┐
                    │              BankAccount                 │
                    ├─────────────────────────────────────────┤
                    │ -nextAccountNumber:Integer               │
                    │ -accountNumber:Integer                   │
                    │ -accountName:String {not null}           │
                    │ -balance:Money = 0                       │
                    │ -overdraftLimit:Money                    │
                    ├─────────────────────────────────────────┤
                    │ +open(accountName:String):Boolean        │
                    │ +close():Boolean                         │
                    │ +credit(amount:Money):Boolean            │
                    │ +debit(amount:Money):Boolean             │
                    │ +viewBalance():Money                     │
                    │ #getBalance():Money                      │
                    │ -setBalance(newBalance:Money)            │
                    │ #getAccountName():String                 │
                    │ #setAccountName(newName:String)          │
                    └─────────────────────────────────────────┘
```

Figure 13.4 BankAccount *class with visibility specified.*

13.3 Interfaces

On occasions a class (or some other component) may present more than one external interface to other classes or the same interface may be required from more than one class. An interface in UML is a group of externally visible (i.e. public) operations. The interface contains no internal structure, it has no attributes, no associations and the implementation of the operations is not defined. Formally, an interface is equivalent to an abstract class that has no attributes, no associations and only abstract operations. Figure 13.5 illustrates two alternative notations for an interface. The simpler of the two UML interface notations is a circle. This is attached by a solid line to the classes that support the interface. For example, in Figure 13.5 the Advert class supports two interfaces, Manageable and Viewable, that is, it provides all of the operations specified by the interface (and maybe more). The circle notation does not include a list of the operations provided by the interface type, though they should be listed in the data dictionary. The dashed arrow from the CreativeStaff class to the Manageable interface circle icon indicates that it uses or needs at most, the operations provided by the interface.

The alternative notation uses a stereotyped class icon. As an interface only specifies the operations and has no internal structure, the attributes compartment is omitted. This notation lists the operations on the diagram. The *realizes* relationship, represented by the dashed line with a triangular arrowhead, indicates that the client class .(e.g. Advert) supports at least the operations listed in the interface (e.g. Manageable or Viewable).[4]

---

[4] The realizes relationship can be used between two classes (rather than between a class and an interface) to mean that the client (at the tail of the arrow) provides at least the operations defined in the supplier class (at the arrowhead), but does not necessarily support any of the structure (i.e. the attributes of the class).

Again the dashed arrow from `CreativeStaff` means that the class needs or uses no more than the operations listed in the interface. The notation used for the realizes relationship (the triangular arrowhead) is deliberately reminiscent of the notation for inheritance, as in a sense `Advert` inherits the `Manageable` interface.

This concept can be implemented using the 'interface' programming language construct in Java (see Chapter 18).

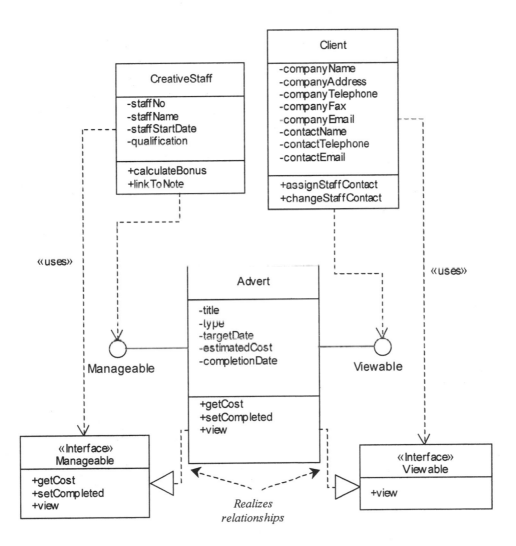

Figure 13.5 *Interfaces for the* `Advert` *class.*

13.4 Criteria for Good Design

13.4.1 Coupling and cohesion

In Chapter 11 we stated that coupling and cohesion are among the criteria for good design. In fact, it is good practice in analysis to prepare models that, as far as possible, satisfy these criteria. Coupling describes the degree of interconnectedness between design components and is reflected by the number of links an object has and by the degree of interaction the object has with other objects. Cohesion is a measure of the degree to which an element contributes to a single purpose. The concepts of coupling and cohesion are not mutually exclusive but actually support each other. Coad and Yourdon (1991) suggested several ways in which coupling and cohesion can be applied within an object-oriented approach. Larman (1998) also considers the application of these criteria. The criteria can be used within object-orientation as described below (adapted from Coad and Yourdon, 1991).

Interaction Coupling is a measure of the number of message types an object sends to other objects and the number of parameters passed with these message types. Interaction coupling should be kept to a minimum to reduce the possibility of changes rippling through the interfaces and to make reuse easier. When an object is reused in another application it will still need to send these messages (unless the object is modified before it is reused) and hence needs objects in the new application that provides these services. This complicates the reuse process as it requires groups of classes to be reused rather than individual classes. (In Chapter 19 we introduce the idea of the *component* as the unit of reuse. Components are groups of objects that together provide a clearly defined service.)

Inheritance Coupling describes the degree to which a subclass actually needs the features it inherits from its base class. For example, in Figure 13.6 the inheritance hierarchy exhibits low inheritance coupling and is poorly designed. The subclass `LandVehicle` needs neither the attributes `maximumAltitude` and `takeOffSpeed` nor the operations `checkAltitude()` and `takeOff()`. They have been inherited unnecessarily. In this example it would appear the base class, `Vehicle`, would perhaps be better named `FlyingVehicle` and the inheritance relationship is somewhat suspect. A land vehicle is not a kind of flying vehicle (not normally anyway). However, many systems developers view designs with a small degree of unnecessary inheritance as being acceptable if the hierarchy is providing valuable reuse and is meaningful. It can be argued that if attributes and operations are inherited unnecessarily it is merely a matter of not using these features in the subclass. However, a subclass with unnecessary attributes or operations is more complex than it needs to be and objects of the subclass may take more memory than they actually need. The real problems may come when the system needs maintenance. The systems maintainer may not realize that some of the inherited attributes and operations are unused and may modify the system incorrectly as a result. Alternatively the systems maintainer may use these unneeded features to provide a fix for a new user requirement, making the system even more difficult to maintain in the future. For these reasons, unnecessary inheritance should be kept as low as possible.

Operation Cohesion measures the degree to which an operation focuses on a single functional requirement. Good design produces highly cohesive operations, each of which deals with a single functional requirement. For example in Figure 13.7 the operation `calculateRoomSpace()` is highly cohesive.

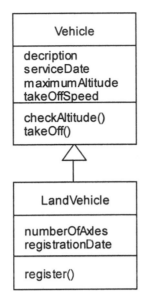

Figure 13.6 *Inheritance coupling.*

Class Cohesion reflects the degree to which a class is focused on a single requirement. The class Lecturer in Figure 13.7 exhibits low levels of cohesion as it has three attributes (roomNumber, roomLength and roomWidth) and one operation (calculateRoomSpace()) that would be more appropriate in a class Room. The class Lecturer should only have attributes that describe a Lecturer object (e.g. lecturerName and lecturerAddress) and operations that use them.

Figure 13.7 *Good operation cohesion but poor class cohesion.*

Specialization Cohesion addresses the semantic cohesion of inheritance hierarchies. For example in Figure 13.8 all the attributes and operations of the Address base class are used by the derived[5] classes: the hierarchy has high inheritance coupling. However, it is neither true that a person is a kind of address nor that a company is a kind of address. The example is only using inheritance as a syntactic structure for sharing attributes and operations. This structure has low specialization cohesion and is poor design. It does not reflect meaningful

---

[5] A derived class is another term for subclass, the superclass is known as the base class.

inheritance in the problem domain. A better design is shown in Figure 13.9, in which a common class `Address` is being used by both the `Person` and `Company` classes.

These design criteria may all be applied at the same time to good effect.

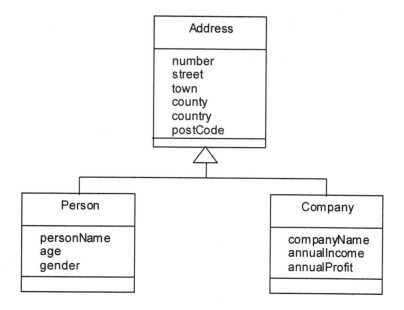

Figure 13.8 *Poor specialization cohesion.*

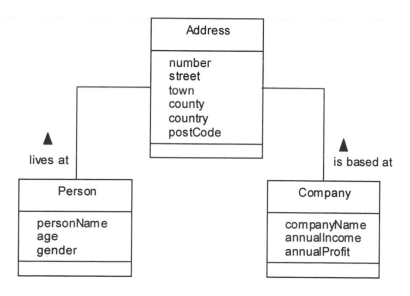

Figure 13.9 *Improved structure using* `Address` *class.*

13.4.2 Liskov Substitution Principle

The *Liskov Substitution Principle (LSP)* is another design criterion that is applicable to inheritance hierarchies. Essentially the principle states that, in object interactions, it should be possible to treat a derived object as if it were a base object. If the principle is not applied then it may be possible to violate the integrity of the derived object. In Figure 13.10 objects of the class MortgageAccount cannot be treated as if they are objects of the class ChequeAccount because MortgageAccount objects do not have a debit operation whereas ChequeAccount objects do. The debit operation is declared private in MortgageAccount and hence cannot be used by any other object. Figure 13.10 shows an alternative structure that satisfies LSP. Interestingly, this inheritance hierarchy has maximal inheritance coupling, and enforcing the LSP normally produces structures with high inheritance coupling.

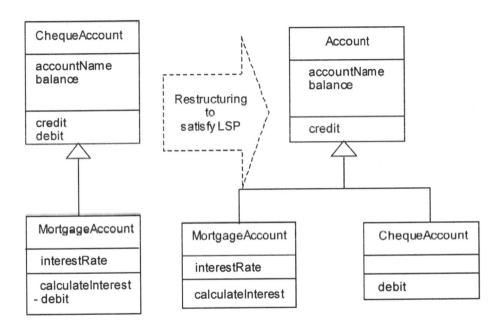

Figure 13.10 *Application of the Liskov Substitution Principle.*

13.4.3 Further design guidelines

Coad and Yourdon (1991) and Yourdon (1994) suggest further design guidelines that are included in the list below.

Design Clarity. A design should be made as easy to understand as possible. This reinforces the need to use design standards or protocols that have been specified.

Don't Over-Design. Developers are on occasions tempted to produce designs that may not only satisfy current requirements but may also be capable of supporting a wide range of future requirements. Designing flexibility (or any other non-functional requirement) into a system has a cost, the system may take longer to design and construct but this may be offset in the future by easier and less expensive modification. However, it is not feasible to design for every eventuality. Systems that are over-designed in first instance are more difficult to extend if the modifications are not sympathetic to the existing structure.

Control Inheritance Hierarchies. Inheritance hierarchies should be neither too deep nor too shallow. If a hierarchy is too deep it is difficult for the developer to understand easily what features are inherited. There is a tendency for developers new to OO to produce over-specialized hierarchies, thus adding complexity rather than reducing it. Yourdon (1994) suggests that hierarchies of up to nine levels are manageable, whereas Rumbaugh et al. (1991) suggest that more than about four or five levels is too many.

Keep Messages and Operations Simple. In general it is better to limit the number of parameters passed in a message to no more than three (of course, a single parameter might be a complete object). Ideally an operation should be capable of specification in no more than one page.

Design Volatility. A good design will be stable in response to changes in requirements. It is reasonable to expect some change in the design if the requirements are changed. However, any change in the design should be commensurate with the change in requirements. Enforcing encapsulation is a key factor in producing stable systems.

Evaluate by Scenario. An effective way of testing the suitability of a design is to role play it against the use cases using CRC cards if appropriate.

Design by Delegation. A complex object should be decomposed (if possible) into component objects forming a composition or aggregation. Behaviour can then be delegated to the component objects producing a group of objects that are easier to construct and maintain. This approach also improves reusability.

Keep Classes Separate. In general, it is better not to place one class inside another. The internal class is encapsulated by the other class and cannot be accessed independently. This reduces the flexibility of the system.

Review Questions

13.1   What levels of visibility may be assigned to an attribute or an operation?

13.2   Why should attributes be private?

13.3   How does the application of the concepts of coupling and cohesion help to produce good object-oriented designs?

13.4   What are the advantages and disadvantages of applying the Liskov Substitution Principle?

13.5 Designing Associations

An association between two classes indicates the possibility that links will exist between the objects of the classes. The links provide the connections necessary for message passing to occur. When deciding how to implement an association it is important to analyse the message passing between the objects tied by the association.

13.5.1 One-to-one associations

In Figure 13.11 objects of the class `Owner` need to send messages to objects of the class `Car` but not vice versa. This particular association may be implemented by placing an attribute to hold the object identifier (some authors prefer to use the term object reference) for the `Car` class in the Owner class. Thus `Owner` objects have the `Car` object identifier and hence can send messages to the linked `Car` object. As a `Car` object does not have the object identifier for the Owner object it cannot send messages to the `Owner` object. The `owns` association is an example of a one-way association: the arrowhead on the association line shows the direction along which it may be navigated.

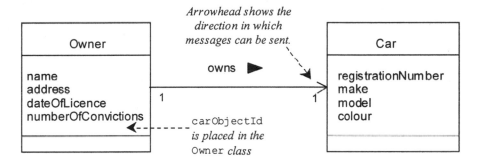

Figure 13.11 *One-way one-to-one association*

So before an association can be designed it is important to decide in which direction or directions messages may be sent. (If messages are not sent in either direction along an association, then the need for its existence should be questioned.) Essentially we are determining the navigability of the association.

In general an association between two classes A and B should be analysed with the questions:

1. Do objects of class A have to send messages to objects of class B?
2. Does an A object have to provide some other object with B object identifiers?

If either of these questions is answered 'yes' then A objects need B object identifiers. However if A objects get the required B object identifiers as parameters in incoming messages, A objects need not remember the B object identifiers. Essentially, if an object needs to send a message to a destination object it must have the destination object's identifier either passed as a parameter in an incoming message just when it is required, or the destination object's identifier must be stored in the sending object. An association that has to support message passing in both directions is a two-way association. A two-way association is indicated with arrowheads at both ends[6]. As discussed earlier, it is important

---

[6] In UML a two-way association may be represented by drawing the association without the navigability arrowheads. However, an association without arrowheads may also represent an uncommitted association, that is, association for which navigability is not yet decided. We suggest that arrowheads are always used to represent navigability.

to minimize the coupling between objects. Minimizing the number of two-way associations keeps the coupling between objects as low as possible.

13.5.2 One-to-many associations

In Figure 13.12, objects of the class Campaign need to send messages to objects of the class Advert but not vice versa. If the association between the classes was one-to-one, the association could be implemented by placing an attribute to hold the object identifier for the Advert class in the Campaign class. However, the association is in fact one-to-many and many Advert object identifiers need to be tied to a single Campaign object. The object identifiers could be held as a simple one-dimensional array in the Campaign object but program code would have to be written to manipulate the array. Another way of handling the group of Advert object identifiers that is more amenable to reuse, is to place them in a separate object, a collection object that has operations to manage the object identifiers and that behaves rather like an index of adverts for the Campaign object. This is shown in the class diagram fragment in Figure 13.13. There will be many instances of the collection

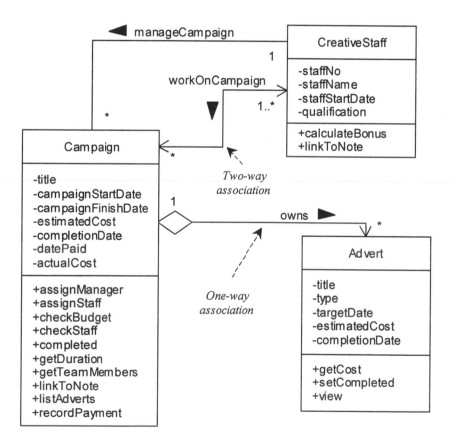

Figure 13.12 *Fragment of class diagram for the Agate case study.*

class, as each `Campaign` object has its own collection of `Advert` object identifiers. Notice that the `AdvertCollection` class has operations that are specifically concerned with the management of the collection. The `findFirst()` operation returns the first object identifier in the list and the `getNext()` gets the next object identifier in the list.

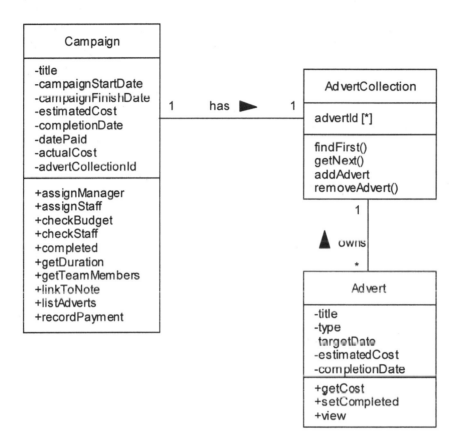

Figure 13.13 *One-to-many association using a collection class.*

When a `Campaign` object wants to send a message to each of its `Advert` objects the `Campaign` object first sends a `findFirst()` message to the class to get the first object identifier. The `Campaign` object can now send a message to its first `Advert` object. The `Campaign` then uses `getNext()` to get the next object identifier from the collection class and sends the message to the next `Advert` object. The `Campaign` object can then iterate through the collection of object identifiers and send the message to each of the `Advert` objects in turn.

Figure 13.14 shows the sequence diagram for the interaction that would enable the `Campaign` object to prepare a list of its adverts with their titles. (The labels on the left hand side of the diagram define the condition on the iteration and the scope of the iteration.) The `Campaign` object holds the object identifier of the collection class so that it can send messages to it. As an `Advert` object does not have the object identifier for the `Campaign` object to which it belongs, it cannot send messages to the `Campaign` object.

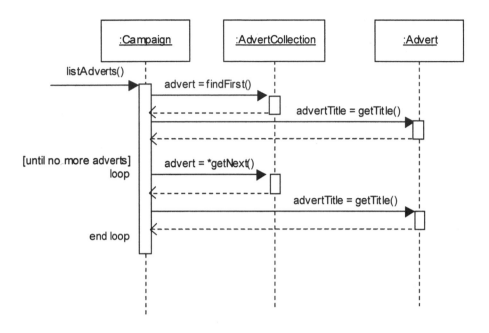

Figure 13.14 *Sequence diagram for* listAdverts().

13.5.3 Many-to-many associations

The design of the many-to-many association workOnCampaign between CreativeStaff and Campaign (see Figure 13.12) follows the principles described above. Assuming this is a two-way association, each Campaign object will need a collection of CreativeStaff object identifiers and each CreativeStaff object will need a collection of Campaign object identifiers. The designed association with the collection classes is shown in Figure 13.15. Both the CreativeStaff and Campaign classes contain an attribute to hold the object identifiers of their respective collection classes.

Collection classes can be designed to provide additional support for object navigation. For example, if there is a requirement to find out if an employee works on a campaign with a particular title, a message may be sent from the CreativeStaff object to each Campaign object the employee works on to get its title until either a match is found or the end of the collection is reached. Two separate messages are required to access each Campaign object. So if an employee works on four campaigns a maximum of eight messages must be sent to find the campaign. In general, if the employee works on N campaigns a maximum of 2N messages must be sent.

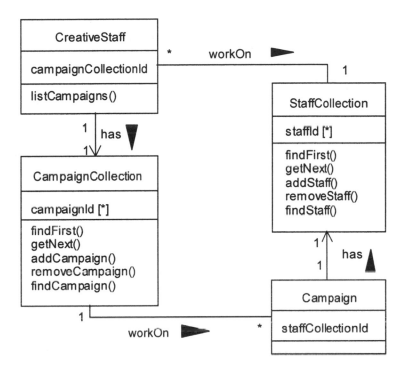

Figure 13.15 *Two-way many-to-many association.*

An alternative search approach uses a findCampaign() operation in the collection class. This operation may be designed either to access an index in the collection object itself or to take responsibility to cycle through the Campaign objects searching for a title match. In the first case only the findCampaign() message is needed to find the campaign as the collection class indexes the campaigns by title. In the second case the collection object may have to send a maximum of N messages (one for each campaign on which the employee works). So in either case the inclusion of the findCampaign() operation in the collection class reduces the message passing required.

The management of object identifiers using collection classes may appear to increase appreciably the development effort required. In fact, the opposite is generally the case, as object-oriented languages normally provide collection classes of various types with standard collection management operations. The standard collection classes may offer various forms of indexing. They may also be subclassed to add additional application functionality. For example, Java provides, among others, standard List, Stack, Hashtable and Dictionary collection classes (Deitel and Deitel, 1997) that may be subclassed to add application specific behaviour.

13.5.4 Keeping classes to a minimum

The association between `Campaign` and `AdvertCollection` is one-to-one (commonly the case with collection classes), and this suggests that one implementation strategy is to place the collection class inside the `Campaign` object. As discussed earlier in Section 13.4.3, this approach generally produces more complex classes and limits extensibility. However, in this case it is likely that collection class behaviour will largely be provided by a library class or by a feature of the development language being used, so the issue of increased complexity may not be so significant. The problem of any reduction in extensibility is also less significant in this case: as only a `Campaign` object would want to know which `Advert` objects are tied to it, and any request to access a `Campaign`'s `Adverts` would be directed to the `Campaign` first. So, on balance, placing the collection class inside the `Campaign` class is a sensible design decision, and using library collection classes where possible maiximizes reuse and reduces development effort. Clearly, if another class, apart from `Campaign`, needs to use the list independently of the `Campaign` class then it is more appropriate to keep the collection class separate.

Review Questions

13.5   How can collection classes be used when designing associations?

13.6   Under what circumstances should a collection of object identifiers be included in a class?

13.6 Integrity Constraints

Systems analysis will have identified a series of integrity constraints that have to be enforced to ensure that the application holds data that is mutually consistent and manipulates it correctly. These integrity constraints come in various forms:

Referential Integrity that ensures that an object identifier in an object is actually referring to an object that exists.

Dependency Constraints that ensures that attribute dependencies, where one attribute may be calculated from other attributes, are maintained consistently.

Domain Integrity that ensures that attributes only hold permissible values.

13.6.1 Referential integrity

The concept of referential integrity as applied to a relational database management system (see Chapter 17) is discussed by Howe (1989). Essentially the same principles apply when considering references between objects. In Figure 13.12 the association `manageCampaign` between `CreativeStaff` and `Campaign` is two-way, and an object identifier called `campaignManagerId` that refers to the particular `CreativeStaff` object that represents the campaign manager is needed in `Campaign`. (`CreativeStaff` needs a collection of `Campaign` object identifiers to manage its end of the association.) To maintain referential integrity the system must ensure that the attribute `campaignManagerId` either is null (not referencing any object) or contains the object identifier of a `CreativeStaff` object that exists. In this particular case the association states that a `Campaign` must have a `CreativeStaff` instance as its manager, and it is not correct to have a `Campaign` with a null `campaignManagerId` attribute. In order to enforce this constraint, the constructor for `Campaign` needs as one of its parameters the object identifier of the `CreativeStaff` object that represents the campaign manager, so that the `campaignManagerId` attribute can be instantiated with a valid object identifier.

Problems in maintaining the referential integrity of a `Campaign` may occur during its lifetime. For instance, the campaign manager, Nita Rosen, may leave the company to move to another job and Nita's `CreativeStaff` object will then be deleted. Referential integrity is maintained by ensuring that the deletion of a `CreativeStaff` object that is a campaign manager always involves allocating a new campaign manager. The task of invoking the operation `assignManager()` is included in the `CreativeStaff` destructor, and it will request the object identifier of the new campaign manager. Similarly, any attempt to remove the current campaign manager from a `Campaign` must always involve allocating the replacement.

The multiplicity of exactly one represents a strong integrity constraint for the system. In the example just discussed, it seems to be appropriate that a campaign should always have a manager, even when it has just been created. However, great care should be taken when assigning a multiplicity of exactly one (or in general a minimum of one) to an association, as the consequences in the implemented system can be quite dramatic. Let us imagine that the campaign manager, Nita Rosen, does leave Agate but that there is no replacement campaign manager available. The strict application of the integrity constraint implied by the `manageCampaign` association means that integrity can only be enforced if all the campaigns that Nita managed are deleted. Of course, because each `Advert` must be linked to a `Campaign`, all the `Advert` objects for each of the `Campaigns` must also be deleted in order to maintain referential integrity. This is an example of a cascade delete: deleting one object results in the deletion of many objects as referential integrity is applied. In the case of Agate, deleting the information about Nita's campaigns and their adverts would be disastrous. There are two solutions: either the constraint on the association is weakened by changing the cardinality to zero or one, or when Nita leaves a dummy `CreativeStaff` object is created and allocated as campaign manager to Nita's campaigns. Although the second solution is a fix, it has the advantage of providing an obvious place-holder, highlighting the problem of unmanaged campaigns but maintaining the integrity constraint. Of course, the minimum of one multiplicity was assigned to the association to reflect company policy that a campaign must always have a manager.

13.6.2 Dependency constraints

Attributes are dependent upon each other in various ways. These dependencies may have been identified during analysis and must now be dealt with during design. A common form of dependency occurs when the value of one attribute may be calculated from other attributes. For instance, a requirement to display the total advertising cost may be satisfied either by storing the value in the attribute `totalAdvertCost` in the `Campaign` class or by calculating the value every time it is required. The attribute `totalAdvertCost` is a derived attribute and its value is calculated by summing the individual advert costs. Including the derived attribute reduces the processing required to display the total advertising cost. On the other hand, whenever the cost of an advert changes, or an advert is either added to or removed from the campaign, then the attribute `totalAdvertCost` has to be adjusted so that it remains consistent with the attributes upon which it depends.

In order to maintain the consistency between the attributes, any operation that changes the value of an `Advert's` cost must trigger an appropriate change in the value of `totalAdvertCost` by sending the message `adjustCost()` to the `Campaign` object. The operation `adjustCost()` is an example of a *synchronizing operation*. The operations that have to maintain the consistency are `setAdvertCost()` and the `Advert` destructor. When a new advert is created the constructor would use `setAdvertCost()` to set the advert cost. This would invoke `adjustCost()` and hence ensure that the `totalAdvertCost` is adjusted. So any change to an `Advert's` cost takes more processing if the derived attribute `totalAdvertCost` is used. Thus one part of the system executes more quickly while another part executes more slowly. Generally it is easier to construct systems without derived attributes, as this obviates the need for complex synchronizing operations. Derived attributes should only be introduced if performance constraints cannot be satisfied without them. If performance is an issue then one of the skills needed in design is how to optimize the critical parts of the system without making the other parts of the system inoperable.

Another form of dependency occurs where the value of one attribute is constrained by the values of other attributes. For example, let us assume that the sum of the total advertising cost, the staff costs, the management costs and the ancillary costs must not exceed the campaign's authorized budget. Any changes to these values must check that the authorized budget is not exceeded by the sum of the costs. If a change to any of these dependent values would cause this constraint to be broken, then some action should be taken. There are two possibilities. Either the system prohibits any change that violates the constraint, and an exception is raised, or it permits the change, and an exception is raised[7]. It is most likely that the violation of the constraint would occur as a result of an attempt interactively to change one of the constrained values and the exception raised would be a warning message to the user. If it is considered permissible for the constraint to be broken then all access to or reporting about these values should produce a warning message to the user.

Dependency constraints can also exist between or among associations. One of the simplest cases is shown in Figure 13.16 where the `chairs` association is a subset of the `isAMemberOf` association.

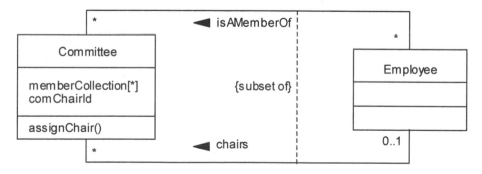

Figure 13.16 *Constraints between associations.*

This constraint is stating that the chair of a committee must be a member of the committee, and it can be enforced by placing a check in the `assignChair()` operation in

---

[7] An exception is a way of handling errors in object-oriented languages. See the data dictionary entries for operations in Case Study A2.

`Committee` to confirm that the `Employee` object identifier passed as a parameter is already in the collection class of committee members. More complex constraints may also exist that require several associations. Derived associations may also be introduced to improve performance if absolutely necessary and, as in the case of derived attributes, synchronizing operations are needed to ensure that the derived links are consistent with the links on which they depend.

13.6.3 Domain integrity

Domain integrity is concerned with ensuring that the values an attribute takes are from the appropriate underlying domain. For instance, the attributes from the `Cost` domain might reasonably be non-negative decimal values with two decimal places. These constraints may be viewed as an extended form of those implied by data types. The necessary integrity checking code is normally placed in the 'set' operations or in any interactive interface that permits the entry of values.

13.7 Designing Operations

The design of operations involves determining the best algorithm to perform the required function. In the simplest case, primary operations require little design apart from the inclusion of code to enforce integrity checks. For more complex operations, algorithm design can be an involved process. Various factors constrain algorithm design:

- the cost of implementation,
- performance constraints,
- requirements for accuracy,
- the capabilities of the implementation platform.

Generally it is best to choose the simplest algorithm that satisfies these constraints, as this makes the operation easier to implement and easier to maintain. Rumbaugh et al. (1991) suggest that the following factors should be considered when choosing among alternative algorithm designs.

- *Computational complexity.* This is concerned with the performance characteristics of the algorithm as it operates on increasing numbers of input values. For example, the bubble sort algorithm has an execution time that is proportional to $N \times N$ where N is the number of items being sorted.
- *Ease of implementation and understandability.* It is generally better to sacrifice some performance to simplify implementation.
- *Flexibility.* Most software systems are subject to change and an algorithm should be designed with this in mind.
- *Fine-tuning the object model.* Some adjustment to the object model may simplify the algorithm and should be considered.

Designing the main operations in a class is likely to highlight the need for lower-level private operations to decompose complex operations. This process is much the same as traditional program design. Techniques such as step-wise refinement (Budgen, 1994) or structure charts (Yourdon and Constantine, 1979) may well be used to good effect. UML offers activity diagrams (see Chapter 10) as a technique both to document and to design operations. In circumstances where high levels of formality are required in operation design, formal specification techniques such as Z or VDM may be used[8].

The new operations that are identified need to be assigned to classes. In general, if an operation operates on some attribute value then it should be placed in the same class as the attribute. On occasions a particular operation may modify attributes in more than one class and could sensibly be placed in one of several classes. In choosing where to locate the operation, one view is that minimizing the amount of object interaction should be a major criterion, while another significant criterion is simplicity. However, in some cases it is not a clear-cut decision. Objectory (Jacobson, 1992) suggests that control classes should be introduced to control the execution of use cases. Such classes may be the best place for operations that have no obvious owning class.

13.8 Normalization

One form of dependency not discussed in Section 13.6.2 is functional dependency. For two attributes A and B, A is functionally dependent on B if for every value of B there is precisely one value of A associated with it at any given time. This is shown notationally as:

$$B \rightarrow A$$

Attributes may be grouped around functional dependencies according to several rules of normalization (see Section 17.3.1) to produce normalized data structures that are largely redundancy free (Howe, 1989). Normalization may be useful when using a relational database management system as part of the implementation platform or as a guide to decomposing a large, complex (and probably not very cohesive) object. Most object-oriented approaches to software development do not view normalization as essential, and structures that are not normalized are considered acceptable. In general however, object-oriented approaches, if applied with suitable quality constraints, will produce structures that are largely redundancy free.

13.9 Summary

Object design is concerned with the detailed design of the system and is conducted within the architectural framework and design guidelines specified during system design. The detailed design process involves determining the data types of the attributes and defining the operation signatures. Interfaces may be specified. Associations have to be designed to support the message passing requirements of the operations. The application of integrity constraints is included in the design of operations. The detailed design process is guided by a series of criteria that incorporate the fundamental principles of coupling and cohesion. Normalization may be applied to complex classes as means of decomposing them to produce redundancy-free structures.

---

[8] Z and VDM are formal languages that can be used to specify a system using mathematical entities such as sets, relations and sequences.

Review Questions

13.7   How can referential integrity be enforced in an object-oriented system?

13.8   Under what circumstances is normalization useful during object-oriented design?

Case Study Work, Exercises and Projects

13.A   Specify the attribute types and the operation signatures for the class `ProductionLine` in the FoodCo case study.

13.B   For each association in which `ProductionLine` participates allocate object identifiers to design the association.

13.C   Show how referential integrity can be enforced for the associations designed in Exercise 13.B.

Further Reading

Rumbaugh et al. (1991) and Booch (1994) provide good advice on object design. Budgen (1994) offers descriptions of various design techniques. Meyer (1997) contains a comprehensive discussion of object-oriented software design and provides many interesting insights. There is an excellent treatment of normalization in Howe (1989) and detailed discussion of integrity constraints from a database perspective in Date (1995).

14

Design Patterns

Objectives
In this chapter you will learn:
- what is meant by the term pattern;
- what types of patterns have been identified in software development;
- how to apply design patterns during software development;
- the benefits and difficulties that may arise when using patterns.

14.1 Introduction

Successful software development relies on the knowledge and expertise of the developer, among other factors. These are built up and refined during the developer's working life. A systems analyst or software engineer applies potential solutions to development problems, monitors their success or failure and produces more effective solutions on the next occasion. It is in the nature of software development that the same problems tend to recur. Different developers may expend a great deal of development time and effort on solving these problems from first principles each time they occur, and the solution that each produces may not be the most appropriate that could be achieved. This can result in information systems that are inflexible, difficult to maintain or have some other disadvantage such as inefficiency. The perpetuation of this cycle of reinventing the wheel has been encouraged partly because there were no effective mechanisms for communicating successful solutions to recurring problems.

Patterns provide a means for capturing knowledge about problems and successful solutions in software development. Experience that has been gained in the past can be reused in similar situations, thus reducing the effort required to produce systems that are more resilient, more effective and more flexible.

In Section 14.2 we consider the areas of software development where patterns have been currently identified. Patterns are documented in various formats and these are considered in Section 14.3. Although patterns have been applied to many aspects of software development we focus our attention on some better known design patterns (Section 14.4) to illustrate their application. Guidelines for using patterns are discussed in Section 14.5. Patterns have many benefits but there are also potential disadvantages associated with their use. These advantages and disadvantages are discussed in the context of the examples used in Section 14.6.

14.2 Software Development Patterns

14.2.1 Pattern definition

The architect Christopher Alexander first used the term 'pattern' to describe solutions to recurring problems in architecture. Alexander identified many related patterns for the development of effective and harmonious architectural forms in buildings. Alexander's patterns address many architectural issues—for example the best place to site a door in a room, or how to organize and structure a waiting area in a building so that waiting can become a positive experience. Alexander argued that his patterns became a design language within which solutions to recurring architectural problems could be developed and described. Alexander's definition of a pattern is as follows.

> Each pattern describes a problem which occurs over and over again in our environment, and then describes the core of a solution to that problem, in such a way that you can use this solution a million times over, without ever doing it the same way twice.
>
> Alexander et al. (1977)

Most particularly a pattern provides a solution that may be applied in different ways depending upon the specific problem to which it is being applied. Alexander's definition makes no reference to buildings or architecture but only to 'our environment'. Although Alexander intended 'environment' to be interpreted as the physical environment in which we live there is clearly an analogous concept appropriate to software development. One definition of a pattern that is appropriate to software systems development is this:

> A pattern is the abstraction from a concrete form which keeps recurring in specific non-arbitrary contexts.
>
> Riehle and Zullighoven (1996)

In fact, this definition could be applied to any domain of human endeavour in which there is some recurring form from which abstractions may be developed. It is not clear how tangible 'concrete' is intended to be but a reasonable interpretation might be 'specific' or 'particular'. Gabriel (1996) gives an alternative and more detailed definition that has some insight into the structure of a pattern.

> Each pattern is a three-part rule, which expresses a relation between a certain context, a certain system of forces which occurs repeatedly in that context, and a certain software configuration which allows these forces to resolve themselves.

This definition focuses on three elements—a *context* that can be understood as a set of circumstances or preconditions, *forces* that are issues that have to be addressed and a software configuration that addresses and resolves the forces.

In this definition the term 'software configuration' might suggest that patterns are limited to software design and construction. In fact, patterns are applied much more widely in systems development. The Layer, Broker and MVC architectures we considered in Chapter 12 are examples of the application of patterns from Buschmann et al. (1996) to the activity of systems design.

Coad et al. (1997) make the distinction between a *strategy* which they describe as a plan of action intended to achieve some defined purpose and a *pattern* which they describe as a template that embodies an example worth emulating. This view of a pattern is slightly different from the views described earlier as it does not emphasize contextual aspects to the same extent. An example of a Coad et al. strategy is 'Organize and Prioritize Features' and relates the need to prioritize requirements (discussed in Chapter 3). A simple example of a pattern from Coad et al. (1997) is the Transaction–Transaction Line Item Pattern (Figure 14.1) as it would be applied to a sales order processing system. The pattern suggests a `SalesOrder` class (the Transaction) and a `SalesOrderLine` class (the Transaction Line Item). This type of structure is commonly used in a wide variety of circumstances (e.g. shipment and shipment line item, payment and payment line item). A novice software developer has to learn this structure, or to reinvent it. The latter is much less efficient. The act of describing it as a pattern highlights it as a useful piece of development expertise and makes it readily available for the novice. An example of a principle would be the desirability of low interaction coupling (Chapter 13).

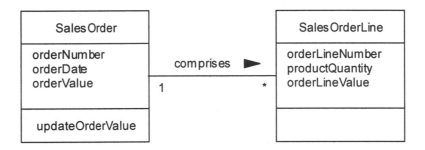

Figure 14.1 *Transaction–Transaction Line Item pattern.*

Coplien (1996) identifies the critical aspects of a pattern as follows.
- It solves a problem.
- It is a proven concept.
- The solution is not obvious.
- It describes a relationship.
- The pattern has a significant human component.

The human component of Christopher Alexander's patterns relates to the aesthetic qualities and the sense of harmony that a building may engender in its users. The analogous concept in software is harder to describe. We do not inhabit software systems—we use them as aids to other activities. But the human component Coplien refers to is not simply a good user interface to a working application, it is concerned with the nature of the software constructs used to build the application. Software patterns should result in structures that are sympathetic to the human perspective. It is argued that a good software pattern not only offers a solution that works but a solution that has an aesthetic quality, that is in some way elegant. This aesthetic quality of patterns is sometimes termed Quality Without A Name (QWAN for short) to avoid the overloading of current terminology. As you might imagine, QWAN is the subject of much controversy. Our discussion of patterns will not address issues of elegance though the reader may judge for themselves whether the solutions developed from the patterns have a sense of elegance.

In the same way that a pattern captures and documents proven good practice, *antipatterns* capture practice that is demonstrably bad. It is sensible to do this. We should ensure not only that a software system embodies good practice but also that it avoids known pitfalls.

Antipatterns are a way of documenting attempted solutions to recurring problems that proved unsuccessful. An antipattern can also include reworked solutions that proved effective (Brown et al., 1998).

For example, Mushroom Management (Brown et al., 1998) is an example of an antipattern in the domain of software development organizations. It describes a situation where there is an explicit policy to isolate systems developers from users in an attempt to limit requirements drift. In such an organization, requirements are passed through an intermediary such as the project manager or a requirements analyst. The negative consequence of this pattern of development organization is that inevitable inadequacies in the analysis documentation are not resolved. Furthermore design decisions are made without user involvement, and the delivered system may not address users' requirements. The reworked solution suggested by Brown et al. (1998) is to use a form of spiral process development model (see Chapter 3). Other reworked solutions include the involvement of domain experts in the development team, as recommended by the Dynamics Systems Development Method (DSDM) (we introduce DSDM in Chapter 21).

Patterns have been applied to many different aspects of software development. Beck and Cunningham (1989) documented some of the earliest software patterns in order to describe aspects of interface design in Smalltalk environments. Subsequently Coplien (1992) catalogued a set of patterns specifically for use in C++ programming (patterns that are related to constructs in a specific programming language are now known as *idioms*). Coad et al. (1997) greatly widened the field by describing a series of analysis and design patterns. Patterns have been applied to software development approaches other than object-orientated ones. For example, Hay (1996) identified a series of analysis patterns for data modelling. Hay discusses patterns relating to concepts such as Party and Contract, which appear commonly in information systems.

Design patterns were popularized by Gamma et al. (1995) in their book 'Design Patterns: Elements of Reusable Object-Oriented Software'[1]. The publication of this book gave significant impetus to the use of patterns in software development, but others authors have identified patterns that are concerned with analysis (Fowler, 1997), organizational issues (Coplien, 1996) and systems architecture using CORBA[2] (OMG, 1995).

Architectural patterns have also been identified that address some of the issues concerning the structural organization of software systems. Definitions have been suggested for some of these pattern categories (Buschmann et al., 1996):

Architectural patterns describe the structure and relationship of major components of a software system. An architectural pattern identifies subsystems, their responsibilities and their interrelationships.

Design patterns identify the interrelationships among a group of software components describing their responsibilites, collaborations and structural relationships

Idioms describe how to implement particular aspects of a software system in a given programming language.

Analysis patterns (Fowler, 1997) are defined as describing groups of concepts that represent common constructions in domain modelling. These patterns may be applicable in one domain or in many domains.

---

[1] The four authors of the book are known as the 'Gang of Four' (GOF) and the book is known as the GOF book.

[2] CORBA (Common Object Request Broker Architecture) is a general definition for a mechanism that uses Object Request Brokers (ORB) to provide interoperability between applications on different machines. We discuss its application in Chapters 17 and 19.

A particular pattern may belong to more than one category. For example, the MVC architecture can be viewed as an architectural pattern (Buschmann et al., 1996) when it is applied to subsystems, as a design pattern when it is concerned with smaller components or individual classes, and as a Smalltalk idiom.

14.2.2 Frameworks

It is important to distinguish between patterns and frameworks. Frameworks are partially completed software systems that may be targeted at a specified type of application, for example sales order processing. An application system tailored to a particular organization may be developed from the framework by completing the unfinished elements and adding application specific elements. This may involve the specialization of classes and the implementation of some operations. Essentially the framework is a reusable mini-architecture that provides structure and behaviour common to all applications of this type.

The major differences between patterns and frameworks can be summarized as follows.

- Patterns are more abstract and general than frameworks. A pattern is a description of the way that a type of problem can be solved, but the pattern is not itself a solution.
- Unlike a framework, a pattern cannot be directly implemented in a particular software environment. A successful implementation is only an example of a design pattern.
- Patterns are more primitive than frameworks. A framework can employ several patterns but a pattern cannot incorporate a framework.

14.2.3 Pattern catalogues and languages

Patterns are grouped into catalogues and languages. A *pattern catalogue* is a group of patterns that are related to some extent and may be used together or independently of each other. The patterns in a *pattern language* are more closely related, and work together to solve problems in a specific domain. For example, Cunningham (1995) documented the 'Check Pattern Language of Information Integrity', which consists of eleven patterns that address issues of data validation. All were developed from his experience of developing interactive financial systems in Smalltalk.

One of these patterns, Echo, describes how data input should be echoed back to the user after it has been modified and validated by the information system (since Cunningham uses the Model View Controller structure he talks about this in terms of changes made by the model). Typically users enter small batches of values and then look at the screen to check that they have been correctly entered. The sequence in which a user can enter data into fields may not be fixed and so validation feedback should be given one field at a time. For example, a user enters a value as 5.236. This might be echoed back by the system as 5.24 (correctly rounded to two decimal places). The user receives direct visual feedback that the value has been accepted and how it has been modified.

14.2.4 Software development principles and patterns

Patterns are intended to embody good design practice and hence are based upon sound software development principles, many of which have been identified since the early days of software development and applied within other development approaches than object-orientated ones. Buschmann et al. (1996) suggest that the following are the key principles that underlie patterns:

- abstraction,
- encapsulation,
- information hiding,
- modularization,
- separation of concerns,
- coupling and cohesion,
- sufficiency, completeness and primitiveness,
- separation of policy and implementation,
- separation of interface and implementation,
- single point of reference, and
- divide and conquer (this means breaking a complex problem into smaller, more manageable ones).

Most of these principles should already be familiar to the reader since they have been discussed in earlier chapters.

14.2.5 Patterns and non-functional requirements

Patterns can address the issues that are raised by non-functional requirements (Chapters 5 and 11). Buschmann et al. (1996) identify these as the important non-functional properties of a software architecture:

- changeability,
- interoperability,
- efficiency,
- reliability,
- testability and
- reusability.

These properties may be required of a complete system or a part of a system. For example, a particular set of functional requirements may be seen as volatile and subject to change. It is important to develop a structure for these requirements that can cope with change. Another requirement may be that a particular aspect of an application must be highly reliable. Again this requirement must be met by the design.

Review Questions

14.1   What is the difference between a pattern and an antipattern, and how is each used?

14.2   Give three example applications of the Transaction–Transaction Line Item pattern.

14.3 Documenting Patterns—Pattern Templates

Patterns may be documented using one of several alternative templates. *Pattern template* determine the style and structure of the pattern description, and these vary in the emphasis they place on different aspects of patterns. The differences between pattern templates may mirror variations in the problem domain but there is no consensus view as to the most appropriate template even within a particular problem domain. Nonetheless it is generally agreed that a pattern description should include the following elements (at least implicitly).

Name. A pattern should be given a meaningful name that reflects the knowledge embodied by the pattern. This may be a single word or a short phrase. These names become the vocabulary for discussing conceptual constructs in the domain of expertise. For instance, the names of three of the Gamma design patterns, Bridge, Mediator and Flyweight, give a strong indication of how they are intended to work.

Problem. This is a description of the problem that the pattern addresses (the intent of the pattern). It should identify and describe the objectives to be achieved, within a specified context and constraining forces. For example, one problem might be concerned with producing a flexible design, another with the validation of data. The problem can frequently be written as a question, for example 'How can a class be constructed that has only one instance and can be accessed globally within the application?'. This question expresses the problem addressed by the Singleton pattern (discussed in Section 14.4.2).

Context. The context of the pattern represents the circumstances or preconditions under which it can occur. The context should provide sufficient detail to allow the applicability of the pattern to be determined.

Forces. The forces embodied in a pattern are the constraints or issues that must be addressed by the solution. These forces may interact with and conflict with each other, and possibly also with the objectives described in the problem. They reflect the intricacies of the pattern.

Solution. The solution is a description of the static and dynamic relationships among the components of the pattern. The structure, the participants and their collaborations are all described. A solution should resolve all the forces in the given context. A solution that does not resolve all the forces fails.

Other aspects of templates

A pattern template may be more extensive than the elements described above. Some other features that have figured in pattern templates are given below:

- an example of the use of a pattern that serves as a guide to its application;
- the context that results from the use of the pattern;
- the rationale that justifies the chosen solution;
- related patterns;
- known uses of the pattern that validate it (some authors suggest that until the problem and its solution have been used successfully at least three times—the *rule of three*—they should not be considered as a pattern);
- a list of aliases for the pattern ('also known as' or AKA);
- sample program code and implementation details (commonly used languages include C++, Java and Smalltalk).

Gamma et al. (1995) use a template that differs from that described above. Although very detailed, this does not explicitly identify the forces. Cunningham's (1995) 'Check Pattern Language of Information Integrity' is described in the Portland Form[3]. Variations in template style and structure make it difficult to compare patterns that are documented in

---

[3] Named after Portland, Oregon where it originated, this is basically free format text.

different templates, and this limits their reusability since it is more difficult to use a pattern that is documented in an unfamiliar template.

14.4 Design Patterns

14.4.1 Types of Design Pattern

In the catalogue of 23 design patterns presented by Gamma et al. (1995) patterns are classified according to their scope and purpose. The three main categories of purpose that a pattern can have are *creational*, *structural* or *behavioural* (these are described in the following sections). The scope of a pattern may be primarily at either the class level or at the object level. Patterns that are principally concerned with objects describe relationships that may change at run-time and hence are more dynamic. Patterns that relate primarily to classes tend to be static and identify relationships between classes and their subclasses that are defined at compile-time. The Gamma patterns are generally concerned with increasing the ease with which an application can be changed, through reducing its coupling and maximizing its cohesion. The patterns are based on principles of good design that include the maximizing of encapsulation and the substitution of composition for inheritance wherever possible.

Changeability involves several different aspects (Buschmann et al., 1996)—maintainability, extensibility, restructuring and portability. Definitions of these terms vary but we use the following.

Maintainability is concerned with fixing errors in the information system.

Extensibility addresses the inclusion of new features and the replacement of existing components with new improved versions. It also involves the removal of unwanted features.

Restructuring focuses on the reorganization of software components and their relationships to provide increased flexibility.

Portability deals with modifying the system so that it may execute in different operating environments, such as different operating systems or different hardware.

14.4.2 Creational patterns

A creational design pattern is concerned with the construction of object instances. In general, creational patterns separate the operation of an application from how its objects are created. This decoupling of object creation from the operation of the application gives the designer considerable flexibility in configuring all aspects of object creation. This configuration may be dynamic (at run-time) or static (at compile-time). For example, when dynamic configuration is appropriate, an object-oriented system may use composition to make a complex object by aggregating simpler component objects. Depending upon circumstances different components may be used to construct the composite and, irrespective of its components, the composite will fulfil the same purpose in the application. A simple analogy illustrates this. A systems development department in an organization will vary in its composition from time to time. When permanent staff are on holiday contract staff may be employed to perform their roles. This enables the department to offer the same service to the organization.

Creating composite objects is not simply a matter of creating a single entity but also involves creating all the component objects. The separation of the creation of a composite object from its use within the application provides design flexibility. By changing the method of construction of a composite object, alternative implementations may be introduced without affecting the current use.

Singleton Pattern

As an example we consider the creational pattern, Singleton, which can be used to ensure that only one instance of a class is created. In order to understand the use of the pattern we need to consider the circumstances under which a single instance may be required.

The Agate campaign management system needs to hold information regarding the company. For example, its name, its head office address and the company registration details need to be stored so that they can be displayed in all application interfaces and printed on reports. This information should only be held in only one place within the application but will be used by many different objects. One design approach would be to create a global data area that can be accessed by all objects, but this violates the principle of encapsulation. Any change to the structure of the elements of global data would require a change to all objects that access them. The creation of a Company class overcomes this problem by encapsulating the company attributes (Figure 14.2). These are then accessible to other objects through the operations of the Company object. But there is still a problem with this proposal. An object that wants to use the Company object needs to know the Company object's identifier so that it can send messages to it. This suggests that the Company object identifier should be globally accessible—but again this is undesirable since it violates encapsulation.

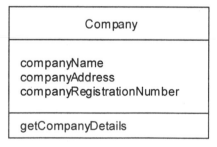

Figure 14.2 Company *class for the Agate case study.*

Some object-oriented programming languages (including Java and C++) provide a mechanism that enables certain types of operations to be accessed without reference to a specified object. These are called *class* (or *static*) *operations*. This offers a solution to the problem of providing global access without the need to globally define the object identifier. For example, a static operation getCompanyInstance() can be defined in such a way that it will provide any client object with the identifier for the Company instance. This operation can be invoked by referencing the class name as shown below.

```
Company.getCompanyInstance()
```

When a client object needs to access the Company object it can send this message to the Company class and receive the object identifier in reply. The client object can now send a getCompanyDetails() message to the Company object.

There is one further aspect to this design problem. It is important that there should only be one instance of this object. To ensure system integrity the application should be constructed so that it is impossible to create more than one. This aspect of the problem can be solved by giving the Company class sole responsibility for creating a Company object. This is achieved by making the class constructor private so that it is not accessible by another object. The next issue that needs to be addressed is the choice of an event that causes the creation of the company object. Perhaps the simplest approach is to create the Company object at the moment when it is first needed. When the Company class first receives the message getCompanyInstance() this can invoke the Company class constructor. Once the Company object has been created, the object identifier is stored in the class (or static) attribute companyInstance so that it can be passed to any future client objects.

So far we have produced a design for the Company class (Figure 14.3) that provides a single global point of access via the class operation getCompanyInstance() and that also ensures that only one instance is created.

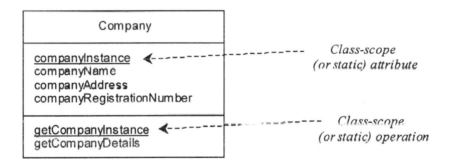

Figure 14.3 Company *class with class-scope operation and attribute.*

The design may need to accommodate further requirements. Since Agate operates as a separate company in each country (each owned by the same multinational), variations in company law from country to country may necessitate different company registration details to be recorded for each country. This suggests a requirement for different types of Company class each with its own variation of the registration details. The creation of a separate subclass for each style of company registration details is a solution to this aspect of the problem (Figure 14.4). When the getCompanyInstance() operation is first called the appropriate subclass is instantiated. If the Company object has not yet been instantiated, its constructor operation can access details of the country (say, held in a CurrentCountry object) to determine which subclass should be instantiated.

This part of the design for the campaign management system now has:
- a class Company that is only instantiated once;
- an instance of this class that is globally accessible;
- different subclasses of Company that are instantiated as needed, depending on run-time circumstances.

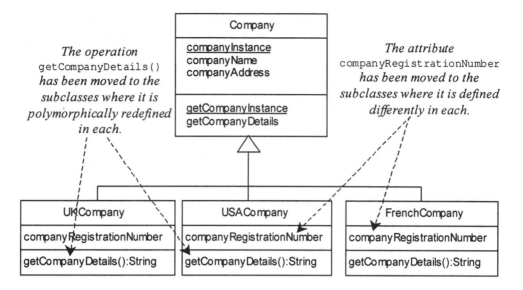

Figure 14.4 Company *class with subclasses.*

This is an application of the Singleton pattern. The pattern is described below in more general language.

Name. Singleton.

Problem. How can a class be constructed that should have only one instance and that can be accessed globally within the application?

Context. In some applications it is important that a class have exactly one instance. A sales order processing application may be dealing with sales for one company. It is necessary to have a Company object that holds details of the company's name, address, taxation reference number and so on. Clearly there should only be one such object. Alternative forms of a singleton object may be required depending upon different initial circumstances.

Forces. One approach to making an object globally accessible is to make it a global variable but in general this is not a good design solution as it violates encapsulation. Another approach is not to create an object instance at all but to use class operations and attributes (called 'static' in C++ and Java). However this limits the extensibility of the model since polymorphic redefinition of class operations is not possible in all development environments (for example C++).

Solution. Create a class with a class operation getInstance(), which, when the class is first accessed, creates the relevant object instance and returns the object identity to the client. On subsequent accesses of the getInstance() operation no additional instance is created but the object identity of the existing object is returned. A class diagram fragment for the singleton pattern is shown in Figure 14.5.

The singleton pattern offers several advantages.

- It provides controlled access to the sole object instance as the Singleton class encapsulates the instance.
- The name space is not unnecessarily extended with global variables.
- The Singleton class may be subclassed. At system start-up user-selected options may determine which of the subclasses is instantiated when the Singleton class is first accessed.

- A variation of this pattern can be used to create a specified number of instances if required.

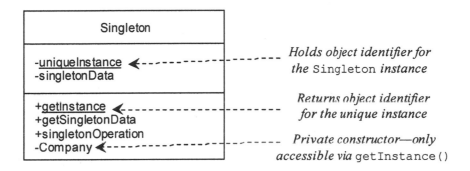

Holds object identifier for
the Singleton *instance*

Returns object identifier
for the unique instance

Private constructor—only
accessible via getInstance()

Figure 14.5 *Creational Patterns—Singleton.*

14.4.3 Structural patterns

Structural patterns address issues concerned with the way in which classes and objects are organized. Structural patterns offer effective ways of using object-oriented constructs such as inheritance, aggregation and composition to satisfy particular requirements. For instance, there may be a requirement for a particular aspect of the application to be extensible. In order to achieve this, the application should be designed with constructs that minimize the side effects of future change. Alternatively it may be necessary to provide the same interface for a series of objects of different classes.

Composite Pattern

It may be appropriate to apply the Composite structural pattern in a design for the Agate case study. In the following example we assume that further work is required to design a multimedia application that can store and play components of an advert.

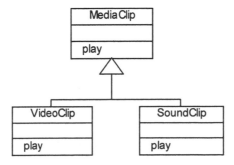

Figure 14.6 MediaClip *inheritance hierarchy.*

Here an advert is made up of sound clips and video clips each of which may be played individually or as part of an advert. The classes SoundClip and VideoClip have attributes and operations in common and it is appropriate that these classes are subclassed from MediaClip (Figure 14.6). But not all advert clips are primitive (that is, made up of only a single MediaClip). Some consist of one or more sequences of clips, such that each sequence is in turn an aggregation of SoundClip and VideoClip objects (Figure 14.7).

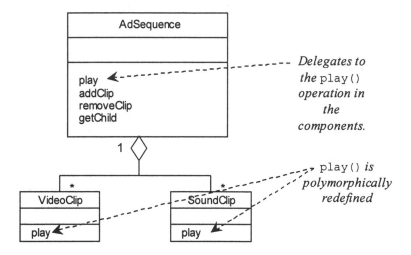

Figure 14.7 AdSequence *aggregation hierarchy.*

These two orthogonal[4] hierarchies can be integrated by treating AdSequence both as a subclass of MediaClip and also as an aggregation of MediaClip objects (see Figure 14.8).

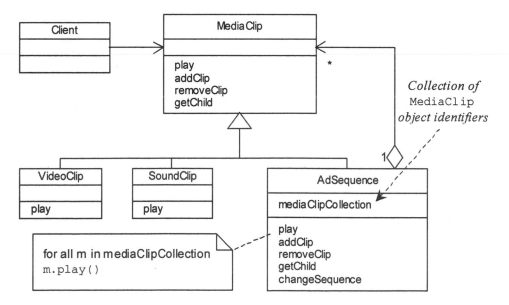

Figure 14.8 *Integrating the two hierarchies for* MediaClip.

---

[4] The term literally means 'at right angles to each other'. It is more loosely used here to describe hierarchies that cannot be directly mapped onto each other.

All the subclasses have the polymorphically redefined operation play(). For the subclasses VideoClip and SoundClip this operation actually plays the object. But for an AdSequence object, an invocation of the play() operation results in it sending a play() message to each of its components in turn. This structure is a straightforward application of the Composite pattern, which is described more generally below.

Name. Composite.

Problem. There is a requirement to represent whole-part hierarchies so that both whole and part objects offer the same interface to client objects.

Context. In an application both composite and component objects exist that are required to offer the same behaviour. Client objects should be able to treat composite or component objects in the same way. A commonly used example for the composite pattern is a graphical drawing package. Using this software package a user can create (from the perspective of the software package) atomic objects like circle or square and can also group a series of atomic objects or composite objects together to make a new composite object. It should be possible to move or copy this composite object in exactly the same way as it is possible to move or copy an individual square or a circle.

Forces. The requirement that the objects, whether composite or component, offer the same interface suggests that they belong to the same inheritance hierarchy. This enables operations to be inherited and to be polymorphically redefined with the same signature. The need to represent whole-part hierarchies indicates the need for an aggregation structure.

Solution. The solution resolves the issues by combining inheritance and aggregation hierarchies. Both subclasses, Leaf and Composite, have a polymorphically redefined operation anOperation(). In the Composite this redefined operation invokes the relevant operation from its components using a simple loop construct (Figure 14.9). The Composite subclass also has additional operations to manage the aggregation hierarchy so that components may be added or removed.

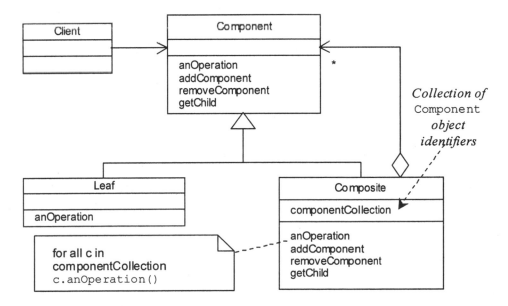

Figure 14.9 *Structural patterns—Composite.*

Further requirements may need to be considered for the Agate example. Perhaps `VideoClip` and `SoundClip` objects must be played in a particular sequence. This can be handled if the aggregate `AdSequence` maintains an ordered list of its components. Each component object can be given a sequence number, and two components that have the same sequence number are played simultaneously. The operation `changeSequence()` allows a component to be moved up or down within the sequence (Figure 14.8).

14.4.4 Behavioural Patterns

Behavioural patterns address the problems that arise when assigning responsibilities to classes and when designing algorithms. Behavioural patterns not only suggest particular static relationships between objects and classes but also describe how the objects communicate. Behavioural patterns may use inheritance structures to spread behaviour across the subclasses or they may use aggregation and composition to build complex behaviour from simpler components. The State pattern, which is considered below, uses both of these techniques.

State Pattern

Let us examine the Agate case study to determine whether it has features that may justify the application of the state pattern. Firstly, are there any objects with significant state dependent behaviour? `Campaign` objects have behaviour that varies according to state; a `Campaign` object may be in one of four main states, as shown in Figures 10.13 and A2.8 (for simplicity we ignore the nested concurrent states of the `Active` state). Clearly a `Campaign` object's state changes dynamically as the campaign progresses, thus necessitating changes in the behaviour of the object.

For example, when the FoodCo campaign is planned a `Campaign` object is created in the `Commissioned` state. It remains in this state until a campaign budget has been agreed and only then does it become possible to run advertisements, although some preparatory work may be done for the campaign in the meantime. Once a `Campaign` object enters the `Active` state all advert preparation and any other work that is done is subject to an agreed billing schedule. Several operations, for example `addAdvert()` and `calcCosts()`, will behave differently depending upon the state of the `Campaign` object. It would be possible to construct a working version of the software using the design for the `Campaign` class that is shown in Figure 14.10. However, this would be a complex class that is further complicated by state dependent operations such as `calcCosts()`, which would need to be specified with a series of case or if-then-else statements to test the state of the object. It would be simpler to subdivide the operations that have state dependent behaviour, which in this case would result in four separate `calcCosts()` operations, one for each state. The inclusion of `calcCostsCommissioned()`, `calcCostsActive()` and so on within `Campaign` would simplify the operation `calcCosts()`, but the class as a whole would become even more complex.

Another possibility is to create additional classes, one for each state so that each holds a state specific version of the operations, and this is how the State pattern works. A class diagram fragment illustrating this application of the State pattern is shown in Figure 14.11. Since the subclasses of `CampaignState` have no attributes specific to a particular `Campaign` object it is possible to have only one instance of each in the system. Thus there will be a maximum of four `CampaignState` objects, one for each state, and the additional overhead of manipulating the objects is unlikely to be significant (Figure 14.12). A variation on the Singleton pattern ensures that only one instance of each subclass can ever exist (this differs from its more usual application of ensuring that there is at most one object for a whole hierarchy).

Figure 14.10 *The* Campaign *class.*

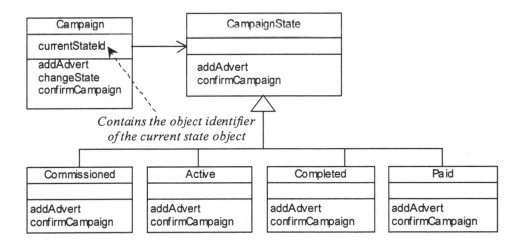

Figure 14.11 *State pattern for Agate.*

The State pattern is described more generally below.

Name. State.

Problem. An object exhibits different behaviour when its internal state changes making the object appear to change class at run-time.

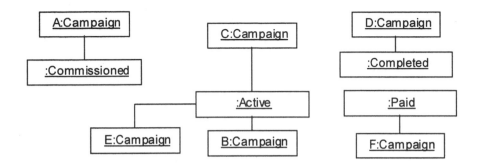

Figure 14.12 *Some State pattern objects for Agate.*

Context. In some applications an object may have complex behaviour that is dependent upon its state. In other words the response to a particular message varies according to the object's state. One example is the `calcCosts()` operation in the `Campaign` class.

Forces. The object has complex behaviour that should be factored into less complex elements. One or more operations have behaviour that varies according to the state of the object. Typically the operation would have large, multi-part conditional statements depending on the state. One approach is to have separate public operations for each state but client objects would need to know the state of the object so that they could invoke the appropriate operation. For example four operations `calcCostsCommissioned()`, `calcCostsActive()`, `calcCostsCompleted()` and `calcCostsPaid()` would be required for the `Campaign` object. The client object would need to know the state of the `Campaign` object in order to invoke the relevant `calcCosts()` operation. This would result in undesirably tight coupling between the client object and the `Campaign` object. An alternative approach is to have a single public `calcCosts()` operation that invokes the relevant private operation (`calcCostsCommissioned()` would be private). However, the inclusion of a separate private operation for each state may result in a large complex object that is difficult to construct, test and maintain.

Solution: The state pattern separates the state dependent behaviour from the original object and allocates this behaviour to a series of other objects, one for each state. These state objects then have sole responsibility for that state's behaviour. The original object, shown as `Context` in Figure 14.13, delegates responsibility to the appropriate state object. The original object becomes an aggregate of its states, only one of which is active at one time. The state objects form an inheritance hierarchy.

The responsibility for transitions from one state to another may either be given to the `Context` object or it may be shared among the State subclasses. If the rules for state changes are volatile and subject to change it may be better for the current State object to be responsible for the next transition. In this way the current state object always knows all the states into which the object may move next. However, this has the disadvantage of producing dependencies between state subclasses.

Use of the State pattern has both advantages and disadvantages, the latter particularly in terms of its possible side effects on system performance.

+ State behaviour is localized and the behaviour for different states is separated. This eases any enhancement of the state behaviour, in particular the addition of extra states.

+ State transitions are made explicit. The state object that is currently active indicates the current state of the Context object.
+ Where a state object has no attributes relevant to a specific Context object it may be shared among the Context objects. This State object is a Singleton!
- If the State objects cannot be shared among the Context objects each Context object will have to have its own State object thus increasing the number of objects and the storage requirements for the system.
- State objects may have to be created and deleted as the Context object changes state, thus introducing a processing overhead.
- Use of the State pattern introduces at least one extra message, the message from the Context class to the State class, thus adding a further processing overhead.

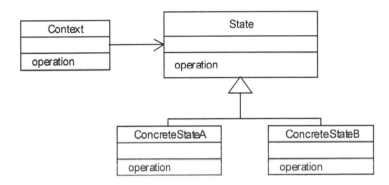

Fig 14.13 *Behavioural patterns—State.*

Review Questions

14.3 What are the advantages of using the Singleton pattern?
14.4 What implementation problems may occur when using the State pattern?

14.5 How to Use Design Patterns

The use of a pattern requires careful analysis of the problem that is to be addressed and the context in which it occurs. Before contemplating the application of patterns within a software development environment it is important to ensure that all members of the team receive appropriate training.

When a developer identifies a part of the application that may be subject to high coupling, a large, complex class or any other undesirable feature, there may be a pattern that addresses the difficulty. The following issues should be considered before employing a pattern to resolve the problem.

• Is there a pattern that addresses a similar problem?
• Does the pattern trigger an alternative solution that may be more acceptable?

- Is there a simpler solution? Patterns should not be used just for the sake of it.
- Is the context of the pattern consistent with that of the problem?
- Are the consequences of using the pattern acceptable?
- Are constraints imposed by the software environment that would conflict with the use of the pattern?

Gamma et al. (1995) suggest a seven-part procedure that should be followed after an appropriate pattern has been selected in order to apply it successfully.

(1) Read the pattern to get a complete overview.
(2) Study the Structure, Participants and Collaborations of the pattern in detail.
(3) Examine the Sample Code to see an example of the pattern in use.
(4) Choose names for the pattern's participants (i.e. classes) that are meaningful to the application.
(5) Define the classes.
(6) Choose application specific names for the operations.
(7) Implement operations that perform the responsibilities and collaborations in the pattern.

A pattern should not be viewed as a prescriptive solution but rather as guidance on how to find a suitable solution. It is quite likely (in fact almost certainly the case) that a pattern will be used differently in each particular set of circumstances. At a simple level the classes involved will have attributes and operations that are determined by application requirements. Often a pattern is modified to accommodate contextual differences. For example, the changeSequence() operation shown in Figure 14.8 represents a variation on the State pattern and may also be a variation on the Singleton pattern. Alternatively a pattern may suggest some other solution to the developer.

It is critical to the use of patterns that pattern catalogues and languages should be made readily available to the developer. Many patterns are documented in hypertext on the Internet or on company Intranets. CASE tool support for patterns is still rather limited but no doubt will emerge. An on-going area of research addresses the provision of suitable browsing capabilities for patterns. It is important to consider the way a pattern is documented so that it is easy for the developer to determine its applicability. We would contend that the minimum information needed is that described earlier in Section 14.3.

Software developers using patterns may wish to capture their experience in the pattern's format thus building up their own pattern catalogue or language. The process of identifying patterns is known as pattern mining and requires careful validation and management to ensure that the patterns that are captured are valuable. Again the pattern elements described earlier provide a checklist against which any candidate pattern can be compared.

If a pattern satisfies these criteria then its quality should be assured via a walkthrough. The most commonly used form of walkthrough for a pattern is known as a pattern writer's workshop. This involves a small of group of pattern authors who constructively comment upon each other's patterns. The focus of a workshop helps the participants describe useful patterns effectively.

14.6 Benefits and Dangers of Using Patterns

One of the most sought after benefits of object-orientation is reuse. Reuse at the object and class level has proved more elusive than was initially expected. Patterns provide a mechanism for the reuse of generic solutions for object-oriented and other approaches. They embody a strong reuse culture. Within the context of design, patterns suggest reusable elements of design and, most significantly, reusable elements of demonstrably successful designs. This reuse permits the transfer of expertise to less experienced developers so that a pattern can be applied again and again.

Another benefit gained from patterns is that they offer a vocabulary for discussing the problem domain (whether it be analysis, design or whatever) at a higher level of abstraction than the class and object level making it easier to consider micro-architectural issues. Pattern catalogues and pattern languages offer a rich source of experience that can be explored and provide patterns that can be used together to generate effective systems.

Some people believe that the use of patterns can limit creativity. Since a pattern provides a standard solution the developer may be tempted not to spend time on considering alternatives. The use of patterns in an uncontrolled manner may lead to over-design. Developers may be tempted to use many patterns irrespective of their benefits, thus rendering the software system more difficult to develop, maintain and enhance. When a pattern is used in an inappropriate context the side effects may even be disastrous. For example, the use of the State pattern may significantly increase the number of objects in the application with a consequent reduction in performance. In extreme cases, especially where the architecture is geographically distributed (as at Agate), this might prevent any successful use of the system. The introduction of any new approach to software development has costs for the organization. Developers need to spend time understanding the relevant pattern catalogues, they need to be provided with easy access to the relevant catalogues and they need to be trained in the use of patterns. Another aspect of the introduction of patterns is the necessary cultural change. Patterns can only be used effectively in the context of an organizational culture of reuse. Ironically, the introduction of a patterns approach may arouse less opposition to the encouragement of a reuse culture than an attempt to introduce procedures for code reuse. Developers need to think about how to apply a pattern to their current context, and thus there are greater opportunities for individual creativity.

These dangers emphasize that the use of patterns in software development requires care and planning. In this respect patterns are no different from any other form of problem solving: they must be used with intelligence. It is also important to appreciate that patterns only address some of the issues that occur during systems development. In no way should patterns be viewed as a 'silver bullet' that conquers all problems in systems development.

14.7 Summary

This chapter has considered how patterns can be used in software development. Patterns have been identified in many different application domains and are applicable at many different stages of the software development process. The most significant aspect of the growth of interest in patterns is the increased awareness of design issues that follows as a consequence. Patterns represent a significant change in the reuse culture in software development. Reuse need no longer be focused solely on elements of code, whether these are individual classes or complex frameworks, but can also realistically include the reuse of analysis or design ideas as described by patterns.

Related patterns are grouped together in catalogues. A pattern language is a group of patterns focused on a particular aspect of a problem domain so that when used together they provide solutions to the problems that arise.

Review Questions

14.5   What are the differences between a pattern language and a pattern catalogue?
14.6   List two general dangers and two general benefits of the use of patterns.
14.7   What seven steps are suggested by Gamma et al. for the effective use of patterns?

Case Study Work, Exercises and Projects

14.A   Read the design patterns Bridge and Façade in Gamma et al. (1995) and rewrite their description using the structure of the template given in Section 14.3.
14.B   In the FoodCo case study the `ProductionLine` class might be a candidate for design using the State pattern. Show how a variation of the State pattern could handle this requirement. What benefits and disadvantages are there in applying this solution?
14.C   Where and how could the Singleton Pattern be used in the FoodCo case study? Prepare a design class specification for a suitable Singleton class.

Further Reading

Gamma et al. (1995) and Buschmann et al. (1996) are two excellent texts that give important advice concerning software construction and should be on the essential reading list of any software developer. Even where the patterns they discuss are not directly relevant there is much to learn from their approach to solving design problems.

Both Fowler (1997) and Coad et al. (1997) are useful sources of information from an analysis perspective.

The 'Pattern Languages of Program Design' (known as the PLOP) books (Coplien and Schmidt, 1995; Vlissides et al., 1996; Martin et al., 1998) catalogue a wide range of patterns for all aspects of software development.

The patterns home page has links to a wealth of patterns and is well worth browsing. This can be found at hillside.net/patterns/patterns.html. Further useful patterns are stored in the Portland Pattern Repository at www.c2.com/ppr.

Some argue that the architectural approach adopted by Shlaer and Mellor (1988 and 1992) is pattern-based, they certainly have interesting ideas.

15

Human–Computer Interaction

Objectives
In this chapter you will learn:

- the importance of good user interface design;
- what is meant by metaphors in human–computer interaction;
- about different approaches to human–computer interaction;
- how to apply the techniques of scenario-based design;
- how standards and the law affect interface design.

15.1 Introduction

Designing the user interface can be critical in the development of an information system. The interface is what the users see. To them it *is* the system. Their attitude towards the entire system can be coloured by their experience of the user interface. Effective design of the interaction between people and the information systems they use is a discipline in its own right—human–computer interaction (HCI)—that combines the techniques of psychology and ergonomics with those of computer science. Before we move on to the design of the classes that make up the user interface in Chapter 16, we will address some of the HCI issues that influence the design of the user interface.

In Chapter 11 we raised the issue of designing the inputs and outputs of a system. This chapter is about the human factors aspects of designing the inputs and outputs. The inputs and outputs can be in the conventional form of data entry and enquiry screens and printed reports, or they can take the form of speech recognition, scanners, touch screens and virtual reality environments. We shall be concentrating on the conventional inputs and outputs used in information systems, although we recognize that the growth in the use of multimedia systems means that even quite conventional business information systems may incorporate multiple media as inputs and outputs.

There are two metaphors that are widely used to represent the user interface: firstly, the idea that the user is conducting a dialogue with the system, and secondly the idea that the user is directly manipulating objects on screen (Section 15.2). Much HCI work in the past has concentrated on producing guidelines for dialogue design, and we include a section on the characteristics of a good dialogue.

In the second part of the chapter, we consider an informal approach to the HCI design of a system and describe some of the activities that may be carried out. We then introduce three different approaches to HCI that reflect different beliefs about the best way to design the user interface of a system (Section 15.3). Examples of some of the techniques used in these approaches are given.

Finally, we make reference to the international standards for the ergonomics of workstation design and, for European readers, to the legal obligations for HCI design that are imposed by the European Union's directive on health and safety requirements for work with display screens (Section 15.4).

15.2 The User Interface

15.2.1 What is the user interface?

Users of an information system need to interact with it in some way. Whether they are users of FoodCo's tele-sales system entering orders made over the telephone by customers, or members of the public using a touch screen system to find tourist information, they will need to carry out the following tasks:

- read and interpret information that instructs them how to use the system;
- issue commands to the system to indicate what they want to do;
- enter words and numbers into the system to provide it with data to work with;
- read and interpret the results that are produced by the system either on screen or as a printed report; and
- respond to and correct errors.

It is important to note that these are mostly secondary tasks: they are concerned with using the system not with the users' primary objectives. In the examples above, the primary tasks are to take a customer order and to find tourist information. If the system has been designed well, the secondary, system-related tasks will be easy to carry out; if it has not been designed well, the secondary tasks will intrude into the process and will make it more difficult for the users to achieve their primary tasks.

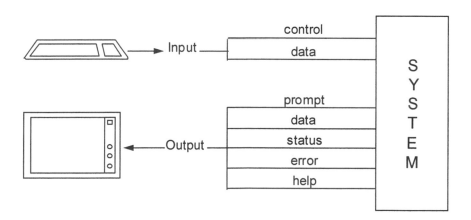

Figure 15.1 *Schematic form of human–computer dialogue.*

Input	control	user directs which way dialogue will proceed
	data	data supplied by user
Output	prompt	request for user input
	data	data from application following user request
	status	acknowledgement that something has happened
	error	processing cannot continue
	help	additional information to user

Figure 15.2 *Types of messages in human–computer dialogue.*

15.2.2 The dialogue metaphor

In the design of many computer systems, interaction between the user and the system takes the form of a *dialogue*. The idea that the user is carrying on a dialogue with the system is a *metaphor*. (A metaphor is a term that is used figuratively to describe something but is not applied literally.) There is no real dialogue in the sense of a conversation between people going on between the user and the computer, but as in dialogues between people, messages are passed from one participant to the other. Figure 15.1 shows the human–computer dialogue in schematic form. Figure 15.2 describes what is meant by each of the types of message that can be found in this dialogue.

Figure 15.3 shows a sample screen layout from FoodCo's existing system which runs on a mini-computer with displays on VT100 terminals. Although this only shows one screen, you can describe it in terms of the dialogue between the user and the system.

- The user may enter a command by selecting an option from a menu (not shown).
- The system responds with this data entry screen and automatically displays the Order Date and next Order No.
- The user enters a Customer Code.
- The system responds with the name and brief address of the customer as a confirmation that the correct number has been entered.

And so on …

The dialogue may not take exactly the same form each time that a user enters data into this screen. Sometimes, the user may not know the Customer Code and may have to use some kind of index look-up facility, perhaps entering the first few characters of the customer name in order to view a display of customers that start with those characters. Sometimes an order may consist of one line, usually it will consist of more, and if it consists of more than eight it will be necessary to clear those that have been entered from the screen and display space for a further eight lines to be entered. It also illustrates elements of the interface that support some of the message types listed in Figure 15.2. These are described in Figure 15.4.

```
CUSTORD1              Customer Order Entry            25/08/1999

Order Date  25/08/1999           Order No.     37291
Customer Code CE102_  Central Stores, Lytham St Annes
Customer Order Ref.   R20716___

   Prod.     Product                              Unit   Line
   Code      Description               Qty        Price  Price
01 12-75___  Sandwich spread 24x250g    _3        18.00  54.00
02 09-103_   Brown sauce 30x500g       _10        24.60 246.00
03 _____                            ___
04 _____                            ___
05 _____                            ___
06 _____                            ___
07 _____                            ___
08 _____                            ___
                                       ___
                                       Total        300.00
                                       VAT           52.50
                                       Order
                                       Total        352.50

F1-Help    F2-Save    F3-Cancel   F4-New    F5-Cust.  F6-Prod.
F10-Exit                          Cust.     Lookup    Lookup
```

Figure 15.3 *FoodCo customer order entry screen layout with sample data.*

In the analysis and design of the new system for FoodCo, there will be a use case for Customer Order Entry, as in Figure 15.5. This will be supported by a use case description, which may be quite brief at the end of the analysis stage of the project. As the design progresses, the use case description will be filled out in more detail. Not all the use cases

Input	control	use of function keys to control dialogue
	data	numbers, codes and quantities typed in by user
Output	prompt	request for user input and labels for automatically generated data, shown in bold, for example Customer Code
	data	automatic display of Order Date and next Order No., automatic calculation of totals and tax (shown in italics to distinguish it from input data)
	status	screen heading; could include display to confirm that a new order has been saved
	error	messages to warn of incorrect data entered, for example if a Customer Code is entered that does not exist or if a negative Quantity is entered
	help	additional information to user in response to the user pressing F1; may be general about the order entry screen or context-sensitive – specific to a particular type of data entry

Figure 15.4 *Examples of types of messages in human–computer dialogue.*

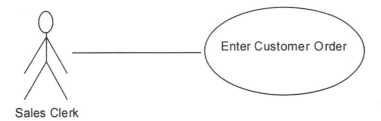

Enter Customer Order

Enter Customer Code or look up customer in index. For each line of the order, enter Product No. and Quantity. Calculate Line Price, VAT and totals automatically.

Figure 15.5 *Use case diagram and description for* Enter Customer Order.

will be for interactive dialogues: some will be for enquiries and some will be for printed reports. Figure 15.6 shows some of the use cases that the FoodCo sales clerks use. For each of these use cases there will be a sequence diagram to show the interaction between the business objects. However, these sequence diagrams will not yet show the details of the interaction between the user and the system at the interface. This will be covered in Section 16.3.

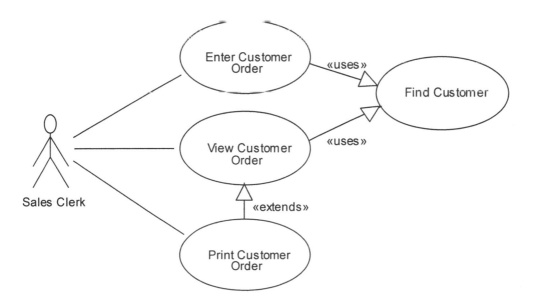

Figure 15.6 *Use case diagram showing use cases used by sales clerks at FoodCo.*

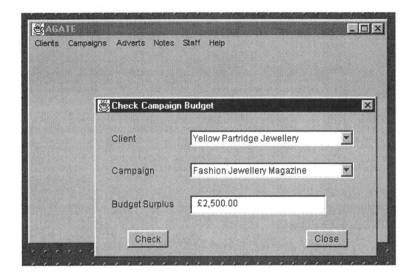

Figure 15.7 *Interface for the use case* Check Campaign Budget *developed in Java.*

15.2.3 The direct manipulation metaphor

The other metaphor for the design of the user interface, which has become more widespread in the last few years, is the *direct manipulation* metaphor. Many people are now familiar with this through the use of GUIs. When you use a software package with this kind of interface you are given the impression that you are manipulating objects on the screen through the use of the mouse. This metaphor is reflected in the concrete nature of the terms that are used. You can:

- drag and drop an icon,
- shrink or expand a window,
- push a button and
- pull down a menu.

Such interfaces are *event-driven*. Graphical objects are displayed on the screen and the window management part of the operating system responds to events. Most such events are the results of the user's actions. The user can click on a button, type a character, press a function key, click on a menu item or hold down a mouse button and move the mouse. The design of user interfaces to support this kind of interaction is more complicated than for interfaces based on the dialogue metaphor. Figure 15.7 shows the interface of a Java program to implement the use case Check Campaign Budget for the Agate case study. In this use case, the user first selects the name of a client from a list box labelled Client. Having selected the client, a list of all active campaigns for that client is placed in the list box labelled Campaign. At this point, no campaign is selected, and the user can click on the arrow at the end of the list box to view the list and select a campaign. When a campaign has been selected, the user can click on the button labelled Check. The program then totals up the cost of adverts in that campaign, subtracts it from the budget and displays the balance as a money value (negative if the campaign is over budget). In this interface design, there is no point in the user selecting a campaign until a client has been selected or clicking the Check button until a client and a campaign have been selected. The designer may choose to disable the Campaign list box until the client has been selected, and disable the button until both client and campaign have been selected. Having checked one campaign, the user may choose a different client, in which case the contents of the Campaign list box have to be changed and the button disabled again until a different campaign has been selected. In Section 16.7 we shall use statechart diagrams to model the state of elements of a user interface like this in order to ensure that we have correctly specified the behaviour of the interface.

Windows like the one in the example above are usually called *dialogue boxes* in GUI environments. In terms of the metaphors that we have discussed, they combine elements of a dialogue with the user with direct manipulation of buttons and lists.

15.2.4 Characteristics of good dialogues

Many authors of books and reports on HCI have produced sets of guidelines to help designers to produce good designs for the user interface. Some such guidelines are specific to certain types of interface. Carlow International Inc. produced a set of guidelines for GUI design (Carlow, 1992) that covers conceptual guidance on factors affecting user interface design, guidelines on overall design considerations and guidelines for the design of screen layouts. Shneiderman (1997) proposes five high-level objectives for data entry dialogue design that date back to his original edition in 1986 when most interfaces were text-based. Other authors such as Gaines and Shaw (1983), also writing at a time when text-based interfaces were predominant, have proposed as many as seventeen.

Regardless of whether a system is being developed for a text-based environment or for a GUI environment, there are a number of important general characteristics of good dialogue design. These include:
- consistency,
- appropriate user support,
- adequate feedback from the system and
- minimal user input.

These are considered in turn below.

Consistency. A consistent user interface design helps users to learn an application and to apply what they know across different parts of that application. This applies to commands, the format for the entry of data such as dates, the layout of screens and the way that information is coded by the use of colour or highlighting. As an example of a command, if the user has to press function key F2 to save data in one part of the system, then they will expect the key to have the same effect elsewhere in the system. If it does not result in data being saved then they will have made an error. The outcome of this could be that pressing F2 does something else that the user did not expect, or that it does nothing, but the user thinks that they have saved the data and then exits without saving it. Whichever is the case, the user is likely to become annoyed or frustrated at the response of the system. Guidelines in corporate style guides or in those from Microsoft and Apple help to prevent this kind of user frustration. Style guides are discussed in Section 15.2.5.

Appropriate user support. When the user does not know what action to take or has made an error, it is important that the system provides appropriate support at the interface. This support can be informative and prevent errors by providing help messages, or it can assist the user in diagnosing what has gone wrong and in recovering from their error. Help messages should be context-sensitive. This means that the help system should be able to detect where the user has got to in a dialogue and provide relevant information. In a GUI environment, this means being able to detect which component of the interface is active (or has the *focus*) and providing help that is appropriate to that part of the interface. The help provided may be general, explaining the overall function of a particular screen or window, or it may be specific, explaining the purpose of a particular field or graphical component and listing the options available to the user. It may be necessary to provide a link between different levels of help so that the user can move between them to find the information they require. The hypertext style of help in Microsoft Windows provides this facility. Help information may be displayed in separate screens or windows, or it may be displayed simultaneously in a status line or using *tooltips* as the user moves through the dialogue or positions the cursor over an item. Many web page designers provide help about elements of their pages by displaying messages in the status line at the bottom of the browser window or by displaying a tooltip-style message in a box as the cursor moves over an item on the page.

Error messages serve a different purpose and require careful design to ensure that they inform rather than irritate the user. An error message that tells the user that he or she has just deleted an essential file and then expects the user to click on a button marked OK when it is anything but OK is likely to annoy the user. Error messages should explain what has gone wrong and they should also clearly explain what the user can or should do to recover the situation. This information should be in language that the user can understand. This may mean using jargon from the user's business that they will recognize and understand rather than using computer jargon. Figure 15.8 shows three different error message boxes for the same situation. Only one is really of any help to the user.

Warning messages can prevent the user from making serious errors by providing a warning or caution message before the system carries out a command from the user that is likely to result in an irreversible action. Warning messages should allow the user to cancel the action that is about to take place. Figure 15.9 shows an example of a warning message.

Figure 15.8 *Example error messages for the same error.*

Figure 15.9 *Example warning message.*

Adequate feedback from the system. Users expect the system to respond when they make some action. If they press a key during data entry, they expect to see the character appear on the screen (unless it is a control command); if they click on something with the mouse, they expect that item to be highlighted and some action from the system. Users who are uncertain whether the system has noticed their action keep on pressing keys or clicking with the mouse, with the result that these further keypresses and clicks are taken by the system to be the response to a later part of the dialogue, with unpredictable results. It is important that users know where they are in a dialogue or direct manipulation interface: in a text-based interface there should be a visible cursor in the current active field; in a GUI environment the active object in the interface should be highlighted. The Yes button in Figure 15.9 is highlighted in this way, and this means that it will respond to the user pressing <Return> on the keyboard.

The system's response time should be appropriate to the type of user action: responses to data entry keypresses should be instantaneous, while responses to commands in menus or by means of buttons may take longer. If a system response is going to take some time, the system should respond first with some kind of feedback indicating that it is busy. This can be changing the cursor to a different form, or it can be displaying a progress monitor that shows what proportion of the task has been completed. If possible, the user should be given the option of cancelling the command. The purpose of this kind of feedback is to reduce the user's uncertainty about whether the system has received the input and is doing something about it or is waiting for the next input.

Minimal user input. Users resent making what they see as unnecessary keypresses and mouse clicks. Reducing unnecessary input also reduces the risk of errors and speeds data entry. The interface should be designed to minimize the amount of input from the user. The user can be helped in this way by:
- using codes and abbreviations,
- selecting from a list rather than having to enter a value,
- editing incorrect values or commands rather than having to type them in again,
- not having to enter or re-enter information that can be derived automatically, and
- using default values.

Some of these have a basis in the psychological aspects of the discipline of HCI. For example, being able to select values from a list rather than having to enter them from memory, allows the user to work by recognizing information rather than having to recall it.

It is also possible to reduce the amount of input as users become more familiar with a system by providing shortcuts or accelerators, key combinations that the user can use instead of selecting a command from a menu. However, these require the user to remember the key combinations and are less useful for new users who will find menus easier to use.

15.2.5 Style guides

In Section 15.2.4 consistency of the interface has been highlighted as one of the characteristics of good dialogue design. Some organizations provide standard guidelines for the design of user interfaces. One way in which standardization of user interface design has come about is through the domination of the PC market by Microsoft. Microsoft produces a book of guidelines "The Windows Interface Guidelines for Software Design" (Microsoft, 1997) that lays down the standards to which developers must adhere if they are to be granted Windows certification. Similar guidelines are available from Apple for the Apple Macintosh operating system—'Macintosh Human Interface Guidelines' (Apple, 1996). The effect of such guidelines is apparent in the similarity of many applications from different sources that make use of toolbars, status bars and dialogue boxes with buttons and other graphical components placed in similar positions. The benefit of this similarity for users is that different applications will look similar and behave in similar ways. This means that users can transfer their knowledge of one application to another and make inferences about the way that it will respond to certain types of user interaction.

Guidelines for user interface design are usually referred to as *style guides*, and large organizations with many different information systems produce their own style guides for the design of systems to ensure that all their applications, whether they are produced in-house or by outside software companies, conform to a standard set of rules that will enable users quickly to become familiar with a new application. Figure 15.3 reflects the use of an existing style guide in FoodCo. The layout of the screen with standard heading information at the top, the use of bold text to highlight prompts and labels, the position of the information about function keys and the use of specific function keys for particular commands are all standards within FoodCo for the design of text-based screens. This is important, as it means that a user can be confident that pressing function key F2 in any application will save the data on the screen.

The use of style guides and the characteristics of a good dialogue relate to dialogue and interface design in general. In the next part of this chapter, we consider how to ensure that the user interface is appropriate to the specific application for which it is being designed.

Review Questions

15.1 Think of a computerized information system that you use regularly. This could be a library system, an automated teller machine (ATM) that you use to get cash, a database that you use in your work or any other system that you are familiar with. Write down which elements of the interface support the five tasks listed at the start of Section 15.2.1.

15.2 For each of the elements of the interface that you have listed in Question 15.1, write down your ideas about how they could be improved.

15.3 What is the difference between the dialogue and direct manipulation metaphors?

15.4 Make a list of direct manipulation metaphors that are used in a GUI that you are familiar with. Are there any metaphors that do not work as you might expect?

15.5 What are the four characteristics of good dialogues described in Section 15.2.4

15.6 Figure 15.9 shows the Yes button in a dialogue highlighted. What do you think is the risk associated with making this the active button by default?

15.3 Approaches to User Interface Design

15.3.1 Informal and formal approaches

There are many different ways of designing and implementing the elements of the user interface that support the interaction with users. The choices that the designer makes will be influenced by a number of factors. These include:
- the nature of the task that the user is carrying out;
- the type of user;
- the amount of training that the user will have undertaken;
- the frequency of use; and
- the hardware and software architecture of the system.

These factors may be very different for different systems. They are listed in Figure 15.10 for the FoodCo tele-sales system and a tourist information system. However, this way of comparing the two systems and identifying factors that affect their design is very informal. More formal and methodical approaches to the analysis of usability requirements have been developed by researchers in the discipline of HCI. These approaches can be categorized under three headings:
- structured approaches,
- ethnographic approaches and
- scenario-based approaches.

These approaches are very different from one another. However, they all carry out three main steps in HCI design:
- requirements gathering,
- design of the interface and
- interface evaluation.

	Tele-Sales System	Touch Screen Tourist Information System
The nature of the task that the user is carrying out	Routine task; closed solution; limited options.	Open-ended task; may be looking for information that is not available.
The type of user	Clerical user of the system; no discretion about use (must use it to do their job).	Could be anyone; discretion about use of system; novice in relation to this system.
The amount of training that the user will have undertaken	Training provided as part of job.	No training provided.
The frequency of use	Very frequent; may be taking an order every few minutes.	Very occasional; may never use it again.
The hardware and software architecture of the system	Mini-computer, VT100 terminals with text screens, keyboard data entry	Touch screen with buttons on screen, no keyboard or mouse.

Figure 15.10 *User interface design factors for two systems.*

Step	Objectives
Requirements gathering	Determine characteristics of the user population: types of user, frequency of use, discretion about use, experience of the task, level of training, experience of computer systems.
	Determine characteristics of the task: complexity of task, breakdown of task, context/environment of task.
	Determine constraints and objectives: choice of hardware and software, desired throughput, acceptable error rate.
Design of the interface	Allocate elements of task to user or system; determine communication requirements between users and system.
	Design elements of the interface to support the communication between users and system in the light of characteristics of the users, characteristics of the task and constraints on design.
Interface evaluation	Develop prototypes of interface designs.
	Test prototypes with users to determine if objectives are met.

Figure 15.11 *Steps in HCI design and objectives in each step.*

Each of these approaches has similar objectives in each of these main steps. Typical objectives are shown in Figure 15.11. However, they differ in the ways that they set out to achieve these objectives. This is described below.

Structured approaches

Structured approaches to user interface design have been developed in response to the growth in the use of structured approaches to systems analysis and design during the 1980s. Structured analysis and design methodologies have a number of characteristics. They are based on a model of the systems development life cycle, which is broken down into stages, each of which is further broken down, for example into steps that are broken down into tasks. Specific analysis and design techniques are used, and the methodology specifies which techniques should be used in which step. Each step is described in terms of its inputs (from earlier steps), the techniques applied and the deliverables that are produced as outputs (diagrams and documentation). These approaches are more structured than the simple waterfall model of the life cycle, as they provide for activities being carried out in parallel where possible rather than being dependent on the completion of the previous step or stage. Typically such structured approaches use data flow diagrams to model processes in the system and take a view of the system that involves decomposing it in a top-down way. Structure charts or structure diagrams are used to design the programs that will implement the system.

Proponents of structured approaches argue that they have a number of benefits.

- They make management of projects easier. The breakdown of the project into stages and steps makes planning and estimating easier, and thus assists management control of the project.
- They provide for standards in diagrams and documentation that improves understanding between the project staff in different roles (analyst, designer and programmer).
- They improve the quality of delivered systems. Because the specification of the system is comprehensive, it is more likely to lead to a system that functions correctly.

Advocates of structured approaches to HCI believe that similar benefits can be brought to HCI by adopting structured approaches. These approaches assume that a structured approach to analysis and design of a system is being used and that a structured approach to the HCI design can take place at the same time and be integrated to some extent into the project life cycle. Two examples of such approaches are given below.

- STUDIO (STructured User-interface Design for Interface Optimization) developed with KPMG Management Consulting in the UK (Browne, 1994).
- The RESPECT User Requirements Framework developed for the European Union Telematics Applications Programme by a consortium of Usability Support Centres (Maguire, 1997).

Structured approaches make use of diagrams to show the structure of tasks and the allocation of tasks between users and the system. They also make extensive use of checklists in order to categorize the users, the tasks and the task environments. Evaluation is typically carried out by assessing the performance of the users against measurable usability criteria. STUDIO is used here as an example of a structured approach.

The overall structure of STUDIO is shown in Figure 15.12. It is divided into Stages, and each Stage is broken down into Steps. The activities undertaken in each of the Stages are

Figure 15.12 *Stages in STUDIO (From Browne, 1994).*

shown in Figure 15.13. Figure 15.14 shows the Steps in Stage 3 as an example. STUDIO uses a number of techniques such as:

- task hierarchy diagrams,
- knowledge representation grammars,
- task allocation charts, and
- statecharts.

It is not possible to provide examples of all of these here. Statecharts are similar to those used in UML and based on the work of Harel (1988). Examples of statecharts applied to user interface design are included in Section 16.7. A sample task hierarchy diagram for 'Take an Order' is shown in Figure 15.15. In this diagram, the boxes with a small circle in the top right-hand corner are selections, only one of which will take place each time an order is taken; the box with an asterisk in the top right-hand corner is an iteration, which will take place usually more than once.

Structured approaches may involve evaluation of the user interface designs in a laboratory situation. This reflects the need to have operational measures of usability that can be tested and used to assess the effectiveness of the design. These operational measures are derived from objectives that were gathered during the requirements analysis phase of the project described in Chapter 5. Examples include measures of how quickly users learn to use the system, error rates and time taken to complete tasks.

The National Physical Laboratory in the UK, one of the partners in the consortium that produced the RESPECT methodology, has a usability laboratory in which the interface designs are tested with users under laboratory conditions. One criticism of this approach is that people do not use systems under laboratory conditions, they use them in busy offices, noisy factories or crowded public spaces. Usability tests conducted in laboratories lack *ecological validity*. This means that they do not reflect the real conditions in the environment in which the systems will be used. As an alternative to laboratory-based usability studies, ethnographic approaches to the establishment of usability requirements, and to the testing of those requirements, have been proposed.

Stage	Summary of activities
Project Proposal and Planning	Decide whether user interface design expenditure can be justified. Produce quality plan.
User Requirements Analysis	Similar to systems analysis, with focus on gathering information relating to user interface design rather than general functionality.
Task Synthesis	Synthesize results of requirements analysis to produce initial user interface design. Produce user support documentation.
Usability Engineering	Prototyping combined with impact analysis to provide an approach to iterative development that is easy to manage.
User Interface Development	Handover of the user interface specification to developers to ensure that usability requirements are understood.

Figure 15.13 *Summary of activities in each Stage of STUDIO.*

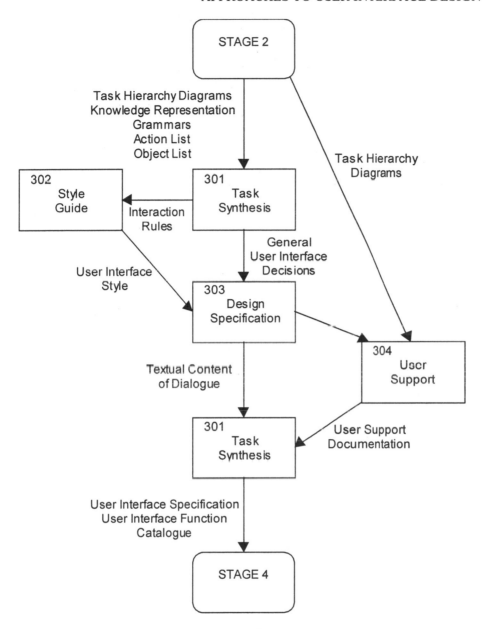

Figure 15.14 *Overview of STUDIO Stage 3 (From Browne, 1994).*

Ethnographic approaches

The term ethnography is applied to a range of techniques used in sociology and anthropology and reflects a particular philosophy about how scientific enquiry should be carried out in these disciplines. Researchers who employ an ethnographic method seek to involve themselves in whatever situation they are studying. They believe that only by doing this and becoming part of that situation can they truly understand and interpret what is happening. Ethnographic methods belong to a wide range of *qualitative* research

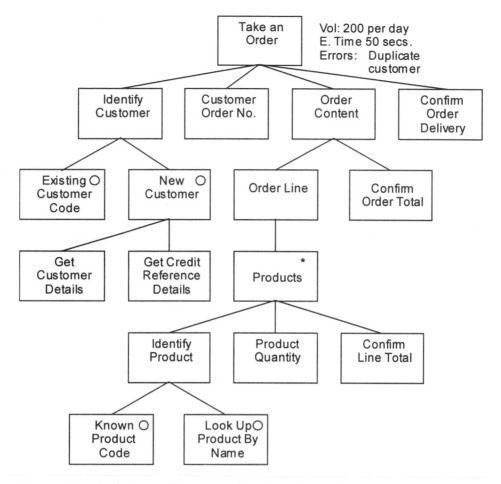

Figure 15.15 *Task Hierarchy Diagram for 'Take an Order'.*

methods. (Qualitative means 'concerned with the quality of something' and is usually taken to be the opposite of *quantitative* which means 'concerned about the quantity of something'. Quantitative methods typically rely on statistics to draw conclusions.) Hammersley and Atkinson (1995) provide a definition of ethnography.

> In its most characteristic form it involves the ethnographer participating, overtly or covertly, in people's daily lives for an extended period of time, watching what happens, listening to what is said, asking questions—in fact, collecting whatever data are available to throw light on the issues that are the focus of the research.

In HCI this means that the professional charged with carrying out the user interface design spends time with the users immersed in their everyday working life. Only by spending time in this way can the real requirements of the users be understood and documented. Ethnographic methods also emphasize that different users interpret their experience of using systems subjectively, and it is this subjective interpretation that the HCI professional must understand rather than assuming that the system can be assessed objectively.

Some HCI methods are criticized for failing to capture information about the context in which people are using systems, by focusing on the user and their tasks. Ethnographic approaches attempt to answer this criticism by the professional experiencing the daily working life of the people who will be the users of the system so that it can be better designed to meet their needs. Some structured approaches have also attempted to respond to the criticism about failing to take context into account, typically by adding some kind of contextual analysis questionnaire to the battery of checklists that they use.

There is no one ethnographic approach that covers the three steps in user interface development: requirements gathering, design of the interface, and interface evaluation. There are a number of approaches that can be classed as ethnographic, although their originators may use particular terms to describe their approaches.

Contextual enquiry is an approach developed by John Whiteside and others at Digital Equipment Corporation (Whiteside et al., 1988). It is used to carry out evaluation of the usability of a system in the users' normal working environment. The aim of contextual enquiry is to get as close to the users as possible and to encourage them to develop their interpretation of the system.

Participative or co-operative design and evaluation involve users actively in the design and evaluation of the user interface of systems (Greenbaum and Kyng, 1991). The social and political issues in the workplace that affect the use of systems are part of the material that is captured by these approaches.

Ethnographic approaches use a range of techniques to capture data: interviews, discussions, prototyping sessions and videos of users at work or using new systems. These data are analysed from different perspectives to gain insights into the behaviour of the users. Video is also used in other approaches, particularly in laboratory-based usability studies, and analysis of video can be particularly time-consuming.

Scenario-based approaches

Scenario-based design has been developed by John Carroll and others (Carroll, 1995). It is less formal than the structured approaches but more clearly defined than most ethnographic approaches. Scenarios are step-by-step descriptions of a user's actions that can be used as a tool in requirements gathering, interface design and evaluation. Use cases are similar to scenarios, and Carroll's book includes chapters by Ivar Jacobson and Rebecca Wirfs-Brock who have developed use cases and responsibility-based approaches to modelling interaction in object-oriented analysis and design.

Pete starts up the word-processor.

He types in a title for the note and changes its style to *Title*.

He types in two paragraphs describing his idea for an advertisement for the Yellow Partridge campaign to be used in fashion magazines in Europe during the summer of 1999.

He types his initials and the date and time.

He uses the short-cut keys to save the file.

The save-as dialogue box appears and, using the mouse, he changes to the *Summer 1999 Campaign* folder in the *Yellow Partridge* folder on the server.

He scrolls to the bottom of the list of files already in the folder and reads the title of the last note to be added, *Note 17*, he calls the new note *Note 18* and clicks on Save.

He exits from the word-processor.

Figure 15.16 *Scenario describing Pete Bywater of Agate adding a new note.*

Scenarios can be textual narrative describing a user's actions or they can be in the form of storyboards (a series of pictures that depict those actions), video mock-ups or even prototypes. Figure 15.16 shows a scenario that describes the actions of Peter Bywater from Agate when he demonstrates how he creates notes following the interview in Chapter 5.

Scenarios can be used like this in requirements gathering to document the actions that a user carries out in their current system. They can also be used to document ideas about how the user would see themselves using the new system. This is called envisioning the design. Alternative scenarios describing different approaches to the design can be compared by the designers and the users. Figure 15.17 shows a scenario describing how a staff member at Agate might use the new system to create a new note about an advert.

The user selects Add a Note from the menu. A new window appears.
From the list box at the top of the window she selects the name of the client.
A list of campaigns appears in the list box below, and she selects a particular campaign.
A list of adverts appears in the next list box, and she selects a specific advert.
She types a few paragraphs into a text box to describe her idea for the advert.
She fills the space on screen and a vertical scrollbar appears and the text in the text box scrolls up.
She enters her initials into a text box, and the system checks that she is allocated to work on that campaign.
The date and time are displayed by the system, and the Save button is enabled.
She clicks on the Save button and the word Saved appears in the status bar.
The text box, the text field for initials and the date and time are cleared.

Figure 15.17 *Scenario describing how a user might add a note in the new system.*

For evaluation of the system more detailed scenarios are prepared so that the actual system can be compared against the expectations that the designer has of how the user will interact with it. Carroll (1995) claims that scenarios can be used in more than just these three ways. He lists the following roles for scenarios:

- Requirements analysis,
- User–designer communication,
- Design rationale,
- Envisionment,
- Software design,
- Implementation,
- Documentation and training,
- Evaluation,
- Abstraction and
- Team building.

Two of these are worth further comment: user–designer communication and design rationale.

In Chapter 5 we pointed out that the diagrams used by systems analysts and designers are used to communicate ideas, among other things. Information systems professionals need to communicate with the end-users of the systems that they are developing. Scenarios provide a means of communication that can be used by professionals and end-users to communicate about the design of the users' interaction with the system. They are simple enough that users can produce them without the need for the kind of training that they would need to understand class diagrams, for example. Scenarios can be used with use cases. The use cases can provide a description of the typical interaction; scenarios can be used to document different versions of the use case, for example, to document what happens when a user is adding a new note but is not authorized to work on the project they try to add it to.

Scenarios can be supported by additional documentation to justify design decisions that have been taken. Carroll (1995) calls these design justifications *claims*. The designer can document the reasoning behind alternative designs, explaining the advantages and disadvantages of each. Figure 15.18 shows some claims for the scenario in Figure 15.17. These usability claims from design can be checked during evaluation of the software or of prototypes.

The Save button is disabled until the user has selected a client and a campaign, entered some text and entered his or her initials. This prevents the user attempting to save the note before all data has been entered and getting an error message.

The initials of the user could be entered automatically from their network login, but observation shows that the creative staff often work together as a group and different people will come up with ideas that they record as notes. It would be inconvenient for them to be logging in and out of the system each time a different person wants to enter a new note. For this reason, they are required to enter their initials.

The initials, date, time and text fields are cleared after a note is saved, but the client, campaign and advert list boxes are left untouched so that the user can enter another note for the same advert or campaign without having to reselect these items.

Figure 15.18 *Claims for the design scenario in Figure 15.18.*

Scenario-based design can result in large volumes of textual information that must be organized and managed so that it is easily accessible. There is a document management task to be undertaken that requires a rigorous approach to control different versions of scenarios and to cross-reference them to claims and feedback from users. Developers run the risk of delaying implementation while they work through and document alternative scenarios for different parts of the system. Rosson and Carroll (1995) present one way to try to prevent this happening. They use a computer-based tool to develop and document their scenarios and to develop working models of the scenarios in Smalltalk as they go along. This allows them to document software implementation decisions at the same time, and they propose that there are benefits to recording design decisions and software implementation decisions together in this way.

These three types of approach have been presented as though they were very separate. However, there are elements that they have in common. Some structured approaches have attempted to take on board the criticisms that they fail to address the context in which people work and use computer systems. Ethnographic methods may use the same data gathering techniques as other approaches, and may be used to provide information that can be used as the basis for drawing up scenarios. What they all share is a concern to enhance the usability of information systems and a recognition that usability issues must be integral to the design of computerized information systems.

15.3.2 Achieving usability

People often talk about how *user-friendly* a piece of software is, but it is often very difficult to tell what it is they mean by this. As a concept it is very vague. Usability may seem like a similar concept, but the HCI community has developed operational definitions of usability that can be used to test a piece of software. Shackel (1990) produced operational definitions of four criteria that were originally developed in the 1980s.

- Learnability—how much time and effort is needed to achieve a particular level of performance.
- Throughput—the speed with which experienced users can accomplish tasks and the number of errors made.
- Flexibility—the ability of the system to handle changes to the tasks that users carry out and the environment in which they operate.
- Attitude—how positive an attitude is produced in users of the system.

In Chapter 5 we mentioned the ISO definition of usability as 'the degree to which specific users can achieve specific goals within a particular environment; effectively, efficiently, comfortably and in an acceptable manner'. These criteria can be used in conjunction with the users' acceptance criteria documented during requirements gathering to assess how easy a software product is to use. Some of these can be quantified, for example, we can count the number of errors made by staff at FoodCo using the new system and compare that with the number of errors made with the old system and the objectives that they have set for the new system.

Sometimes conflicts will exist between different criteria and between usability criteria and other design objectives, and the designers will have to make compromises or trade-offs between different objectives. In particular, increasing flexibility is likely to conflict with the objective of developing the system at a reasonable cost.

15.4 Standards and Legal Requirements

In Section 15.2.5 we discussed style guides, which set standards for the design of user interfaces. Style guides like these determine the use of standard layouts, colour and function keys and the overall appearance of the system. The International Standards Organization (ISO) has produced standards that have a broader impact on the use of computer systems. ISO 9241 is an international standard for the ergonomic requirements for work with Visual Display Terminals, including both hardware and software. The standard covers physical aspects of the user's workstation (including positioning of equipment and furniture), the design of the computer equipment and the design of the software systems. These standards are intended to ensure the quality of systems and to prevent local standards becoming barriers to free trade.

In the European Union (EU), this has been taken one step further and the EU Council issued a directive on 29th May 1990 that has the force of law for member states. In the United Kingdom, for example, this directive has been implemented in the Health and Safety (Display Screen Equipment) Regulations 1992. Under these regulations all workstations must now comply with certain minimum requirements, and employers have a duty in law to ensure the health and safety of employees using display screen equipment.

The regulations provide a number of definitions:

- display screen equipment—any alphanumeric or graphic display screen;
- user—an employee who habitually uses display screen equipment as a significant part of his or her normal work (see table of criteria in the Health and Safety Executive guidance document);
- operator—self-employed person as above;
- workstation—display screen equipment, software providing the interface, keyboard, optional accessories, disk drive, telephone, modem, printer, document holder, work chair, work desk, work surface or other peripheral item and the immediate work environment around the display screen equipment.

The definition of display screen equipment excludes certain types of equipment, such as equipment in the cab of a vehicle, cash registers and some portable equipment.

As well as covering the physical equipment that is used by the user, the regulations cover environmental factors such as position of equipment, lighting, noise, heat and humidity in the workplace. Employers are required to:

- analyse workstations to assess and reduce risks,
- take action to reduce risks identified,
- ensure workstations meet the requirements of the regulations by the necessary dates,
- plan the work activities of users to provide breaks,
- provide eyesight tests for users,
- provide corrective appliances for eyes if required,
- provide training relevant to health and safety issues and workstations, and
- provide information to employees about health and safety risks and measures taken to reduce them.

The analysis of workstations in order to reduce risks includes analysis of the software, and the guidelines published to assist employers to meet their responsibilities state the following requirements.

Employers must take into account the following principles in designing, choosing, commissioning and modifying software and in designing tasks for people that require them to use display screen equipment.

- The software that is used must be suitable for the task.
- The software must be easy to use and able to be adapted to the level of knowledge or experience of the operator or user.
- The employer is not allowed to use any kind of quantitative or qualitative checking facility without the knowledge of the operators or users.
- Systems must give feedback to operators or users about the performance of the systems that they are using.
- Systems must display information for users both in a format and at a pace that are adapted to the operators or users.
- The principles of software ergonomics must be applied, particularly to the way that people process data.

Clearly the effect of this is to require employers, and so also software developers, to demonstrate that they are applying good HCI practice in the way that they design software.

15.5 Summary

System designers must take account of the requirements of the people who will use their software if they are to reduce errors and maximize the satisfaction of the users with the system. The user interface can be viewed as part of a dialogue between the user and the system and there are a number of characteristics of good dialogue design that can be used to ensure that the user is supported by the interface and assisted in carrying out their primary task.

It is possible to apply an informal approach to determining characteristics of the users, the task and the situation that will affect the interface design, or to apply a more formal approach using either structured, ethnographic or scenario-based techniques or some combination of these. The main aim of this is to produce software that can be demonstrated to meet the usability requirements of the people who will use it. This may be done in order to ensure compliance with international standards or it may be to meet legal requirements in some countries.

Review Questions

15.7 For the system that you wrote about in Question 15.1, note down information relevant to the design factors in Figure 15.10.

15.8 List as many differences as you can think of between structured, ethnographic and scenario-based approaches.

15.9 Make your own list of what you think the advantages and disadvantages could be of structured, ethnographic and scenario-based approaches.

Exercises, Case Study Work and Projects

15.A Using a GUI that you are familiar with as an example, try to identify features that you think might be part of the style guidelines for that GUI.

15.B Using the four criteria for good dialogues discussed in Section 15.2.4 evaluate an application that you use regularly. Identify the ways in which it meets these criteria and the ways in which it does not meet the criteria. Suggest ways in which it could be improved.

15.C Write a scenario to describe what is done when Rik Sharma of FoodCo starts to plan staff allocation, based on the interview transcript in Exercise 5.B (Chapter 5). (Make sure that you concentrate on what he does and not on what is done by other staff at other times.)

15.D For the system that you wrote about in Question 15.1, identify measurable objectives that could be used to measure how usable that system is. (You may like to start by thinking about how long it takes you to use it and how many errors you make.)

15.E Find out whether there are any legal requirements on software designers to comply with legislation that covers ergonomics or HCI in your country. Write a short report to summarize these requirements as though you were an analyst reporting to your manager on this legislation.

Further Reading

Many computer science and information systems courses now include HCI as a subject. If you have not come across HCI before and want to find out more, there are a number of suitable textbooks, such as Booth (1989), Preece et al. (1994) and Dix et al. (1998). A classic text in this area is Shneiderman (1997) which has been updated to cover graphical user interfaces in more detail since its first publication.

Style guidelines are available from some of the largest companies in the industry. Links are available to the on-line versions of these in the book's web site.

For a structured method for user interface design Dermot Browne's book (Browne, 1994) provides a step-by-step approach to user requirements analysis, task analysis, usability and interface design. Browne also uses Harel statecharts to model the behaviour of the interface in a more thorough way than many authors. John Carroll's book (1995) provides a good coverage of scenario-based methods, and is very practical in approach.

In the UK, Her Majesty's Stationery Office (HMSO) publishes a booklet that explains the requirements of the Display Screen Regulations

16

Designing Interface Objects

Objectives
In this chapter you will learn:
- how to add interface objects to the class diagram;
- how prototyping is used in interface design;
- how to extend sequence diagrams to include interface objects;
- how design patterns can be applied to the design of the user interface;
- how to model the human–computer interface using statechart diagrams.

16.1 Introduction

In Chapters 11 and 12 we introduced the three-tier system architecture. The presentation layer in this architecture contains the classes that handle the interface with the user—usually windows and reports. We can use the techniques and diagrams of UML to model the user interface by adding user interface classes to the class diagram. These classes handle the user interface and allow us to design business classes that do not contain details of how they will be presented. This enables the reuse of the business classes.

Practical example window layouts are used to show how the interaction between the user and the system is designed. These window layouts can be modelled as classes if required, but this is not always necessary unless the behaviour of the interface or of graphical objects in the user interface is the subject of the application being developed. The UML notation for packages and for package dependency can be used to show how class diagrams can reference classes from reusable class libraries (Section 16.2).

There is rarely one right solution to the problem of designing the user interface for a particular application, and prototyping can be used to try out different interface designs (Section 16.3).

The sequence diagrams developed in Chapter 8 can be extended to include the new user interface classes (Section 16.4). As we add more classes to the sequence diagrams, we need ways of simplifying sequence diagrams to prevent them becoming complex and difficult to read. Sequence diagrams are used to model on-screen dialogues and reports.

The new interface classes can be represented in the class diagram and we introduce an inheritance hierarchy to represent the structure of the interface classes and their associations (Section 16.5). The UML notation for stereotypes can be used to provide a high-level classification of classes, and the stereotype of «user interface» can be added to the class diagram together with the stereotype icon used originally by Jacobson et al. (1992) for user interface classes.

Some design patterns apply to the user interface. Many systems written in Smalltalk use the Model-View-Controller (MVC) architecture, which separates the model (domain or business classes) from classes that handle the interaction between user and system:

- view classes that present information to the user, and
- controller classes which receive events from the user such as mouse clicks and keypresses and dispatch them to the relevant model or view classes.

In Section 16.6 we examine the MVC architecture in terms of patterns.

The dynamic behaviour of the user interface is modelled with statechart diagrams (Section 16.7). UML statecharts were used in Chapter 10 to model the response of objects to events that take place during their lifetimes and to show how they change state as time passes. The same notation can be used to show the state of the user interface and how it responds to events such as mouse clicks on buttons or the entry of text into data entry screens.

16.2 The Presentation Layer

In Chapters 11 and 12 a layered model of the system was introduced. The three-tier architecture is a common way to separate out user interface classes from the business and application logic classes and from mechanisms for data storage. There are a number of reasons for doing this, and these are shown in Figure 16.1.

This is not to say that classes should contain no means of displaying their contents to the outside world. It is common practice to include in each class a print() method that can be used to test the classes before the presentation layer has been developed. Such methods typically take an output stream as a parameter and produce a string representation of their attributes on that stream. This enables the programmer to check the results of operations

Logical design	The project team may be producing analysis and design models which are independent of the hardware and software environment in which they are to be implemented. For this reason, the business classes which provide the functionality of the application will not include details of how they will be displayed.
Interface independence	Even if display methods could be added to classes in the application, it would not make sense to do so. Object instances of any one class will be used in many different use cases: sometimes their attributes will be displayed on screen, sometimes on a printer. There will not necessarily be any standard layout of the attributes that can be built into the class definition, so presentation of the attributes is usually handled by another class.
Reuse	One of the aims is to produce classes that can be reused in different applications. For this to be possible, the classes should not be tied to a particular implementation environment or to a particular way of displaying the attributes of their instances.

Figure 16.1 *Reasons for separating business and user interface classes.*

carried out by classes without needing the full system in place. (See Chapter 18 for more on testing.)

Coad and Yourdon (1991) call the presentation layer the Human Interaction Component and keep it separate from what they call the Problem Domain Component which contains what we have called the business classes. Larman (1998) uses the three-tier architecture (Schulte, 1995) that was discussed in Chapter 12. Developers using Smalltalk to implement systems have for many years adopted a similar approach using the Model-View-Controller (MVC) approach that was described in Chapter 12. In the (MVC) approach a system is divided into three components:

• Model—the classes that provide the functionality of the system;
• View—the classes that provide the display to the user;
• Controller—the classes that handle the input from the user and send messages to the other two components to tell them what operations to carry out.

Whatever approach is chosen in a particular project, all these approaches share the objective of keeping the behaviour of the interface separate from the behaviour of the classes that provide the main functionality of the system. To use the anthropomorphic style of some authors about object-oriented systems, the business classes 'know' nothing about how they will be displayed. We shall return to this later in the chapter when we consider how the objects in the presentation layer communicate with the instances of the business classes.

Without knowing in detail how this communication is going to work, the design of the presentation layer can begin. The use case for Check Campaign Budget is used as an example. At the simplest level, there will be an object that provides an interface onto the functionality of this use case. This could be a dialogue window like the one shown in Figure 16.2 (in the foreground). This can be added into the sequence diagram. The messages that previously came from the user can now be shown in the sequence diagram as originating from this dialogue window. (Some CASE tools have the convention of showing objects that are part of the interface to the left of a boundary line in sequence diagrams to clearly distinguish them from business objects.)

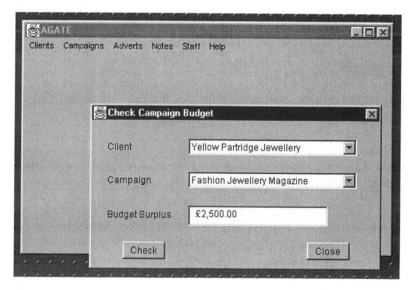

Figure 16.2 *Dialogue window for the use case* Check Campaign Budget.

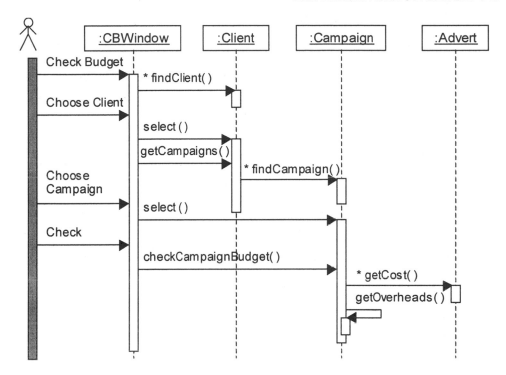

Figure 16.3 *Sequence diagram for* Check Campaign Budget *including user interface.*

Figure 16.3 shows the sequence diagram with this new object (:CBWindow) added to it. In producing this diagram, the designer does not need to worry about the individual objects that make up the CBWindow; it can be treated as a single object. In reality, this window may well be an instance of a subclass of a class such as Dialog that is available in a library of user interface classes, and it may contain a number of components: buttons, labels, listboxes (Choice is the Java term for a listbox) and a textbox. This can be shown in a class diagram, as in Figure 16.4. The composition associations represent the fact that the

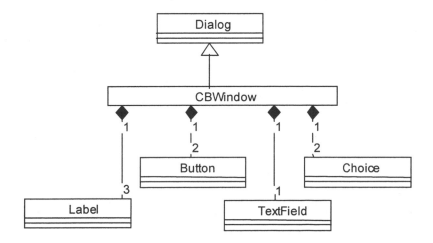

Figure 16.4 *Class diagram showing dialogue components.*

CBWindow is made up of instances of the other classes. (Alternatively, this can be represented as a class with attributes for each of the components. This is shown in Figure 16.5, and makes it easier to draw a class diagram for the user interface classes.) The component classes that are used here all come from the Java Abstract Windowing Toolkit (AWT). The CBWindow class is dependent on the classes in the AWT, and this can be shown using packages in a class diagram, as in Figure 16.6.

Figure 16.5 *Class for dialogue window showing dialogue components as attributes.*

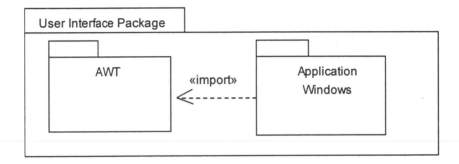

Figure 16.6 *Class diagram to show dependency between classes in different packages.*

The «import» stereotype on the dependency shows that it will be necessary to import the classes from this AWT package in order to make them available to be used with the classes in the Application Windows package. How this is done will depend on the language that is used for implementation. In Java it is done simply with a line of code:

```
import java.awt.*
```

Figure 16.7 illustrates how classes from other packages can be shown in the class diagram by adding the pathname of the package, followed by two colons, to the name of the class. The Java AWT is only one example of this. Most object-oriented programming languages or development environments are provided with *class libraries* which contain many of the classes that are needed to build a working system. Microsoft, for example, provides the Microsoft® Foundation Classes (MFC) which include all the classes such as buttons and text fields that are required to build a Windows interface to an application. These classes are grouped together and provided in what UML terms packages. This is an example of the reuse that is claimed as a benefit of object-oriented systems. These user interface classes, whether in Java or Visual C++® or another language, have been implemented by other developers and can be reused in many new applications.

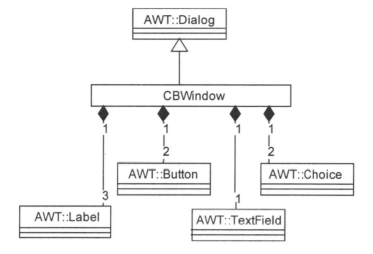

Figure 16.7 *Class diagram showing AWT components with their package name.*

If the application being designed is mainly concerned with the behaviour of the objects in the interface itself, for example a drawing package, a CASE Tool or an application with a strong visual element, then it may be advisable to model the user interface using a class diagram as shown in Figure 16.7. In most applications in which the user interface will display text and numbers, it is not necessary to produce a model of the classes which make up the interface. The simplest way to design the interface is to build prototypes of the interface using a prototyping tool or a visual programming environment

16.3 Prototyping the User Interface

Prototyping is a technique that can be used to produce a model of the user interface. The models we have produced up to now have been analysis and design diagrams; they are rather like an architect's drawings—they represent what the finished product will be like but they do not really show how it will look. A prototype is a model that looks, and to some extent behaves like the finished product, but is lacking in certain features; it is more like an architect's scale model of a new building.

There are different kinds of prototype that can be built. In Chapter 3, prototyping approaches were discussed in much more detail in relation to the different life cycles that can be adopted for a project. A prototype that only provides a model of the user interface is one example of a *horizontal prototype*. It is horizontal because it deals with one layer of the layered architecture of the system (see Figure 16.8). A *vertical prototype* takes one subsystem of the whole system and develops it through each layer: user interface, business classes, application logic and data storage. A horizontal prototype need not only deal with the user interface; there may be circumstances where it is more important to prototype the middle tier in order to test whether an innovative aspect of the system functionality works correctly.

Another distinction is made between those prototypes that are developed further and eventually by an iterative process become part of the finished system, and those prototypes that are simply used to test out design ideas and then thrown away after they have served their purpose. This second kind of prototype is known as a *throwaway prototype*. A

Figure 16.8 *The three-tier architecture showing where the interface prototype fits in.*

throwaway prototype can be built using any programming language that is suitable for the purpose. Figure 16.9 shows a prototype of the user interface for Agate that was created in a matter of minutes using Microsoft Visual Basic. Although this application will be developed in Java, a visual programming environment such as Visual Basic or Delphi can be used to produce prototypes of the user interface. In fact, the Java examples used in the book take no more time than the Visual Basic examples, as they are also produced in a visual programming environment—Symantec Visual Café™ for Java.

Visual programming environments can be used to develop prototypes of the user interface to applications. These can be shown to the users, and used, for example, in conjunction with techniques from scenario-based design (described in Chapter 15) to agree with users how they will interact with the user interface.

Figure 16.9 *Visual Basic prototype of the* Check Campaign Budget *interface.*

Figure 16.10 *Prototype developed in Visual Basic showing separate look-up window.*

Because visual programming environments are so easy to use, developers are often tempted to develop applications from the outside in: starting with the interface and linking the functionality to the visual components that appear on the screen. Without thorough analysis of the business requirements for the system, this can lead to a blurring of the distinction between the presentation layer and the business classes and application logic. Applications developed in this way often have a large amount of program code associated with interface objects such as buttons. This program code should be an operation of a controller or of one or more business classes. If it is linked to a button it cannot be reused in the same way as if it is carefully encapsulated in a class. Typically, the programmer then needs to reuse the code in another window and copies and pastes the code to another button in the new window, then when a change is made to the code linked to one button it may not be copied to the code linked to the second button, and discrepancies creep into the system. This is not to say that it is not possible to develop good applications in visual environments; the important thing is that a thorough analysis should have been carried out first and the business objects and application logic should be kept separate from the visual components.

Prototyping can be used to try out alternative approaches to the same use case. In the example screen layout shown in Figure 16.9 we have assumed that the users will select first a client and then a campaign from a listbox. There are many possible alternatives to this. Three of these are:

- to use a separate look-up window for each class;
- to allow the user to enter part of a name (for example of a client) and for a list of close matches to be returned; and
- to use a tree structure which shows the instances of clients and campaigns in a tree-like hierarchy.

Prototyping allows us to experiment with these approaches and build models which the users can try out for themselves. Figures 16.10 and 16.11 show screen shots of prototypes based on two of these different ways of handling the look-up process.

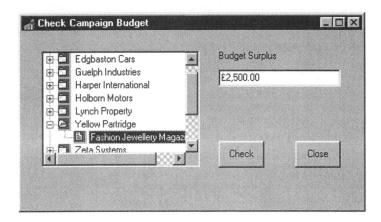

Figure 16.11 *Prototype developed in Delphi using TreeView control.*

The choice of how the look-up in this use case is handled on screen will be determined by the style guidelines that were discussed in Chapter 15. It is important that style guidelines are agreed before development starts; prototyping can be used during design to try out various different interface styles and to get the users' agreement on which style will be followed.

Whichever style is adopted here will be adopted in other use cases in which the user needs to be able to look up clients from a list. The same approach will also be adopted in all use cases in which the user looks up any kind of class. For example, Figure 16.12 shows the same method as in Figure 16.10 being used to look up campaigns for a particular client in the use case for Add a Concept Note.

Figure 16.12 *Use of the same style of look-up as in Figure 16.10 in a different use case.*

16.4 Updating the Sequence Diagram

Using additional windows in this way requires changes to the sequence diagram. Figure 16.13 shows a sequence diagram for Check Campaign Budget with two new interface classes: CLWindow a client look-up window and CAWindow a campaign look-up window. There are a number of features of this sequence diagram worth noting.

- New instances of CLLookup and CALookup are created to handle the look-up of clients and campaigns.
- These instances are destroyed once they are no longer needed.
- Both look-up dialogue boxes are modal. This means that the user cannot do anything with the main window until he or she has closed the look-up window.
- In using this design approach the responsibility for finding a list of clients belongs to the CLLookup window rather than the main window. Similarly, the responsibility for finding a list of campaigns for a particular client now belongs to the CALookup window, and a reference to the client must be passed to the instance of CALookup as a parameter.
- We have used a select() operation as a shorthand for whatever operation or operations are necessary to bring a particular instance into memory. This could be handled by a collection class (Chapter 13) or by a database broker (Chapter 17).

However, the diagram is now becoming more cluttered. We can simplify it by using the *probe* notation of Jacobson et al. (1992). The associated use case can then be shown in a separate sequence diagram, and this makes it possible to keep the diagrams simpler. Figure 16.14 shows part of the sequence diagram in Figure 16.13 with the probe notation. Figure

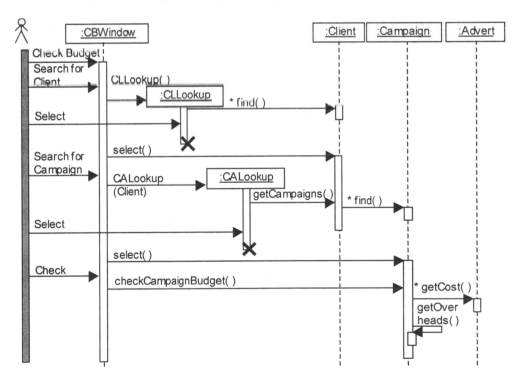

Figure 16.13 *Sequence diagram revised to show interface objects to handle look-up.*

16.15 shows the separate sequence diagram for Search for Client. Figure 16.16 shows how this is reflected in the use case diagram. However, this notation is not part of UML, even though it can be useful. An alternative was suggested in Chapter 8.

Figure 16.14 *Simplified sequence diagram showing probe notation.*

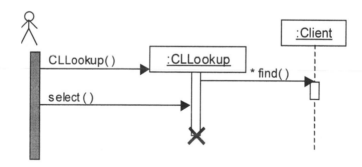

Figure 16.15 *Sequence diagram for* Search for Client.

It is possible to achieve the same effect using UML notation by choosing not to show the detail of the interaction which occurs after the look-up windows have been instantiated. The main diagram includes the message to create each window (see Figure 16.17), but the detail is shown in a separate diagram as in Figure 16.15 for the probe notation. In this way, we can reuse the use case description and sequence diagram for all the use cases that handle look-up throughout the system. The advantage of the probe notation used by Jacobson over this UML notation is that it is not necessary to show the window objects that receive the messages in the sequence diagram, and this helps to keep it simpler and clearer.

Interface objects include reports as well as screen displays, and report objects can be shown in sequence diagrams as well. The simplest form of report is produced by opening an output stream to a device such as a printer. In this case each object that is to be printed

Figure 16.16 *Use case diagram reflecting use of probe.*

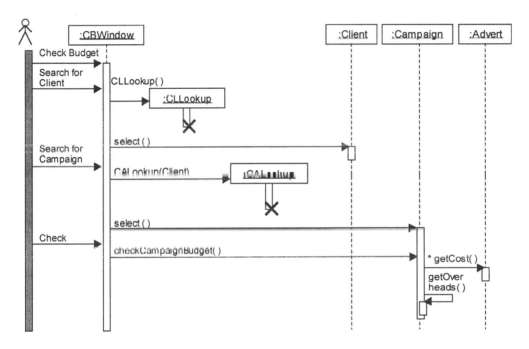

Figure 16.17 *Sequence diagram revised to hide detail of look-up.*

out can be sent the message to print itself with a reference to the output stream as a parameter. This is shown in Figure 16.18 for a simple report of all clients.

There are two alternatives to this approach. The first is to create a class for each type of report that is required and to give these classes the necessary operations to obtain the data that they require from instances of other classes. The second is to leave the reports with relatively limited functionality (perhaps just the ability to format the output) and to introduce separate control classes that handle the task of obtaining the data for the report from other objects. The decision as to which of these approaches is best will depend on the development environment being used and factors such as whether any dialogue is required before the report is printed. If a dialogue on the screen is required, then the dialogue interface object can be allowed to handle the control of the report; if there is no dialogue on the screen, then either the report object will have to do more work in obtaining the data or a

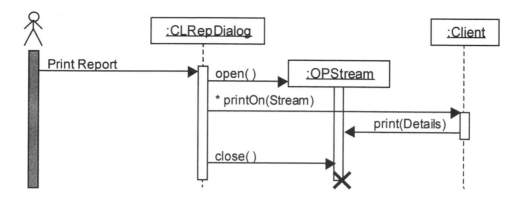

Figure 16.18 *Report shown as an output stream.*

control class will have to be designed and an instance of that class will control the report when it is run. Figure 16.19 shows a sequence diagram that uses a control object for the report of all clients.

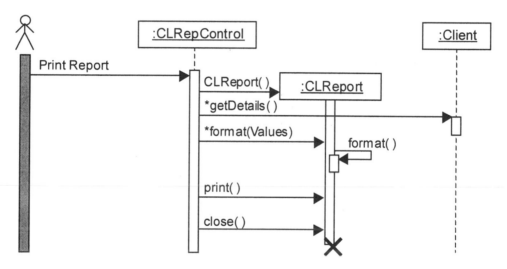

Figure 16.19 *More complex version of* Report of All Clients *using a control object.*

Control objects can be used in any use case. They separate the control of the use case from the presentation. Control objects are not always necessary. In Java applets, the Applet class can handle both the presentation and the control. In some languages, and in applications where there are many different windows that could be displayed, it may be better to adopt a design approach that separates control from presentation in the same way as presentation is separated from the business classes or problem domain component. This strategy has its origins in the Smalltalk Model-View-Controller style of developing applications, and is discussed in Section 16.6.

Collaboration diagrams can be used instead of sequence diagrams to model the interaction between interface classes and the business classes. Figure 16.20 shows an example equivalent to the relatively simple sequence diagram of Figure 16.18.

Figure 16.20 *Collaboration diagram for* `Report of All Clients`.

Review Questions

16.1 Why should the user interface classes be kept separate from the business classes and application logic?

16.2 Explain the difference between vertical and horizontal prototyping.

16.3 What is meant by a throwaway prototype?

16.5 Extending the Class Diagram

The interface classes can be added to the class diagram. These can be shown in a single diagram for the Interface package. The buttons and other classes that are used to make these interface classes need not be shown in this diagram. However, all the interface classes in the application will have a package dependency to the package where the buttons and other classes are held. Figure 16.21 shows the various dialogue classes that have been added in the sequence diagrams shown in the last section.

From the classes for the campaign and client look-up dialogues we can generalize to a `LookupDialog` class which will be the parent of all look-up dialogues in the system. This is in turn a sub-class of the `Dialog` class from the Java AWT package. The `Dialog` class is also the superclass of `ReportDialog`, the parent of all dialogues used to run reports, and of `DataWindow` which will be the parent of all windows used for data entry.

UML notation allows classes to be marked with stereotypes. A stereotype is a high-level classification of different types of classes. For example, Jacobson et al. (1992) classify all classes as interface classes (user interface), control classes and entity classes. A stereotype is displayed in the top compartment of the class in the class diagram. It can either be displayed as a word or words in guillemets (angle brackets) thus «user interface», or as an icon. Figure 16.22 shows a user interface class in this way. You may recognize the notation from the «uses» and «extends» notation in use case diagrams. In use case diagrams, the stereotypes classify the two different kinds of associations between use cases.

Figure 16.21 *Interface classes for Agate system based on Figure 16.13 sequence diagram.*

Figure 16.22 *User interface class showing stereotype and stereotype icon.*

Unless you are following Jacobson's approach to drawing classes, it is not necessary to use this approach, but you may see it in diagrams.

In the example here, we have proposed that there should be a separate dialogue for each class that we may want to look up. This is only one design approach. It is possible instead to have a single look-up dialogue box and for the window that invokes it to pass all the necessary data to it for it to display and for the user to select from. This simplifies the system in one way, as it means that there are fewer interface classes. However, it moves the responsibility for building a list of names of whatever is being looked up—clients, campaigns, adverts etc.—from the specialized look-up dialogue box to the class that invokes it. The operation to do this will then be replicated in a number of places in the system and we run the risk of different versions getting out of step with one another. The solution to this is to take the operation to look up each class and encapsulate it in a specialized interface class to handle the look-up of that business class ... which brings us back to the solution we started with of a look-up dialogue for each class, as in Figure 16.21.

16.6 User Interface Design Patterns

The Model-View-Controller architecture has been mentioned more than once in this chapter and was described in detail in Chapter 12. Jacobson's three types of class—entity, interface and control—reflect the MVC architecture. The Model class is the business object, the View is its presentation to the user and the Controller defines how the user interface responds to the user's actions. In Figure 16.19 a controller class was introduced into the sequence diagram to separate the control of the use case from the presentation. Figure 16.19 is a simple example of the MVC architecture. The class CLRepControl handles the control of the use case: it creates an instance of the CLReport class, it obtains data from the instances of the business class Client and it passes that data to the CLReport object which is responsible for formatting it appropriately and presenting it to the user. The control class synchronizes the behaviour of the other two types of class. The use of this architecture in Smalltalk systems is more sophisticated. Model, View and Controller are all available as classes from which developers can subclass the classes they need for their applications. These subclasses then inherit methods which provide for communication between the three components. When the controller detects interaction from the user it sends a message that is received by the relevant model class and tells it to modify itself; when a model class changes, the change is notified to the controller and the relevant view classes; the views then check with the model class to find out what has changed and update the data that is presented to the user. This sequence of interaction is shown in Figure 16.23 as a collaboration diagram. (Compare this with Figure 12.12.)

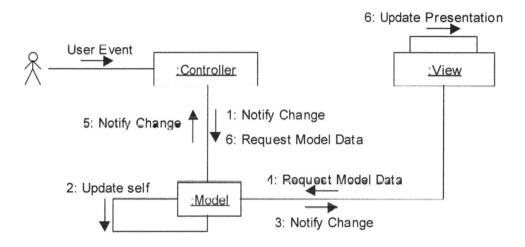

Figure 16.23 *Model-View-Controller response to an external event.*

In the Smalltalk implementation of MVC, each subclass of the Model class has an attribute which refers to a Controller class. For each instance of a Model subclass, this attribute can refer to an instance of a subclass of Controller, and that subclass can be any subclass designed by the developers of an application. When an instance of a particular Model subclass is created, it can be given a reference to the Controller subclass which it must notify whenever any of its data changes. Similarly the Controller subclass is told which View or Views it must notify when a Model subclass changes. Once the mechanism by which this operates is understood, and the links between the instances of the different classes are set up, this is relatively transparent in operation.

Java provides two similar mechanisms: firstly the Observer interface and Observable class, and secondly the EventListener interface and Event class. The latter is used in the user interface and there are a number of subclasses of Event for different types of event, such as mouse events, and different subinterfaces of EventListener to handle them. We shall illustrate this by means of the ActionListener interface. In Java the term interface does not mean the user interface but a definition of a set of methods that a class must provide if its class definition states that it implements that particular interface (see Chapter 7). The use of interfaces is a way of avoiding the problems of multiple inheritance. In Java a class can only inherit from a single superclass. However, its behaviour can be extended through the use of interfaces.

Any class that implements the `ActionListener` interface must implement the method `actionPerformed()`; this means that it must include code so that its instances can respond appropriately to the message `actionPerformed(ActionEvent)` whenever they receive it. Any class that implements `ActionListener` can register its interest in the events that affect an instance of certain user interface components, such as the `Button`, the `Menu` and the `TextField` class. This is done by sending the message: `AddActionListener(ActionListener)` to an instance of one of these classes. It adds the new instance of `ActionListener` to a list it keeps of all the `ActionListeners` it must notify of any events. Then when the Button is clicked or the `TextField` is changed, it sends the message `ActionPerformed()` to each of the objects in its list and passes them an `ActionEvent` object as a parameter. Each of the objects that receives the message can then inspect the `ActionEvent` object and decide whether it is interested in that particular event and whether it needs to take some action in response. The action could be to notify other classes of the event, particularly if the class that is implementing the `ActionListener` interface is a control class. This is shown in Figure 16.24. The use of the interface mechanism in Java means that the user interface component does not need to know the actual class of the class that implements `ActionListener`; as long as it implements the interface it will know what to do when it is sent the `ActionPerformed()` message.

Gamma et al. (1995) describe the MVC architecture in terms of three patterns: the Observer, the Composite and the Strategy design patterns.

• The Observer pattern provides a mechanism for decoupling business objects from their views. It is based on a publish-subscribe model. There are two main types of objects: Observers which are the View objects and Subjects which are the business objects. Observers can subscribe to a subject and ask to be notified of any changes to the subject. When a change takes place in the subject, it publishes information about the changes to

Figure 16.24 *Java* `ActionListener` *response to an external event.*

all the observers that have subscribed to it. This is the basis of the Observer–Observable approach adopted in Java 1.1 (and later versions of Java). The Observer pattern is the core of the MVC architecture, but two other patterns also apply.

- The Composite pattern provides a means to structure objects into whole-part hierarchies. Most windowing interfaces use this approach to form composite views made up of smaller components. The class diagram of Figure 16.7 shows this kind of composite structure. The Composite pattern is a way of structuring views to include nested views. For example, a graphical display may represent the same model in two different views, a graph and a table of numbers that may be nested within another view.
- The Strategy pattern offers a way of defining a family of algorithms, encapsulating each one, and making them interchangeable. This allows the strategy to vary independently of client objects that use it. In MVC, one controller can be replaced by another. For example, an on-screen calculator could be designed and built initially to respond to mouse clicks on the buttons on the screen. The controller for the calculator detects mouse clicks and notifies the internal application objects and the view objects about these mouse events. Later the calculator could be modified to respond instead to keypresses on the computer numeric keypad, or to both mouse clicks and keypresses. The controller object can be replaced without having any impact on the application objects.

Design patterns provide a way of reusing design experience and best practice. Separating the concerns of the user interface and the control of interaction from the business application objects is a proven technique to produce good designs and maximize the opportunities for reuse of the classes that are designed and developed.

16.7 Modelling the Dynamic Behaviour of the Interface

In Figure 16.7 we showed the classes that make up the user interface for the use case Check Campaign Budget in the Agate project. That diagram shows the static structure of the interface, but it provides no information about how it will behave in response to the user. The sequence diagram in Figure 16.3 and the prototype in Figure 16.9 provide additional information, but they tell us nothing about the permitted states of the interface. The sequence diagram shows the sequential view of the user working through the fields on the screen from top to bottom, but it is in the nature of GUI interfaces that the user can click on interface objects out of sequence. So what happens if the user clicks on the Check button before a client and a campaign have been selected? The user may choose to check more than one budget. What happens if they select a different client—how does that affect the other fields where data has already been selected? All these issues can be modelled using a statechart diagram. Statecharts were introduced in Chapter 10, and were used there to model the way that events affect instances of a class over its lifetime. They can also be used to model the short-term effects of events in the user interface. Browne (1994) uses statecharts in this way to model the user interface as part of the STUDIO methodology.

For the example that follows, we are using the original design for the user interface with listboxes for Client and Campaign, as in Figure 16.9.

In the Check Campaign Budget use case, we may decide to prevent the user making mistakes by disabling the Campaign listbox until a client has been selected and disabling the Check button until both a client and a campaign have been selected. We can build up

Figure 16.25 *Initial statechart for Client listbox.*

the statechart diagrams for different interface components step by step to show this. Figure 16.25 shows the initial two possible states for the Client listbox. (Initially, we are using English descriptions of events, but as the design progresses, it will be possible to relate these to specific operations of the objects involved. In later diagrams, we use a Java style syntax for the operations.)

Once a client has been selected, the listbox for campaigns can be enabled. It is also necessary first to find all the campaigns for that particular client and load their names into the listbox. For this to work, we need to consider two things: firstly, when a client is selected a message needs to be sent to the Campaign listbox, and secondly, while the list of campaigns is loading, we may want to change the cursor to give an indication of the need for the user to wait. Initially the Campaign listbox is disabled, then it is loading data, then it is enabled and ready for a campaign to be selected. Figure 16.26 shows a first attempt at the statechart for the Campaign listbox. Note that some events result in messages being sent to other objects (for example ^checkButton.enable() to enable the Check button so that the user can check the budget.

Figure 16.26 *Initial statechart for Campaign listbox.*

This statechart highlights two things: firstly, we need to model the state of the cursor; secondly, we should perhaps have included a Loading state in the statechart for the Client listbox. We can add the necessary state to the Client listbox statechart later. The cursor statechart seems quite simple and is shown in Figure 16.27.

The statechart for cursor shows how the cursor can move from one state to the other and back again. It raises the question of whether this is the case for other statecharts. There is

Figure 16.27 *Statechart for cursor.*

no reason to disable the Client listbox, but if the user selects a different client, the list of campaigns must be updated and reloaded. This means that the event `Select Client from List` can happen again when the listbox is in the `Client Selected` state, and it will need to send a message to the Campaign listbox. The Campaign listbox will then go back into the `Loading` state. Figures 16.28 and 16.29 show the changes to these two statecharts. Figure 16.28 shows the following additions:

- `Loading` state added and cursor changed in line with Campaign listbox;
- message `clientSelected(aClient)` sent from the Client listbox to the Campaign listbox;
- the event `Select Client from List` permitted if the listbox is already in the `Client Selected` state to allow the client to be changed.

Figure 16.29 shows the addition of the loop back to the `Loading` state.

In the same way as the Campaign listbox is disabled until a client has been selected, the Check button is disabled until both a client and a campaign have been selected. It also needs to be disabled if the client is changed until a new campaign has been selected. Figure 16.30 shows the first attempt at the statechart for the Check button.

However, the whole point of a button is that the user presses it (or clicks on it with the mouse) and something happens. When the button is pressed in the Check Campaign Budget dialogue box, then the work of the use case gets done. At this point the message `checkCampaignBudget()` needs to be sent to the selected instance of `Campaign` in order to get a result back. However, it is not enough just to send the message and return the button to the `Enabled` state; the cursor should be changed while the calculations take

Figure 16.28 *Revised statechart for Client listbox.*

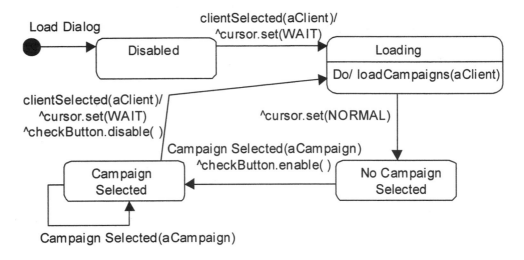

Figure 16.29 *Revised statechart for Campaign listbox.*

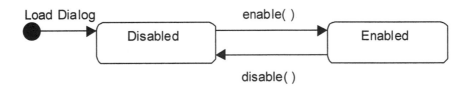

Figure 16.30 *Initial statechart for Check button.*

place, and something in the interface has to set the returned value into the textbox for display. Remember the business classes know nothing about the interface, so the `Campaign` instance will not set the value into the textbox itself. To achieve this we need a further state for the button: it needs to send the message to the instance of `Campaign` and then set the returned value into the Budget textbox. Figure 16.31 shows this revised version of the statechart for the Check button using a Java style syntax.

At this point, the messages that are being sent have become more specific, and some can be recognized from the sequence diagram, for example `checkCampaignBudget()`. However, the `checkButton` object will not know which `Campaign` instance has been selected; it needs to get the selected `Campaign` from the `campaignTextbox`. Note that it is not really the button that will carry out the operation to send the message to the `Campaign`. This will almost certainly be done by a controller object. When the button is pressed, an event occurs and the controller recognizes the event and acts accordingly. For now, however, we can focus on the state of objects in the interface and show the relevant messages that will be sent to business objects in the application logic layer of the three-tier model. Showing these messages helps us to understand the interaction between the interface and the rest of the objects in this use case.

A further addition needs to be made: we need a statechart for the textbox. The textbox will be cleared whenever the client is reselected or the campaign is reselected. Having recognized that the message is going to the selected `Campaign` object to check the budget,

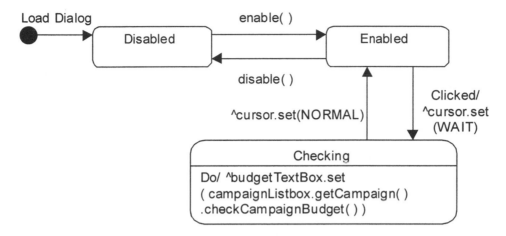

Figure 16.31 *Revised statechart for Check button.*

we should recognize that there are other messages going from the interface to the business objects in the sequence diagram, but that these need not be included in the statecharts. Figure 16.32 shows the statechart for the textbox.

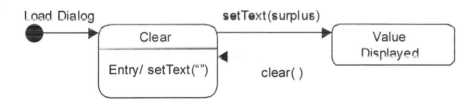

Figure 16.32 *Initial statechart for Budget textbox.*

Finally, we need to recognize that all these states of objects in the interface are part of a composite object. Combined together they represent the state of the whole interface—the Check Campaign Budget Dialogue. All the statecharts that have been developed over the last few pages can be shown in a single diagram. Figures 16.33(a) and (b) show these statecharts for all the different components.

Where messages are sent from one object to another, it is possible to show this in statechart diagrams by means of dashed arrows from the transition that results in the sending of the message to the object boundary of the object that is the target of the message. This would make the diagrams of Figures 16.33(a) and (b) very cluttered, so we have not included this part of the UML notation.

You might want to add a statechart for the dialogue as a whole and for the Close button, so that events that close the dialogue such as an onClick event for the closeButton can be modelled. The user could potentially exit from the dialogue by clicking on the Close button in the dialogue or by closing the window. It may be necessary to consider whether it should be possible to close the window while one of the operations loadClients(), loadCampaigns() or Campaign.checkCampaignBudget() is taking place. If the programming language to be used allows multiple threads of control, then these can all be run on separate threads. If the Close button is pressed, a message will have to be sent to

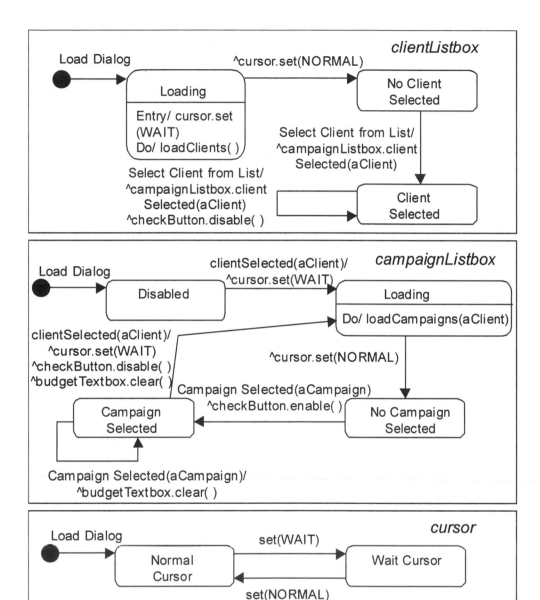

Figure 16.33(a) *Statecharts for components of the* Check Campaign Budget *dialogue.*

stop the separate thread running. If the language does not allow this, then the user will have to wait until the action is completed and the cursor changes back to normal before closing the window.

It is worth commenting that this diagram (Figures 16.33(a) and (b)) is relatively complicated for a simple interface with only five active components. For more complex interfaces, it may not be so easy to combine all the separate statecharts into a single diagram. However, having a clear specification of the behaviour of the interface is an important part of the design process.

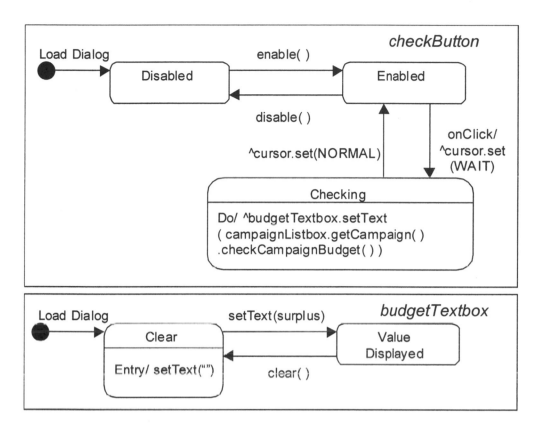

Figure 16.33(b) *Statecharts for components of the* Check Campaign Budget *dialogue.*

16.8 Summary

Designing the interface objects requires us to model three important aspects of the interface. Firstly, we need to determine the classes of the objects that will participate in the interaction with the user and decide on how we will reuse interface classes that are available in class libraries. The choice of interface objects will depend on the style guidelines that have been adopted for the system. These classes are shown in class and package diagrams. Secondly, we need to model the interaction with the user in sequence or collaboration diagrams. The way that the interaction is modelled will depend on the architecture that has been chosen for the system. The three-tier architecture and the Model-View-Controller architecture separate the interface objects from the business and application logic objects using well-defined methods. Thirdly, we need to model the state of the interface to ensure that we understand how the interface will respond to events and what sequences of events are permitted. We use statechart diagrams to do this.

While carrying out these three modelling tasks, we can draw on design patterns to inform the way in which we select the interface classes and design the interaction between them and the business classes. Prototyping can be used to build models of the interface and test how it will work. Users can be involved in this process to ensure that the interface meets their requirements and to validate the analysts' understanding of their requirements for how tasks should be carried out.

Review Questions

16.4   What role does each element of the Model-View-Controller architecture play?
16.5   What else do we use statechart diagrams for, apart from modelling the state of interface objects?
16.6   Convert the collaboration diagram of Figure 16.21 into a sequence diagram.
16.7   Convert the collaboration diagram of Figure 16.22 into a sequence diagram.
16.8   What are the differences between the MVC and Java `EventListener` approaches?

Case Study Work, Exercises and Projects

16.A   Decide how you will handle the interaction between the user and the system for the use case `Record Problem on Line` for FoodCo. Draw a sequence diagram to include the interface objects that are needed.
16.B   Draw a class diagram to show the classes that are used in your sequence diagram.
16.C   Extend your class diagram from Exercise 16.5 to show the superclasses of the interface classes.
16.D   If you are familiar with a class library such as the Java AWT, Java Swing™, Microsoft Foundation Classes or Borland Object Windows Library, then try to determine how your interface classes relate to classes in that class library.
16.E   Produce a prototype for the use case `Record Problem on Line` using a language or visual programming environment with which you are familiar.
16.F   Draw a statechart diagram for the interface to the use case `Record Problem on Line` to model the behaviour of your prototype developed in Exercise 16.E.

Further Reading

The MVC architecture is explained in a number of books, particularly books which use Smalltalk as a programming language. Gamma et al. (1995) specifically explain it in terms of patterns.

The Java EventListener event model was introduced into Java in Version 1.1. (Version 1.0 used a less efficient model.) There are a number of these EventListener interfaces for different kinds of events. Most introductory Java programming books explain them.

Few books on object-oriented analysis and design provide much detail on statecharts, and those that do often provide simple models of telephone systems as examples rather than user interfaces. Browne (1994) is one of the few authors who seriously applies statechart diagrams to user interface design. Although he does not use the UML notation, his diagrams in Chapter 3 are similar enough to provide a clear idea of how they can be used to model the detail of user interface interaction.

17

Data Management Design

Objectives
In this chapter you will learn:
- the different ways of storing persistent objects;
- the differences between object and relational databases;
- how to show data management objects in the class diagram;
- how to extend sequence diagrams to include data management objects.

17.1 Introduction

Real information systems require persistent data: data that continues to exist even when the system is not active. In this chapter we discuss how persistent data can be modelled using UML. Persistent data can be stored in files or in databases (Section 17.2). Using a database management system (DBMS) offers a number of advantages over files. The use of a relational DBMS or an object DBMS will affect the design of the data management.

Many object-oriented systems are constrained by organizations' existing investments in hardware and software and have to use relational database management systems (DBMS) to store data (Section 17.3). *Normalization* can be used to design tables for a relational database, and these can be modelled in a storage object model.

Designing for an object DBMS will have a different impact on the design model (Section 17.4). Depending on the object DBMS, making objects in the class diagram persistent can be straightforward. CORBA (Common Object Request Broker Architecture) can be used to separate the persistent objects from the interface objects running on a client machine.

Database objects can be reflected in the UML models that are produced in design (Section 17.5). ODBC (Open Database Connectivity) and the specific Java implementation of this—JDBC (Java Database Connectivity)—can be used to link an object-oriented system to a relational database. Design patterns may be used to show how persistence can be designed into a system using the experience of many previous developers.

17.2 Storage Mechanisms—an Overview

17.2.1 The requirement for persistence

For some applications the data that is created or used while the application is running is not required when the application terminates. This applies mainly to simple applications: an example would be the on-screen calculators provided with GUI operating systems. Such data is called *transient data*.

Most applications, however, need to store data between one execution of the program and the next. In some cases, the data that is stored is secondary to the operation of the application. When you use a browser one of the first things that happens as it loads is that it reads data from one or more files that describes the user's preferences and records the last web sites visited. The ability to store user settings and to configure applications in this way is an important factor in their usability, but it is not their primary purpose.

In the case of information systems, storing data is a primary requirement. Businesses and other organizations rely on their information systems to record data about other organizations, people, physical objects and business events and transactions. The data entered into such a system today will be required in the future; operations being carried out on data today rely on data that was stored in the past. Computerized information systems have replaced systems based on paper in files and ledgers, and must provide the same relatively permanent storage that is provided by paper-based systems. In most organizations, it is also important that data can be shared between different users. Data in the memory of a particular computer is not normally accessible to multiple users. It must be written away to some kind of shared data storage system so that other users can retrieve it when they require access to it.

This is what we mean by *persistent data*. It is data that must be stored in a secondary data storage system, not just in computer memory, that must be stored after the program that creates or amends it stops running, and that usually must be available to other users. Information systems also use transient data: for example, the results of calculations or lists of objects that are required for a particular purpose such as printing a report, but that are not required permanently and can be destroyed after they have been used.

In an object-oriented system, we are concerned with both *persistent objects* and *transient objects*. Persistent objects are those which must be stored using some kind of storage mechanism, while transient objects will be erased from memory after they have been used.

17.2.2 File systems

The simplest means of persistent storage available for computerized information systems uses files. Most users of personal computers are familiar with the idea of files. Word-processors and spreadsheets store their data in files; browsers download them from web sites where they are stored. At the simplest level, a file is a stream of bytes of data stored on some physical medium. In most cases, files are stored magnetically by adjusting the magnetic properties of the surface layer of a disk or tape. Files can also be stored optically on CD-ROMs and other optical storage systems, and electronically in special kinds of memory, such as the flash memory used in palmtops. However, the user is normally shielded from the physical implementation of the file system by the operating system of the computer or by a programming language, which provides high-level functions to create files, store data in them and retrieve that data.

Programming languages, and in some cases operating systems, also impose a structure on files. This structure breaks a file up into individual records, each of which groups together a number of fields representing the data that is to be held in the file. In the same way as each object contains a number of attributes, each one of which holds a particular kind of data about the object, each field in a record holds a particular kind of data about whatever it is that the record describes. Records in files can take different forms, described below.

- Fixed-length—Each record is made up of a number of fields, each of which has a fixed length in bytes. If the data in a particular field does not fill it, the field is padded out with special characters (often spaces). Each record is of the same, fixed length, and it is possible to skip from the beginning of one record to the beginning of the next by jumping that fixed number of bytes ahead.
- Variable length—Each record is made up of a number of fields, each of which may have a maximum length but has a minimum length of zero bytes. Fields are usually separated or delimited by a special character that would not appear in the data. Records may also be delimited by a special character. The length of each record may also be stored at the start of the record, making it possible to skip to the beginning of the next record by jumping that variable number of bytes ahead.
- Header and detail—Records may be of two types: each transaction recorded consists of a header record, followed by a variable number of detail records. This approach can be used with many business documents, such as orders, invoices and delivery notes that have a variable number of lines on them. Each record will contain a `record type` field. The number of detail records may be held in the header so that it is possible to tell where the next header record starts.
- Tagged data—The data may have a complex structure, as in object-oriented systems, and it may even be necessary to hold objects of different classes in the same file. Every object and attribute may be tagged with some kind of description that tells a program reading the file what the type of each item is. This approach is used for data in files that use Hypertext Markup Language (HTML) or Extensible Markup Language (XML).

Some systems store information about the structure of the file in a data dictionary, and this may be held in a separate file or at the start of the data file itself. This makes it possible to write programs which can read the data out of any file that uses this format: the program first reads the data dictionary information and configures itself to read the appropriate data structures from the rest of the file.

As well as having alternative ways in which the data can be structured within files, files can have different types of organization, can be accessed in different ways and can serve different purposes in a system.

17.2.3 File organization

There are three ways in which files can be organized: serial, sequential and random.

Serial organization. Each record in the file is written onto the end of the existing records in the file. If a record is to be deleted, the file must be copied from the start up to the deleted record, which is skipped, and then the rest of the file is copied back to the disk.

Sequential organization. In the basic form of sequential organization, each record is written to the file in some pre-determined order, usually based on the value of one of the fields in each record, such as an invoice number. Records must be deleted in the same way as for serial files. Each record must be added to the file in its appropriate place, and if it is necessary to insert a record into a file, the file is copied up to the point where the record is to be inserted, the new record is written to the file and then the rest of the file is copied.

Random organization. The term random is a poor way of describing the organization of random files, as the organization is really anything but random. The records are added to the file by means of precise algorithms that allow records to be written and read directly without having to read through the rest of the file. What this means is that if you choose any record *at random*, it should be possible to access it more or less straightaway without searching through the file. The algorithm usually converts a key field of each record into an address in the file that can be reached directly.

17.2.4 File access

Depending on the file organization chosen, different ways of accessing the data in the files are available to the designer. The main ones are serial, indexed sequential and direct.

Serial access. Serial and basic sequential files can only be accessed serially. To find a particular record, it is necessary to read through the file, record by record, until the required record is located.

Indexed sequential access. Access to sequential files can be improved by maintaining an index on the field that is used to order the data in the file. Records are stored sequentially within blocks (areas of the disk that have a defined size). Enough blocks are allocated to the file for the total anticipated number of records. Records are written into blocks in key order, but the blocks are not filled up from the start, rather records are distributed evenly across the blocks, leaving space for new records in each block. Blocks are grouped together in some way. Typically, this used to be done by grouping together all the blocks in a particular cylinder on the hard disk. Hard disks typically have several disk platters within them, and each disk holds data in concentric tracks on these platters. A cylinder is the set of tracks, one track per platter, a particular number of tracks in from the outside of the disk. All the tracks in a cylinder can be read without having to move the read–write heads. It is the tracks that are divided up into blocks. This is shown in Figure 17.1. In reality, the hard disk in a PC may have a few thousand cylinders, and each track may be split into many sectors or blocks (typically 17 or 63). However, not all computer operating systems make it possible to allocate space to files in this way, so some other way of grouping blocks may be necessary.

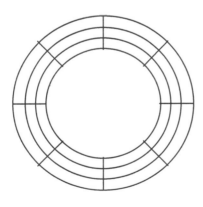

Each track is divided into blocks.

All the tracks a certain distance from the centre of each disk platter form a cylinder and can be accessed together without needing to move the read–write heads over the disk platters.

Figure 17.1 *Schematic diagram of tracks and cylinders on a hard disk.*

Access to the records in an indexed sequential file is achieved by maintaining two levels of indexes. A master index holds the value of the key field of the last record on each cylinder. A cylinder index on each cylinder holds the value of the key field of the last record on each block in that cylinder. To find a record by its key, the master index is read until a key value is found that is greater than or equal to the key of the record being sought. This makes it possible to go to the cylinder index for that record. The cylinder index is then read until a key value is found that is greater than or equal to the key of the record being sought. This makes it possible to go to the block in which the record is held. The records in the block are then read sequentially until the desired record is found. (This sounds slow, but will take a matter of milliseconds.) A similar approach is taken in order to add records to an indexed sequential file. At the last stage, the records in the block will be read into memory and copied to the disk up to the point that the new record is to be inserted, the new record will be written into the block, and then the rest of the records in the block will be copied back to the block. Eventually some blocks will fill up, and it will be necessary to write some records into special overflow blocks. The addresses of the overflow blocks will be held in the blocks that have overflowed. Figure 17.2 shows the organization of data and indexes in an indexed sequential file.

Indexed sequential files have the advantage over sequential files that records can be read and written more quickly, although there is a storage overhead associated with maintaining the indexes. Compared to direct access, which is described next, there is also the overhead of the time taken to access the indexes before the data is reached. However, indexed sequential files have the advantage that records can be accessed either sequentially or more or less directly.

Master Index

Record key	Cylinder address
Feng	1023
Patel	1024
Zarzycki	1025

Cylinder Index (Cylinder 1024)

Record key	Block address
Finlayson	64450
Gomez	64451
Hanson	64452
Jacobson	64453
Khan	64454
....	....

Data Records in blocks (for simplicity only the keys are shown)

Block address	Records				
64450	Fern	Finch	Finlayson		
64451	Finn	Firmin	Ford	Gangar	Gomez
64452	Gordon	Govan	Hamer	Hanson	
64453	Ho	Ibrahim	Jacobson		
....	....				

Figure 17.2 *Schematic of indexes and data in an indexed sequential file ordered by surname.*

Direct access. Direct access methods rely on the use of algorithms to convert the values of key fields in the records to addresses in the file. The first and simplest of these is *relative addressing*. This access method requires the use of fixed length records and successive positive integers as keys. If each record is 200 bytes long, then record 1 will start at byte 1 of the file, record 2 at byte 201, record 3 at byte 401 and so on. It is possible to calculate the position of any record in the file by subtracting 1 from its key, multiplying the result by the size of the records and adding 1. Each record can be read directly by reading from that point in the file.

Hashed addressing is the second approach. This can use keys of any form. As with indexed sequential files a fixed number of blocks is initially allocated to the file. This is usually a prime number of blocks, as this helps to achieve a more even spread of records into blocks. The key is then hashed to determine which block a particular record will be allocated to. The hashing function is an algorithm that takes an ASCII string and converts it to an integer. There are many approaches. A simple approach is to take the characters in the string and convert them to their ASCII values (for example 'A' is 65). These ASCII values are summed together. The sum is divided by the number of blocks in the file and the remainder or modulo gives the number of the block in the file into which that record will be placed, starting at block 0. If a block fills up, then an additional block will be used as overflow, and its address will be held in the block that the record would have been stored in. Figure 17.3 shows the organization of a hashed direct access file and the calculation of a simplified version of the hashing algorithm based on just three characters of the key.

Improving access. There are a number of ways of improving access to data in files. Files with a random organization can normally only be accessed directly or serially (for example to copy the file), so it is not possible to read through them sequentially in an order based on

Records hashed on first three characters of key field

Khan → Kha
ASCII Values = 75, 104, 97
75 + 104 + 97 = 276
276 divided by 7 leaves a modulo of 3
So Khan will be added in Block 3.

Data records in blocks (for simplicity only the keys are shown)

Block no. in file	Records			
0	Hao			
1	Ford	Farmer		
2	Firmin	Firth		
3	Harris			
4	Hastings	Gomez	Jacobson	
5	Ibrahim	Finch	Fern	Gangar
6	Hanson			

Figure 17.3 *Organization of a sample hashed direct access file.*

key fields. However, it is possible to add two extra fields to each record containing the key values of the next record and the previous record in sequence. This makes it possible to read through the records sequentially, but adds a considerable overhead when a record is added or deleted.

A common way of improving access to a file is to add an index. This is a similar approach to that used in adding an index to an indexed sequential file. It is used when there is a requirement either to access records in a file based on the values in some field other than the key field, for example to find records by postal code or zipcode, or to provide sequential access to a random file, by building an index of sequential keys. A separate file is created in which the keys are the values from the indexed field (for example the postal code) in all the records in the main file. Each record also contains either the keys or the block addresses of each of the records that contain that indexed field. This kind of index is known as an *inverted file*.

17.2.5 File types

We have seen that as well as files that hold data, there can also be files that hold indexes to the data in the main files. Other types of files may be required in a file-based system.

Master files. Master files are the files that hold the essential, persistent data records for the system. In transaction-processing systems the master files are updated with details of transactions that are recorded in transaction files. Master files usually require some kind of direct access so that records can be updated quickly.

Transaction files. Transaction files record business transactions or events that are of interest to the organization and that are used to update records in master files. In a banking system, transactions that take place when customers withdraw cash from an automatic teller machine (ATM) may be recorded in a transaction file. At the end of the day, the transaction file is processed in order to update all the accounts of customers who have withdrawn cash.

Index files. Indexes are used to speed up access to records in master files, as described above. There are many index file structures that can be used, and the choice of index structure will depend on the nature of the data and the type of access required.

Temporary files or work files. During the course of processing data, it may be necessary to create a temporary file that is used to hold the results of one process before it is used in another process. For example, it may be necessary to sort records into a particular order so that a report can be produced, a work file containing the records in the correct order would be produced and then deleted once the report had been created. If you use a PC that prints your work in the background (while you get on with another task), then it is using spool files to hold the printed output that is sent to the printer. These spool files are deleted when they have been printed.

Backup files. Backup files may be direct copies of master files or transaction files that are held on the system and that allow the data to be restored if the originals are destroyed or corrupted. Alternatively, they may be files with a special structure that can be used to reconstruct the data in the system.

Parameter files. Many programs need to store information about items of data of which there is only one record. These are typically system settings, such as the name and address of the company using the software, and configuration information, such as the currency format to be used to print out money values or the format to be used for printing the date. Parameter files hold this kind of information.

In a project that uses files to store data, part of the role of the designer is to choose the appropriate file organization and access method for the storage of all the objects in the system. In most cases, objects will need to be stored and retrieved using some kind of object identifier, and direct access files will be required. However, there may be requirements for all the objects of a particular type to be retrieved from the file in sequence, which may indicate the need for an organization that supports sequential access, or for the addition of an index. Some object-oriented languages provide mechanisms for streaming objects out to a serial file. This is the case with both Smalltalk and Java. Java, for example, contains two classes called `ObjectOutputStream` and `ObjectInputStream` that can be used to write objects together with information about their class to a disk file and read them back again.

Many organizations use database management systems to hold data. In their systems, there will be no need for designers to make decisions about the physical file structures used to provide persistent storage for objects.

17.2.6 Database management systems

Files are appropriate for simple programs and for storing data that does not need to be shared by many users. During the 1960s systems were built using files to store data; since the 1970s most large commercial systems have used databases of some sort to hold their data, and more importantly database management systems to organize and manage the tasks associated with storing and providing effective access to large volumes of data.

Using files to store data can result in a number of problems.

- As the number of applications grows, the number of different files grows. Some of these files may hold the same data for different applications in different formats. So, data is duplicated, taking up unnecessary storage space. This is known as redundancy.
- There is the risk that the updates to data in different applications will not be synchronized. For example, a customer address may be changed in one file but not in another, leaving the data inconsistent.
- Each application must contain its own mechanisms for storing the data in its set of files, if the data changes or the way that it is stored has to be changed, then each program within an application that accesses that data must be amended. This makes it difficult to add new programs to an application that use some of the same data but also need to store additional data.
- As business requirements change, users may want to access the data in new ways, for example to produce a report combining data from different applications. This cannot be implemented without considerable programming effort.

The first step towards resolving these problems is to analyse the storage requirements of different applications across the organization and to build a corporate database that contains all the data from different applications. Each application then uses a subset of this database for its own requirements. The second step is to use a database management system to organize and manage the data and to decouple the storage mechanisms from the application programs. The aim of using a DBMS is to separate the details of the way that data is stored physically from the way that it is used in application programs. This is achieved by

External Schema	The view on data used by application programs.
Conceptual Schema	The logical model of data that is separate from how it is used.
Internal Schema	The physical storage of data in files and indexes.

Figure 17.4 *The Three-schema Architecture.*

producing a logical model of the data that is independent of the demands of applications and that could be stored in different ways in the database. This is known as the Three-schema Architecture. Figure 17.4 shows the Three-schema Architecture. The external schema represents the ways in which data is used in application programs. The conceptual schema is a logical model of the data and is independent both of the external schema and of the details of how the data will be stored. The physical organization of the data is to be found in the internal schema, which defines the files used to store the data. The aim of this approach is to isolate the application programs from the details of how any particular item of data is stored. This is central to the way that relational DBMS work. Relational DBMS are described in Section 17.3.

DBMS provide more than just a means of storing data that can be shared across many applications. They provide tools and features that can be used to manage the data.

- Data definition language (DDL). The DDL is used to specify the data that is held in a database management system and the structures that are used to hold it
- Data manipulation language (DML). The DML is used to specify updates and retrievals of the data in the DBMS.
- Integrity constraints. Constraints can be specified to ensure that the integrity of the data is maintained.
- Transaction management. Updates to the database can be specified as transactions in which all of the component updates must succeed, otherwise the entire transaction is *rolled back*, which means that it is not committed to the database.
- Concurrency. Many users can simultaneously use the database and update its contents.
- Security. Access to the data in the database can be controlled, and permissions granted to different users for different levels of access (for example in SQL, select, insert, update and delete).
- Tuning of storage. Tools can be used to monitor the way that data is accessed and to improve the structures in the internal schema in order to make the access more efficient.

These changes can then be made without having any impact on the application programs. As mentioned above, these structures in the internal schema will be files. An important feature of a DBMS is that the kind of file used, the access methods and the indexes that are held on the file are hidden from users and can be changed without affecting the programs that use that data.

These changes to the database do not happen automatically. A Database Administrator (DBA) must be employed to manage the database and to ensure that it is running efficiently. The DBA will be responsible for controlling the data dictionary that defines the conceptual schema of the database, for controlling access to data and for tuning the performance of the database.

In summary, the use of a DBMS based on the Three-schema Architecture has a number of advantages over the use of files to store data for an information system.

- The use of a conceptual schema can eliminate unnecessary duplication of data.
- Data integrity can be ensured by the use of integrity constraints and transaction management techniques.
- Changes to the conceptual schema, the logical model, should not affect the application programs, provided the external schema used by the application programs does not have to be changed.
- Changes to the internal schema, the physical storage of the data, have no impact on the conceptual schema and should not affect the application programs, except perhaps positively in enabling them to access data more efficiently. Compromises may have to be made between the needs of different application programs.
- Tools are available for tuning the performance of the database.
- Facilities are provided for the back-up and recovery of data.
- Security and access to data by multiple simultaneous users can be controlled.

However the use of DBMS may also have disadvantages for organizations that decide to go down this route.

- There is a cost associated with investing in a large DBMS package.
- There is a running cost involved in employing staff to manage the DBMS.
- There will be a processing overhead in converting data from the database to the format required by the application programs.

The most widely used type of DBMS is the relational DBMS. For object-oriented systems one would hope to be able to use an object DBMS. However, in many situations, organizations have an existing relational DBMS and new object-oriented applications must share that corporate database.

17.2.7 Types of DBMS

The three main types of database that we are concerned with here are relational, object and hybrid object-relational. Hierarchical and network databases also exist. They are less widely used now than they were ten years ago, but still exist on mainframe systems. For an example of a network database model and implementation, see Howe (1989).

Relational databases. The idea of relational databases was first suggested by Codd (1970). His proposal was followed by considerable research effort that led to the development of commercial relational database management systems (RDBMS) during the 1970s. However, it was not until 1986 that the American National Standards Institute published the first SQL standard based on this work (ANSI, 1986). SQL (Structured Query Language) is now the standard language for relational databases and provides both DDL and DML capabilities.

Relational databases have a theoretical foundation in set theory, and their operations are defined in terms of the *relational algebra*, a mathematical specification of the operations that can be carried out on *relations*. The essence of the relational model is to eliminate redundancy from data and to provide the simplest possible logical representation of that data. This is achieved by means of a series of steps that can be applied in analysing the data and that result in normalized data. This normalized data is held in relations or tables. This process simplifies a complex data structure until it can be held in a series of tables. Each table is made up of *rows* of data. Each row contains attribute values that are

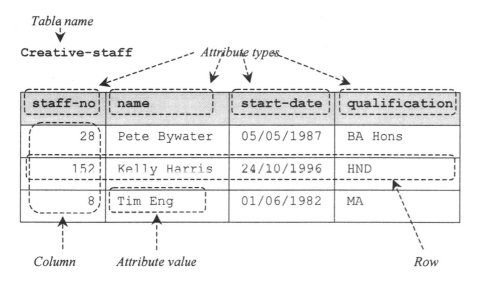

Figure 17.5 *Conventional representation of a table.*

organized in *columns*. Each column contains data values of the same attribute type. The data in each row must be distinct and can be uniquely identified by some combination of attribute values in that row. Each attribute value in the table must be atomic, that is, it may not contain multiple values or be capable of being broken down further. Figure 17.5 shows the conventional form for representing tables on paper. In existing RDBMS, all data structures must be decomposed into this kind of two-dimensional table[1].

The weakness of current implementations of RDBMS lies in the fact that objects in object-oriented systems are not flat two-dimensional objects. They can be broken down into tables, as is shown in Section 17.3, but there is a processing overhead associated with breaking them down and reconstructing them. References to other objects (represented by associations in the class diagram) must also be maintained when an object is stored in a relational database, and restored when it is retrieved, and even if the associated object is not itself in memory some mechanism must be created to allow it to be referenced and to be sent messages. Finally, RDBMS do not handle multimedia objects such as video and audio clips well. Most now offer facilities to store them as BLOBs (Binary Large Objects), but information about the type of such objects is lost when they are stored.

Relational DBMS are currently the most widely used type of DBMS. They are based on a sound mathematical theory, they have been developed over a number of years, they are robust and they are efficient and flexible for the kind of data that they are designed to handle. The best known is probably Oracle; others include DB2®, Informix®, Ingres®, Progress® and Sybase®.

Object databases. In contrast, objects in an object-oriented system are not flat, two-dimensional structures. Each object may contain other objects nested within it. For example, a SalesOrder object could contain an Address object that contains its own attributes, and a List of OrderLine objects, each of which is made up of two attributes.

---

[1] Date and Darwen (1998) argue that the relational model can handle complex data types, but it is the way that relational DBMS have been implemented that leads to their inability to handle these data types.

```
salesOrderNo:        37291
orderDate:           25/08/1999
customerID:          CE102
customerOrderRef:    R20716
deliveryAddress:

    line1:       23 High Street
    line2:       Lytham St Annes
    line3:       Lancashire
    postcode:    LA43 7TH

orderLineList:

    productID:   12-75
    quantity:    3

    productID:   09-103
    quantity:    10
```

(a) *Complex object.*

```
              SalesOrder

salesOrderNo: Integer
orderDate: Date
customerID: String
customerOrderRef: String
deliveryAddress: Address
orderLineList: OrderLine[ ]
```

(b) *Class with classes of nested objects as attributes.*

Figure 17.6 *Composite* SalesOrder *object with equivalent UML class.*

An example of this is shown in Figure 17.6 with a SalesOrder class in UML notation. (Nested objects can be shown as attributes like this or alternatively by using the composition notation.)   Object database management systems (ODBMS) have been developed to handle complex objects of this sort.   Part of the motivation for the development of ODBMS has been the growth in the number of applications that use complex data structures.   These include multimedia applications in which objects such as sounds, images and video clips are not easily represented in tables, and applications such as computer-aided design packages in which the designer may want to deal with different levels of abstraction, for example treating a sub-assembly in terms of its behaviour as a sub-assembly, in terms of the individual chips or in terms of the components such as logic gates that make up those chips.   ODBMS provide services that make it possible to store complex objects of this type.

Examples of ODBMS include GemStone®, O₂ System®, ObjectStore®, Ontos® and Versant ODBMS™.

Object-relational databases.   Object-relational databases combine the simplicity and efficiency of relational databases with the ability of object databases to store complex objects and to navigate through the associations between classes.   The SQL3 standard, due for completion in 1999, details how the relational model can incorporate many features of object-oriented systems such as user definable abstract data types, inheritance and operations.   Illustra™ is probably the most well-known hybrid DBMS.   Oracle *8i*™ includes some hybrid features.   In what follows we shall focus on relational and object databases.

Review Questions

17.1   Give one example each of a persistent and a transient object.

17.2   Explain the difference between different types of file organization and file access.

17.3   Of the different kinds of record type listed in Section 17.2.2 suggest which would be most appropriate for storing complex nested objects. Explain the reasons for your choice.

17.4   Outline the advantages and disadvantages of using a database management system over developing an application using files.

17.5   What is the key difference between a relational DBMS and an object DBMS?

17.3 Designing for Relational Database Management Systems

RDBMS have been in use since the 1970s. They use mature technology and are robust. Many organizations already have a relational DBMS. It is, therefore, common for an object-oriented system to be built to use a relational database. Relational databases hold data in flat two-dimensional tables whereas classes may have complex nested structures with objects embedded within other objects. If it is necessary to use an RDBMS to provide the storage for a system built using an object-oriented programming language, then it will be necessary to 'flatten' the classes into tables in order to design the storage structures. When the system requires an instance of a class from the database, it will have to retrieve the data from all the tables that hold parts of that object instance and reconstruct it into an object. When a complex object instance has to be stored, it will have to be taken apart and parts of it will be stored in different tables. The designer of such a system has to decide on the structure of the tables to use to represent classes in the database. It should be emphasized that it is only the attribute values of object instances that are stored in an RDBMS; operations are implemented in the programming language used.

There are two ways in which classes can be converted to tables in a relational database. The first, normalization, is suitable for decomposing complex objects into tables. It is used in systems that are not object-oriented to design the structure of tables in databases. It can also be used during object design to simplify large complex objects that are not cohesive (see Chapter 13). The second approach is based on a series of rules of thumb that can be applied to classes in a class diagram to produce a set of table structures. In this section, we describe these two approaches, and in Section 17.5 we discuss the impact that this will have on the design of classes in the system.

17.3.1 Data modelling and normalization

In order to store the objects from an object-oriented system in a relational database, the objects must be flattened out. *Normalization* is an approach that is also used to convert the complex structures in business documents into tables to be stored in a relational database. A typical example of its use would be to design a set of tables to hold the data in a sales order like the FoodCo example in Figure 15.3. How then do we apply normalization? Normalization is based on the idea of *functional dependency*. This kind of dependency was mentioned in Section 13.8. It was not discussed in detail there and is explained here.

For two attributes, A and B, A is functionally dependent on B if for every value of B there is exactly one value of A associated with it at any given time. This is shown as:

$$B \rightarrow A$$

Attributes may be grouped around functional dependencies according to the rules of normalization to produce normalized data structures that are largely free of redundancy. There are five *normal forms* of normalized data. The data is free of redundancy in *fifth normal form*. For practical purposes it is usually adequate to normalize data into *third normal form*. Normalization is carried out in a number of steps, and we shall apply these to an example from the Agate case study.

Analysis has identified a class called `InternationalCampaign`, and the attribute values of two instances of this class are shown in Figure 17.7. As stated earlier, we may wish to decompose this class into simpler classes because it is not cohesive (as suggested in Chapter 13) or we may need to decompose it into table structures for storage using an RDBMS. The same approach is used in both cases. Here we are applying normalization as part of the design for a relational database.

The first step is to remove any calculated values. There are no examples of this in the `InternationalCampaign` instances.

We have now created a relation that is said to be in unnormalized form. Each `InternationalCampaign` is uniquely identified in this system by its `campaignCode`. This is the *primary key* attribute. Figure 17.8 shows the data from these instances in a table. Each `InternationalCampaign` is represented in a single row. Note that there are multiple values in some of the columns in each row. These relate to the locations where the campaign will run and the manager in each location.

```
SMGL:InternationalCampaign

campaignCode = SMGL
campaignTitle =  Soong Motors
                 Granda Launch

locationsList =
[
 locationCode =  HK
 locationName =  Hong Kong
 locationMgr =   Vincent Sieuw
 locationMgrTel = ext. 456

 locationCode =  NY
 locationName =  New York
 locationMgr =   Martina Duarte
 locationMgrTel = ext. 312

 locationCode =  TO
 locationName =  Toronto
 locationMgr =   Pierre Dubois
 locationMgrTel = ext. 37
]
```

```
YPSC:InternationalCampaign

campaignCode = YPSC
campaignTitle =  Yellow Partridge
                 Summer Collection

locationsList =
[
 locationCode =  HK
 locationName =  Hong Kong
 locationMgr =   Jenny Lee
 locationMgrTel = ext. 413

 locationCode =  NY
 locationName =  New York
 locationMgr =   Martina Duarte
 locationMgrTel = ext. 312
]
```

Figure 17.7 *Example InternationalCampaign objects.*

InternationalCampaign

campaign Code	campaign Title	location Code	location Name	locationMgr	location MgrTel
SMGL	Soong Motors Granda Launch	HK	Hong Kong	Vincent Sieuw	456
		NY	New York	Martina Duarte	312
		TO	Toronto	Pierre Dubois	37
YPSC	Yellow Partridge Summer Collection	HK	Hong Kong	Jenny Lee	413
		NY	New York	Martina Duarte	312

Figure 17.8 *Table for sample International Campaigns.*

A table is in *first normal form* (1NF) if and only if all row/column intersections contain atomic values. The table in Figure 17.8 does not conform to this criterion and must be redrawn as in Figure 17.9. These multiple values are often known as *repeating groups*.

The `campaignCode` no longer uniquely identifies each row in the table. Each row is identified by a combination of `campaignCode` and `locationCode`. These attributes form a candidate *primary key* for the table.

InternationalCampaign-1

campaign Code	campaign Title	location Code	location Name	locationMgr	location MgrTel
SMGL	Soong Motors Granda Launch	HK	Hong Kong	Vincent Sleuw	456
SMGL	Soong Motors Granda Launch	NY	New York	Martina Duarte	312
SMGL	Soong Motors Granda Launch	TO	Toronto	Pierre Dubois	37
YPSC	Yellow Partridge Summer Collection	HK	Hong Kong	Jenny Lee	413
YPSC	Yellow Partridge Summer Collection	NY	New York	Martina Duarte	312

Figure 17.9 *Revised table for International Campaigns without repeating groups.*

The data values have been flattened out into a two-dimensional table and could now be stored in a relational database as they are. However, there is redundancy that we want to eliminate from the data. If redundant data is held in the database, there is the risk that values will not be updated correctly. For example, if Martina Duarte's telephone extension number changes, the system must ensure that it is correctly updated in every row in which it appears.

The next step is to convert these relations to *second normal form* (2NF). A relation is in 2NF if and only if it is in 1NF and every non-key attribute is fully dependent on the primary key. Here the attribute `campaignTitle` is only dependent on `campaignCode`, and `locationName` is only dependent on `locationCode`. (These are sometimes called part-

InternationalCampaign-2

campaign Code	location Code	locationMgr	location MgrTel
SMGL	HK	Vincent Sieuw	456
SMGL	NY	Martina Duarte	312
SMGL	TO	Pierre Dubois	37
YPSC	HK	Jenny Lee	413
YPSC	NY	Martina Duarte	312

Campaign

campaign Code	campaignTitle
SMGL	Soong Motors Granda Launch
YPSC	Yellow Partridge Summer Collection

Location

location Code	location Name
HK	Hong Kong
NY	New York
TO	Toronto

Figure 17.10 *2NF tables.*

key dependencies.) The other attributes are dependent on the whole primary key. (Remember A is dependent on B if for every value of B there is exactly one value of A associated with it at a given time.) Figure 17.10 shows the creation of two new relations `Campaign` and `Location`.

The next step is to convert the tables to *third normal form* (3NF). A relation is in 3NF if and only if it is in 2NF and every attribute is dependent on the primary key and not on another non-key attribute. `Campaign` and `Location` are in 3NF, but in `InternationalCampaign-2` `locationMgrTel` is dependent on `locationMgr` and not on the primary key. Figure 17.11 shows the tables in 3NF with the addition of a new table called `LocationManager`.

InternationalCampaign-3

campaign Code	location Code	locationMgr
SMGL	HK	Vincent Sieuw
SMGL	NY	Martina Duarte
SMGL	TO	Pierre Dubois
YPSC	HK	Jenny Lee
YPSC	NY	Martina Duarte

LocationManager

locationMgr	location MgrTel
Vincent Sieuw	456
Martina Duarte	312
Pierre Dubois	37
Jenny Lee	413

Campaign

campaign Code	campaignTitle
SMGL	Soong Motors Granda Launch
YPSC	Yellow Partridge Summer Collection

Location

location Code	location Name
HK	Hong Kong
NY	New York
TO	Toronto

Figure 17.11 *3NF tables.*

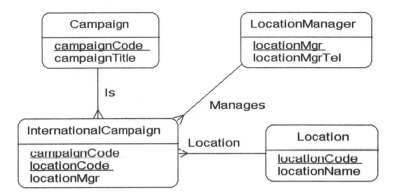

Figure 17.12 *Entity-relationship diagram for tables of Figure 17.11.*

These relations can be shown in a diagram using the notation of entity-relationship diagrams, which are often used to represent the logical structure (conceptual schema) of relational databases. Note that this is not part of the UML notation. Figure 17.12 shows the relations of Figure 17.11 as an entity-relationship diagram. This diagram was produced using the SELECT Enterprise CASE tool as a Storage Class Diagram from which it can generate the SQL statements to create these tables.

If we examine the attributes of these relations, we may come to the conclusion that InternationalCampaign was not a very well analysed class in the first place. It should perhaps be a subclass of Campaign. LocationManager appears to be nothing more than CreativeStaff with an association to InternationalCampaign. We also require a new class called Location with an association to InternationalCampaign. One possible design is shown in Figure 17.13. Note that in this class diagram we have included

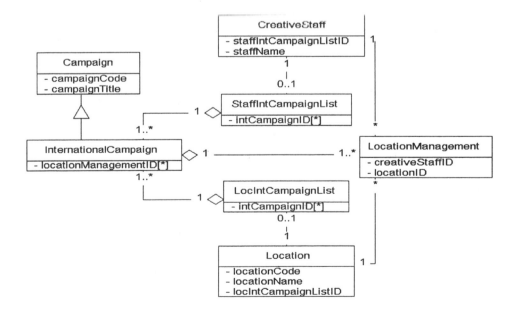

Figure 17.13 *Excerpt from design class diagram for* InternationalCampaign.

the collection classes that were introduced during object design (Chapter 13). There is presumably also an association between StaffMember and Location, but we should not rely on that to find out which Location an InternationalCampaign is running in, as the member of staff who is LocationManager could move offices.

The classes in Figure 17.13 raise a number of questions that should have been addressed during the analysis of requirements. Some of these are listed below.

- Does each campaign have a location? Is this the main location for an international campaign?
- Does each member of staff work in a location?
- Are there different versions of adverts for different locations?
- How are international campaigns costed? What currencies are used?

You can probably think of others. In an iterative life cycle, it may be acceptable to be raising these issues at this stage. In a traditional waterfall life cycle, these issues should have been resolved during the analysis stage.

17.3.2 Mapping classes to tables

An alternative approach to that provided by normalization is to follow a set of guidelines for how to map the classes and multiplicities in the class diagram to tables in a relational database design. A summary of the patterns that can be applied to this mapping can be found in Brown and Whitenack (1996). The following guidelines are derived from Rumbaugh et al. (1991) and Brown and Whitenack (1996).

- Classes with a simple data structure. These classes become tables.
- Object identifiers become primary keys. Each class is allocated an object identifier attribute that can be used as a primary key in the relational table for that class.
- Classes that contain an instance of another class as an attribute. A separate table should be created for the embedded class. Objects of the embedded class should be allocated a unique object identifier. The object identifier should replace the embedded object in the table for the container class as a foreign key.
- Classes that contain collections. Allocate an object identifier to the class held in the collection. This class will be represented by a table. Create a separate table that contains two columns. The first holds the object identifiers of the objects that contain the collection; the second holds the object identifiers of the objects that are held in the collection.
- One-to-many associations can be treated like collections.
- Many-to-many associations become separate tables. Create a table that contains two columns. Each row contains a pair of object identifiers, one from each object participating in the association. (These are like two collections.)
- One-to-one associations are implemented as foreign key attributes. Each class gains an extra attribute in which to hold the object identifier of the object with which it is associated.

(A *foreign key* is used in relational databases to create the relationships between tables. locationCode in the InternationalCampaign table in Figure 17.11 is an example of a foreign key. Objects do not have keys, and this is why object identifiers are allocated to them. It may be possible to use an attribute that will have a unique value in each instance of a class as an object identifier.)

When a relational database is used, collection classes that exist only to provide access to a set of objects of the same class are not part of the data that is stored in tables. If it is necessary to iterate through every instance of a particular class, this can be done by selecting every row from the table.

Inheritance poses more of a problem. There are three alternative ways of mapping an inheritance hierarchy to relational database tables.

- Only implement the superclass as a table. Attributes of subclasses become attributes of the superclass table and hold null values where they are not used. This approach is most appropriate where subclasses differ from their superclass more in behaviour than in attributes.
- Only implement the subclasses as tables. The attributes of the superclass are held in all the subclass tables. This only works if the superclass is abstract and there will be no instances of it.
- Implement all the classes (both superclass and subclasses) as separate tables. To retrieve the data for a subclass both its own table and the table of its superclass must be accessed.

The solution that is chosen may depend on the requirements of the particular application or may be constrained by the use that will be made of the data in the database by other applications.

This brings us to a further aspect of relational databases: data is added to and retrieved from them using SQL statements. SQL provides both the DDL and DML for relational databases. Figure 17.14 shows the SQL statements necessary to create the tables of Figures 17.11 and 17.12 in Oracle 7™ generated by the CASE tool from the storage class diagram.

Figure 17.15 shows an SQL statement that finds all the international campaigns with the locationCode 'HK' (i.e. running in Hong Kong). There is a design decision to be made in deciding where to place the responsibility for this kind of requirement.

- This SQL statement could be executed and only data for those objects that are required would be returned from the database. This replaces the interaction modelled in a sequence diagram with functionality provided by the DBMS.
- Alternatively, the data from all these tables could be retrieved from the database and used to instantiate the objects in the system. Each InternationalCampaign object could then be sent a message to check whether it includes the Location 'Hong Kong'. This will involve it sending a message to each associated Location object. This is more object-oriented, but will take a longer time to execute.
- The third alternative is to retrieve data from each table in turn, as though navigating through the structure of the class diagram, first the InternationalCampaign then each of the Locations for that InternationalCampaign. This approach requires use of indexes on the tables to make access possible in a reasonable time.

This kind of design decision trades off the pure object-oriented approach against the efficiency of the relational database. In order to retrieve this data into objects in a programming language such as Java, the SQL statements must be embedded in the program. In Chapter 18, we present examples of how this can be done using JDBC (Java Database Connectivity). During design, we have to decide which classes have the responsibility for accessing the database. In Section 17.5 we describe two different approaches to this design decision and show how they can be modelled in UML. However, before we address how we can model the database management responsibilities of the system, we need to consider object DBMS and what they have to offer the designer as an alternative to relational DBMS.

```
CREATE TABLE Campaign (
      campaignCode VARCHAR(6) NOT NULL,
      campaignTitle VARCHAR(50) NULL,
      PRIMARY KEY (campaignCode)
      );

CREATE TABLE InternationalCampaign (
      campaignCode VARCHAR(6) NOT NULL,
      locationCode VARCHAR(2) NOT NULL,
      locationMgr VARCHAR(30) NULL,
      PRIMARY KEY (campaignCode, locationCode)
      );

CREATE TABLE Location (
      locationCode VARCHAR(2) NOT NULL,
      locationName VARCHAR(20) NULL,
      PRIMARY KEY (locationCode)
      );

CREATE TABLE LocationManager (
      locationMgr VARCHAR(30) NOT NULL,
      locationMgrTel INT NULL,
      PRIMARY KEY (locationMgr)
      );
```

Figure 17.14 *SQL statements to create tables of Figures 17.11 and 17.12.*

```
SELECT campaignTitle FROM Campaign c, InternationalCampaign ic,
Location l
      WHERE c.campaignCode = ic.campaignCode
      AND ic.locationCode = l.locationCode
      AND l.locationName = 'Hong Kong'
```

Figure 17.15 *SQL statement to find campaigns running in Hong Kong.*

17.4 Designing for Object Database Management Systems

17.4.1 Object databases

Object DBMS differ from current relational DBMS in that they are capable of storing objects with all their complex structure. It is not necessary to transform the classes in the design model of the system in order to map them to storage objects. As you might expect, using an object database maintains the seamlessness that is claimed for object-oriented systems right through to the storage of objects in the database. Designing for an object database will have a minimal impact on the design of the system.

The standard for object databases has been set by the Object Data Management Group (ODMG) and is currently available in Version 2.0 (Cattell and Barry, 1997). The standard defines both the Object Definition Language (ODL™) and the Object Manipulation Language (OML™) for object databases. The ODL is similar to the DDL elements of SQL for relational databases but allows objects to maintain their complex structure: objects can

```
interface  InternationalCampaign
(extent    international_campaigns
  key      (campaignCode, locationCode))
{
          attribute      String campaignCode;
          attribute      String locationCode;
}
interface  StaffMember
(extent    staff_members
  key      staffNo)
{
          attribute      Short staffNo;
          attribute      String staffName;
          attribute      Date staffStartDate;
}
interface  CreativeStaff: StaffMember
(extent    creative_staff
  key      staffNo)
{
          attribute      String qualification;
          attribute      List<InternationalCampaign>
                                    staffIntCampaignList
          relationship Client isContact;
          inverse Client::hasContact;
}
```

Figure 17.16 *ODL for* InternationalCampaign, StaffMember *and* CreativeStaff.

contain other objects, including collections, as attributes. Figure 17.16 shows the ODL definition of the InternationalCampaign, StaffMember and CreativeStaff classes based on Figure 17.12 but with the StaffIntCampaignList as an embedded attribute of CreativeStaff. Note also that the client contact association with Client is shown as a one-to-one association to illustrate the syntax.

However, individual ODBMS do not necessarily conform to the standard. Eaglestone and Ridley (1998) describe both the ODMG standard and the syntax of O_2. Blaha and Premerlani (1998) describe how to map an object-oriented design to ObjectStore. Some object databases include a DML similar to SQL that can be used to query data in the database. Statements in this OML are usually embedded in the program code of an object-oriented language. Unlike SQL statements, which are usually built as strings and then executed by a call to a system function or passed to a database server, OML queries are extensions to the programming language that is being used. Figure 17.17 shows a C++ operation to query the collection of InternationalCampaign objects (abbreviated in the figure to IntCampaign for simplicity) associated with a particular CreativeStaff instance in order to find one that contains a particular campaignCode. The campaignCode would be passed as a parameter to the operation and not hard-coded as in the SQL of Figure 17.15.

Object databases provide much better facilities for navigating the structure of the classes in the class diagram. Figure 17.18 shows an excerpt from a class declaration and implementation in C++ that illustrates how the associations are created between classes using ObjectStore's macros. This excerpt declares an attribute of CreativeStaff called isContact, the type of which is an os_Set of pointers to objects of type Client. The inverse association from Client is called hasContact. This technique makes it

```
IntCampaign * CreativeStaff::findIntCampaign ( string campaignCode )
{
   IntCampaign * intCampaignPointer;
   intCampaignPointer = staffIntCampaignList.getValue().query_pick(
      "IntCampaign*",
      "campaignCode == this->campaignCode",
      os_database::of(this));
   return intCampaignPointer;
}
```

Figure 17.17 *Example query (highlighted in bold) in C++ for ObjectStore.*

straightforward to implement the associations, and it is not necessary to manually insert object identifiers into the classes to handle the associations. This would make the collection classes of Figure 17.13 unnecessary.

Where an object contains a reference to another object as an attribute, it is possible to directly invoke its public methods or access its public attributes even if the object in question is not in memory. The great benefit of object databases is that they transparently materialize objects from the database into memory without the need for special action to be taken by the developer. Typically the only requirement is for some part of the application to make a connection to the database and for operations that read from or write to the database to be bracketed by commands to start and finish a transaction. This approach is used, for example, in ObjectStore for Java and C++.

The structure of the class diagram will require minimal changes to be used with an object database. Minor changes may be required. For example, if you use ObjectStore Persistent Storage Engine (PSE Pro™) with Java you will find that some of the Java collection classes cannot be made persistent. However ObjectStore provides persistent equivalents for these classes. For example, wherever you have used a class such as a Hashtable and require it to be persistent, it must be replaced with an instance of the class OSHashtable.

Although object DBMSs make it easier to map an object-oriented design straight to a database, there is one area where the seamlessness breaks down. Apart from very simple operations, such as those to insert new values or to do simple arithmetic, ODL and hence object databases do not support the storage of operations in the database. Operations must still be implemented in an object-oriented programming language such as Java or C++. In Chapter 18 we consider some of the issues surrounding implementation using an object DBMS.

```
os_relationship_1_m(CreativeStaff, isContact, Client, hasContact,
      os_Set<Client*>) isContact;

...

os_rel_1_m_body(CreativeStaff, isContact, Client, hasContact);
```

Figure 17.18 *Example association declaration and implementation in C++ for ObjectStore.*

17.4.2 Distributed databases

Before we consider how to design the classes that handle data management for a database that is not object-oriented, it is worth considering the issue of design for distributed databases.

Up to this point, we have been assuming that wherever and however objects are stored, they will be brought into the memory of the user's machine and will be sent messages to invoke their operations before being transferred back to the database and removed from memory. This need not be the case. The objects that participate in a particular use case need not be on the same machine as the user interface, indeed they may be distributed on different machines. Several ways of doing this exist, for example Remote Procedure Calls (RPC) in languages such as C or C++ and Remote Method Invocation (RMI) in Java. The new object-oriented industry standard, however, is CORBA (the Common Object Request Broker Architecture). CORBA Version 2.0 is defined as a standard by the Object Management Group (OMG, 1995), a different group from the ODMG.

CORBA separates the interface of a class (the operations that it can carry out) from the implementation of that class. The interface can be compiled into a program running on one computer. An object instance can be created or accessed by name. To the client program it appears to be in memory on the same machine. However, it may actually be running on another computer. When the client program sends it a message to invoke one of its operations, the message and its parameters are converted into a format that can be sent over the network (known as marshalling), at the other end the server unmarshals the data back into a message and parameters and passes it to the implementation of the target object. This object then carries out the operation and, if it returns a value, that value is marshalled on the server, unmarshalled on the client, and finally provided as a return value to the client program. The marshalling and unmarshalling process makes it possible to pass objects with a complex structure over the network by flattening them out and reconstructing them at the other end.

CORBA achieves this by means of programs known as ORBs (Object Request Brokers) that run on each machine. The ORBs communicate with each other by means of an Inter-ORB Protocol (IOP). Over the Internet, the protocol used is IIOP (Internet IOP). To use this facility, the developer must specify the interface for each class in an Interface Definition Language (IDL). IDL defines the interface of the class in terms of its public attributes and operations. Figure 17.19 shows a sample IDL file for the Location class. The IDL file is then processed by a program that converts the interface to a series of files in the target language or languages. In the case of Java, this program is called IDL2JAVA and produces several files. These files include:
- a file that defines the interface in Java,
- a stub file that provides the link between the client program and the ORB,
- a file that provides a skeleton for the implementation on the server.

There are a number of other files that may be generated, but these are the important ones. The stub file implements the interface on the client and is compiled into the client program; the skeleton file also implements the interface (but on the host) and is amended by the developer to provide the implementation and compiled on the host.

Systems developed using CORBA can be set up so that the remote objects are located on a named machine and accessed by name. However, CORBA also provides services for locating objects by name when it is not known where they are running. It also includes sophisticated services for locating objects that implement a certain interface and for

```
module Agate
{  interface Location
     { attribute string locationCode;
       attribute string locationName;
       void addIntCampaign( in IntCampaign campaign );
       void removeIntCampaign( in string campaignCode );
       int numberOfCampaigns( );
     };
};
```

Figure 17.19 *CORBA IDL for* Location *class.*

interrogating an object to determine its interface (operations, parameter types and return types) in order to dynamically invoke its operations. However, for the vast majority of applications, these services are not required.

CORBA is known as *middleware*, as it acts as an intermediary between clients and servers. As such it enables the implementation of a three- or four-tier architecture that isolates the user interface and client programs from the implementation of classes on one or more servers. CORBA also provides interoperability between different languages: a Java client program can invoke operations on a C++ or Smalltalk object that exists on a separate machine. CORBA also makes it possible to encapsulate pre-existing programs written in non-object-oriented languages by wrapping them in an interface. To the client it looks like an object, but internally it may be implemented in a language such as COBOL.

17.5 Designing Data Management Classes

17.5.1 The data storage layer

In designing the presentation layer one of our aims was to isolate the business classes in the system from the way that they are presented on screen and in reports and documents. We did this in order to maximize the reusability of the classes. Our aim is to do the same with the data storage layer. Classes are less reusable if they are tightly coupled to the mechanisms by which instances are stored in some kind of file system or database. What are the options for locating the operations that handle the tasks of storing and retrieving objects?

(1) We could add operations to each class to enable objects to save and store themselves.
- This reduces reusability. Each class must now contain code to implement the operations that couple the class to the particular data storage mechanism used.
- This breaches the idea of cohesion. Each business class should contain operations that are relevant to the behaviour of instances of that class, data storage methods belong in a data storage class.
- If an object is not currently instantiated, how can we send it a message to invoke an operation to load itself?

(2) We can get around this last problem by making the storage and retrieval operations class-scope methods rather than instance-scope methods (static methods in Java or static member functions in C++).

- This still suffers from the first two problems listed above for option 1. The class is less reusable and lacks cohesion.

(3) All persistent objects in the system could inherit methods for storage from an abstract superclass—PersistentObject for example rather than Object in Java.
- This has the effect of strongly coupling existing classes to the PersistentObject superclass, so all business classes end up inheriting from a utility class.

(4) Where we have introduced collection classes into the design to manage collections of objects, we could make these collection classes responsible for storing and retrieving object instances.
- This is closer to a solution. The collection classes are design artefacts—not part of the business classes in the system. However, we may wish to reuse the design, and we are coupling it to the storage mechanisms.

(5) We could introduce separate classes into the system whose role is to deal with the storage and retrieval of other classes. This is the database broker approach.
- This solution fits the layered architecture. These classes are part of the data storage layer.
- The data storage classes are decoupled from the business classes. The business classes will contain nothing that indicates how they are to be stored. The same business classes can be reused unchanged with different storage mechanisms.

(6) We could limit the number of new data storage classes to one. Different instances of this class would be created with attributes to hold the names of tables or files that are to be used to store and retrieve instances of their associated classes.
- This parameterized version is more difficult to set up and more difficult to implement.
- It requires some part of the system outside the database broker class to know what parameters to set for each instance that is created.

Option 5 is the approach that is favoured by most developers of object-oriented systems. It involves the use of a number of patterns. Larman (1998) describes it in some detail as a *persistence framework* the main feature of which is the use of *database brokers*, which mediate between the business classes and the persistent storage and which are responsible for storing and retrieving objects. However, we shall first describe the use of option 3— inheritance from a persistent superclass—before looking at option 5.

17.5.2 PersistentObject superclass

A simple approach to the design of data storage classes is to design an abstract superclass PersistentObject that encapsulates the mechanisms for an object of any class to store itself in and retrieve itself from a database. Eriksson and Penker (1998) use this approach in order to keep their example case study application simple and independent of any vendor's DBMS. The PersistentObject superclass implements operations to get an object by object identifier, store, delete and update objects and to iterate through a set of objects. These operations are implemented in terms of two abstract operations, to write and read objects, that must be implemented by each subclass that inherits from the

PersistentObject superclass. This is shown in Figure 17.20 (adapted from Eriksson and Penker, 1998). This approach also uses an aspect of option 2, as the getObject() operation is a class-scope method rather than an instance-scope method (as are the others underlined in the class diagram of Figure 17.20).

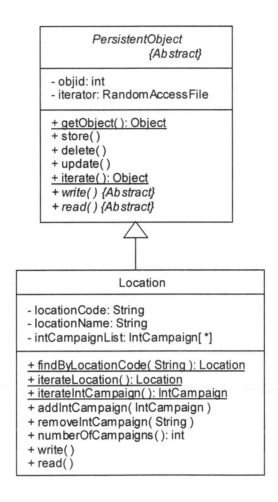

Figure 17.20 *Excerpt from class diagram to show inheritance from* PersistentObject.

The PersistentObject hides some of the detail of the implementation of the data storage from the business objects in the application. However, they must implement the write() and read() operations, and this will limit their reusability.

This approach does have the benefit of limiting the changes that will be made to sequence diagrams. Messages that have been shown being sent to object instances to select an instance or to iterate through a set of instances can be shown as being sent to the class rather than the instances. Figure 17.21 shows an example of this for the use case Get Number of Campaigns for Location.

However, the use of a persistent superclass is unlikely to be robust enough for business applications, and a more sophisticated approach, such as the database broker must be used.

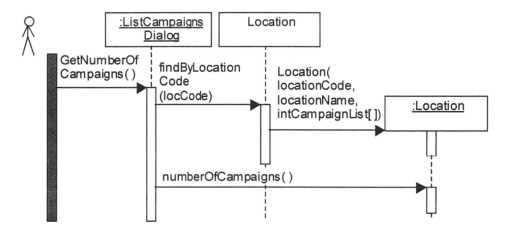

Figure 17.21 *Sequence diagram for* Get Number of Campaigns for Location *showing* Location *retrieving an instance.*

17.5.3 Database broker framework

The database broker framework separates the business objects from the data storage implementation. The classes that provide the data storage services will be held in a separate package. Coad and Yourdon (1991) talk about a Data Management Component, but then expect operations to be added to their Problem Domain Component (business) classes to allow them to save themselves in persistent storage. Our objective here is to separate the data storage mechanisms completely from the business classes. For each business class that needs to be persistent, there will be an associated database broker class that provides the mechanisms to materialize objects from the database and dematerialize them back to the database. A simple form of this is shown in Figure 17.22 for the Location class. The LocationBroker is responsible for the storage and retrieval of Location object instances. In order to ensure that there is only ever one LocationBroker instance, we can use the Singleton pattern (see Chapter 14). This means that we use a class-scope operation, but only to obtain an instance of the LocationBroker that can be used subsequently to access the database. The sequence diagram involving the LocationBroker is very similar to that of Figure 17.21, and is shown in Figure 17.23.

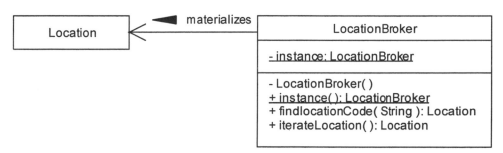

Figure 17.22 Location *and* LocationBroker *classes.*

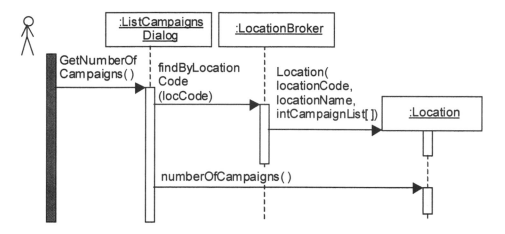

Figure 17.23 *Sequence diagram for* Get Number of Campaigns for Location *showing* LocationBroker *retrieving an instance of* Location.

(Note that this diagram does not show the creation of the instance of LocationBroker.)

Each persistent class in the system will require a broker class, so it makes sense to create a superclass that provides the services required by all these broker classes. Larman (1998) suggests two levels of generalization. At the top of his hierarchy is an abstract Broker class that provides the operation to materialize an object using its object identifier. This is then subclassed to produce different abstract classes of brokers for different kinds of storage, for example, one for a relational database and one for a file system. Finally, the appropriate broker is subclassed into the concrete classes for each persistent class in the system. A simplified version of this inheritance hierarchy is shown in Figure 17.24.

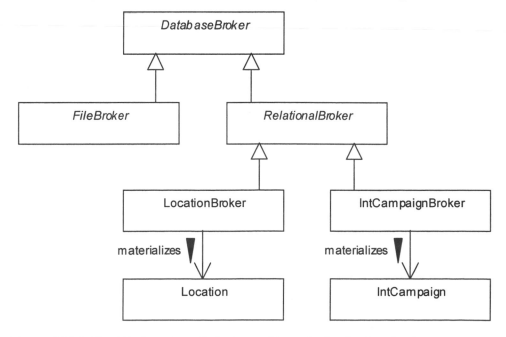

Figure 17.24 *Simplified version of inheritance hierarchy for database brokers.*

In the Agate case study, we could use JDBC (Java Database Connectivity), the Java version of Open Database Connectivity (ODBC), to link the Java programs to a relational database. ODBC is a standard means for client programs to connect to databases.

This will require the use of classes from the `java.sql` package, in particular `Connection`, which is used to make a connection to the database, `Statement`, which is used to execute SQL statements, and `ResultSet`, into which the results of SQL Select statements are placed (we can then iterate through the `ResultSet` retrieving each row in turn and extracting the values from each column). An appropriate driver will also be required. Figure 17.25 shows the associations between the `RelationalBroker` abstract class and these other classes. The figure shows the Oracle JDBC driver; to access a database such as Access via a link from JDBC to ODBC, the appropriate driver would be `sun.jdbc.odbc::JdbcOdbcDriver`.

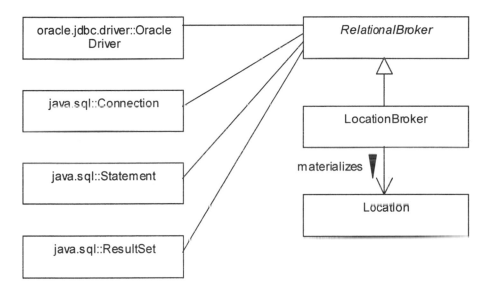

Figure 17.25 `RelationalBroker` *class and classes from other packages.*

The dependencies between the application classes and those in other packages can be shown using packages in a class diagram, as for the classes in the presentation layer in Chapter 16, Figure 16.6. This is shown in Figure 17.26.

This simple framework using database broker classes can be extended to deal with some of the problems that remain.

- The most important problem concerns the way in which persistent objects maintain references to other objects. If the `LocationBroker` retrieves an instance of `Location`, what happens when an operation of the `Location` requires it to send a message to one of its `IntCampaigns`. The `IntCampaign` will not necessarily have been retrieved from the database. The same applies to the many other operations that involve collaboration between objects.
- The second problem concerns the ability to manage transactions in which a number of objects are created, retrieved from the database, updated and deleted.

Two extensions to the database broker framework can be used to resolve these problems. The first uses the Proxy pattern to provide proxy objects for those objects that have not yet been retrieved from the database. The second uses caches to hold objects in memory and

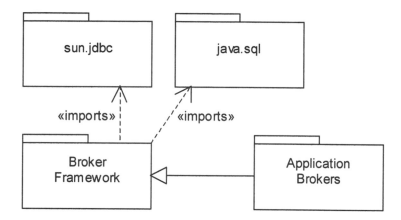

Figure 17.26 *Class diagram showing packages for database brokers package.*

keep track of which have been created, updated or deleted.

The Proxy pattern (Gamma et al., 1995) provides a proxy object as a placeholder for another object until it is required. In this case, we can use proxies for each business class to link to where there is an association with another object or objects. If no message is sent to the associated objects, then the proxy does nothing, if a message is sent then the proxy asks the relevant database broker to retrieve the object from the database, and once it has been materialized, the proxy can pass the message directly to it. Subsequently, messages can be sent directly to the object by the proxy, or the proxy can replace the reference to itself in the object that sent the message with a reference to the real object. For this to work, the proxy must hold the object identifier of the object that it is a placeholder for. When the object itself is retrieved from the database, the object identifier is effectively transformed into a reference to the object itself. The proxy class must also implement the same interface as the real class so that it appears to other objects as if it is the real thing.

Caches can be combined with this approach. The database broker can maintain one or more caches of objects that have been retrieved from the database. Each cache can be implemented as a Hashtable, using the object identifier as the key. Either a single cache is maintained and some mechanism is used to keep track of the state of each object, or six caches can be maintained:

- new clean cache—newly created objects,
- new dirty cache—newly created objects that have been amended,
- new deleted objects—newly created objects that have been deleted,
- old clean cache—objects retrieved from the database,
- old dirty cache—retrieved objects that have been amended,
- old deleted objects—retrieved objects that have been deleted.

As objects are changed, the broker must be notified so that it can move them from one cache to the other. This can be achieved using the Observer–Observable pattern: the object implements Observable, and the broker inherits from Observer.

When the transaction is complete, the broker can be notified. If the transaction is to be committed, the broker can process each object according to which cache it is in:

- new clean cache—write to the database,
- new dirty cache—write to the database,
- new deleted objects—delete from the cache,
- old clean cache—delete from the cache,

- old dirty cache—write to the database,
- old deleted objects—delete from the database.

The cache or caches can be used by the proxy object to check whether an object is already available in memory. When it receives a message, the proxy can ask the broker for the object, if it is in a cache, the broker will return a reference to it directly, if it is not in the cache, the broker will retrieve it. Figure 17.27 shows the associations between the broker class, the caches and the proxy.

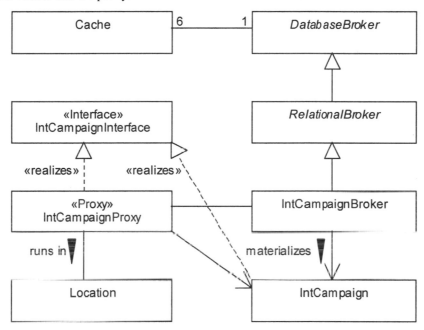

Figure 17.27 *Extension of the database broker framework to include caches and proxies.*

The collaboration between these classes can be seen in the collaboration diagram in Figure 17.28, which represents the following interaction.

A Location is in memory with an IntCampaignProxy as a placeholder for the real IntCampaign object that runs in that Location. In order to print a list of campaigns, the title of the IntCampaign is required.

1.1 The Location sends the message getTitle() to the IntCampaignProxy.
1.2 The IntCampaignProxy checks whether the IntCampaign is in memory.
1.3 It is not, so it then requests the object by object identifier from the broker, IntCampaignBroker.
1.4 The IntCampaignBroker checks if the object is in a Cache.
1.5 It is not, so the IntCampaignBroker retrieves the object from the database and returns it to the IntCampaignProxy.
1.6 The IntCampaignProxy sends the getTitle() message to the IntCampaign object and returns the result to the Location.

This may appear to be overkill, but it enables us to maintain our objective of decoupling the business classes from the data storage mechanisms that are provided by the classes in the persistence framework. This should make it possible to migrate the data storage to a different platform without having an impact on the business classes or the application logic, simply by replacing the database broker class with the broker for a different database.

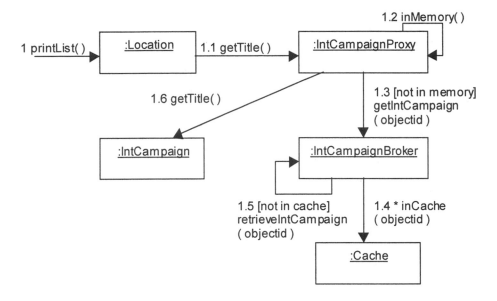

Figure 17.28 *Collaboration diagram showing proxy, broker and cache objects collaborating to retrieve an object instance.*

17.6 Summary

The design of persistent storage for object-oriented systems is not straightforward. For simple systems, it is possible to use files to store the objects. However, commercial systems require a more robust and sophisticated approach so that objects can be shared between applications and users. Database management systems provide the facilities to build robust, commercial strength information systems and offer a number of advantages. Object DBMS can be used and will have a less significant impact on the design of the classes in the system than if a relational DBMS is used. However, many organizations have an existing investment in a relational DBMS, and it may be necessary to build the system to use this database. In this case it is necessary to design tables, either by normalizing object instances or by following a set of guidelines for mapping classes and associations to tables. To decouple the business objects from the persistent storage mechanism, a persistence framework can be designed that can be extended to handle the resolution of object identifiers into references to real objects and that can use caches to manage transactions involving multiple objects.

The design of the persistent data storage mechanisms should ideally be carried out in conjunction with the object design activities of Chapter 13. If an object DBMS is being used, it may provide mechanisms by which to implement collection classes to handle associations (as discussed in Section 13.5), and class diagrams and interaction diagrams should reflect this. On the other hand, if a relational DBMS is being used, then the broker classes will play the role of collection classes and provide operations such as findFirst(), getNext() and findObject(), shown for example in Figure 13.15. If the object design is carried out before the data management design, then the collection classes that are produced as part of an implementation-independent design model can be converted to database broker classes at a later stage.

Review Questions

17.6 List in your own words the three steps used in going from an unnormalized relation to a relation in third normal form.

17.7 What are the three ways of mapping the classes in an inheritance hierarchy to tables?

17.8 What is meant by OML and ODL?

17.9 What is the difference between ODBC and CORBA as ways of connecting to databases?

17.10 Explain what is meant by (i) a broker and (ii) a proxy.

Case Study Work, Exercises and Projects

17.A Normalize the data in the Agate invoice in Figure 5.1. (Remember to remove the calculated values first.)

17.B Normalize the data in the FoodCo sales order entry screen of Figure 15.3. (Remember to remove the calculated values first.)

17.C Use the guidelines in Section 17.3.2 to decide on the tables necessary to hold the classes of Figure 13.12.

17.D Find information about a relational DBMS and an object DBMS and write a short report comparing the features they offer.

17.E Extend the sequence diagram of Figure 16.15 to show the use of a proxy class and database broker.

17.F Redraw your answer to 17.E as a collaboration diagram.

17.G Draw collaboration diagrams similar to Figure 17.27 to show what happens (i) when the IntCampaign is already in memory, and (ii) when it is in one of the caches.

Further Reading

Codd's (1970) paper on relational databases was reprinted in the 25th Anniversary issue of the Communications of the ACM, which is more likely to be in your library (1983, 26(1) pp.64–69). This 25th Anniversary Issue is well worth looking at for other papers by some of the greats of computer science.

Silberschatz et al. (1996) provide a good overview of database theory.

Loomis (1995) deals with background to object databases and the functionality they offer, whereas Eaglestone and Ridley (1998) present the ODMG standard and provide a worked example case study using O_2.

There are links to the web pages of the standards bodies—ANSI, ODMG and OMG—and to the web pages of various database vendors in the web site for this book.

A3

Agate Ltd Case Study—Excerpt from Design Specification

Agate Ltd

A3.1 Introduction

In this chapter we bring together some models (diagrams and supporting textual information) from different chapters in the third part of the book. In Chapters 11 to 17 we have
introduced or further developed the following UML diagrams:
- design class diagram
- behaviour diagrams
  - o statechart diagram
  - o interaction diagrams
    - – sequence diagram
    - – collaboration diagram
- implementation diagrams
  - o component diagram
  - o deployment diagram

As in the Requirements Specification (Chapter A2), we do not have the space to present
either a complete or a final Design Specification. However, we have included a sample of
the diagrams and other information in this chapter firstly to illustrate the kind of material
that should be brought together in a design specification at the end of the design phase of a
project, and secondly to allow the reader to see more easily how the elements of different
models can be cross-referenced and cross-checked with one another.

A3.2 Design Class Diagram

A revised analysis class diagram is included in Figure A3.1 for the Campaign class and
some of the classes with which it has associations. The Campaign class appeared to have
grown very large, and its responsibilities were no longer cohesive. We have reviewed it
from an analysis perspective and removed some of the responsibilities from the Campaign
class itself and given them to three new classes: CampaignBudget, CampaignSchedule
and CampaignSurvey. These new classes are more cohesive.

In Figure A3.2, we show a design class diagram for the classes in Figure A3.1. We have
added object identifiers to classes and, on the associations, we have shown the navigability

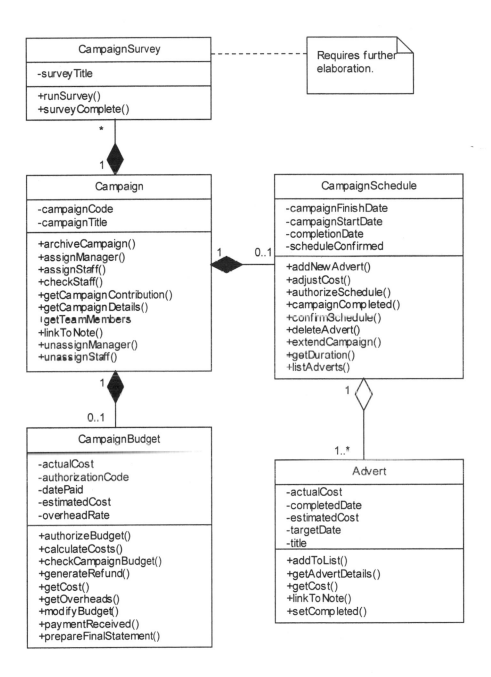

Figure A3.1 *Revised analysis class diagram for* Campaign *and related classes.*

that is supported by these identifiers. The complex state dependent behaviour that led to the use of the State pattern in Figure 14.12 is no longer a feature of the Campaign class. Much of that behaviour is now the responsibility of the three new classes, so we do not need to use the State pattern here.

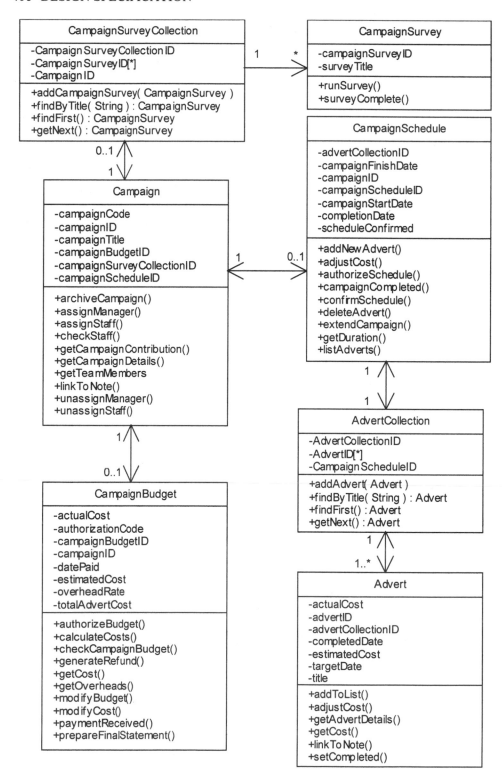

Figure A3.2 *Design class diagram for* Campaign *and related classes.*

Collection classes have also been added, and we have included the operation signatures of the operations of these collection classes.

A new attribute, `totalAdvertCost`, has been added to `CampaignBudget` in order to improve the performance of the `Check Campaign Budget` use case. This new attribute means that the `Campaign` does not have to obtain the cost of all its adverts, but can access the total cost directly. The result of this is that we require the synchronizing operation `adjustCost()` in `Advert`, which sends a `modifyCost()` message to the `CampaignBudget` when the `Advert`'s cost is changed or a new `Advert` is added.

One result of breaking the `Campaign` class down is that we have broken the encapsulation of the `Campaign` and exposed its inner workings. Classes that require services from `Campaign` will now have to deal directly with its component classes. One solution to this problem is to add the operations of the component classes back into `Campaign`. These operations will each be implemented by only a couple of lines of code, as they will delegate responsibility to the operations of the component classes. An example of a similar solution, using the Façade pattern, is shown in Chapter 19.

One other class diagram has been included as an example to show classes in other packages that provide the user interface (Figure A3.3). Figure A3.4 shows a class diagram using the package notation to show the package dependencies of the business and user interface classes.

We are assuming that the Agate campaign management system will be implemented using CORBA to provide distributed access to an ODBMS, so there is no need for the brokers and proxies of Figure 17.27. The collection classes will be implemented using the collection classes available in the ODBMS, as for example in the ODL of Figure 17.16 or the sample ObjectStore code of Figure 17.18.

The use of CORBA means that instances of the interface classes will run on client machines and that the actual business classes will reside on different servers, perhaps one in each office with that office's local `Campaigns` etc. Only the interfaces to these classes (defined in CORBA IDL) will be present on the client machines. Chapter 19 includes more detail on the use of CORBA.

A class specification has been included for CampaignBudget, and two attribute specifications that show more detailed format information. Two operation specifications have also been included for the new synchronizing operations (see Figure A3.5).

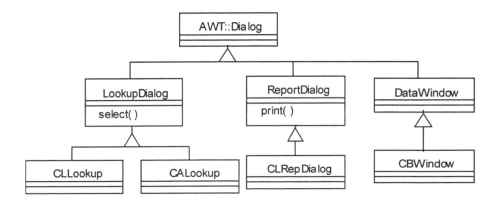

Figure A3.3 *Design class diagram showing interface classes.*

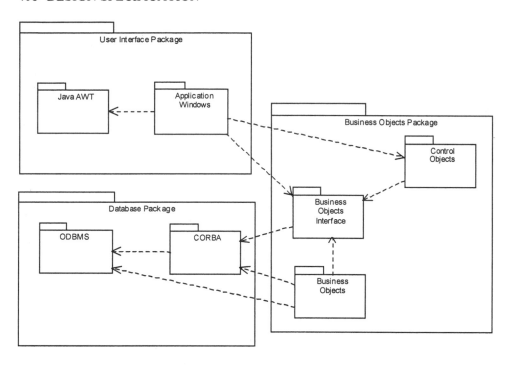

Figure A3.4 *Class diagram showing packages.*

Class description: **CampaignBudget**

Description: Each campaign has a budget associated with it. The budget is set up when the campaign is authorized. This class is responsible for all data concerned with the planned and actual cost of the campaign.

Synonyms: Budget.

Persistent: Yes. Using ODBMS on server. Stored on same server as associated Campaign.

Notes: The issue of whether a campaign should have more than one budget over time has been resolved. Historical information about previous versions of the budget is not required.

Attribute description: **datePaid**

Description: The date on which the final payment for a campaign is received. At this point a refund may be generated if an over-payment is made.

Format: 4-digit year. Should conform to local date format, e.g. dd/mm/yyyy in UK, mm/dd/yyyy in USA. Allow entry as 2-digit year, but validate and redisplay as 4-digit year. Defaults to today's date.

Constraints: `datePaid <= date`

Figure A3.5 *Selected class, operation and attribute specifications. (Continues on next page.)*

Attribute description: **totalAdvertCost**

Description: Sum total of all adverts for the campaign. Must be updated automatically whenever an advert is added or the cost of an advert is amended.

Format: Money value in local currency.

Constraints: totalAdvertCost =
Campaign.CampaignSchedule.AdvertCollection.getCost()->sum

Operation specification: **modifyCost**

Operation intent: Update value of totalAdvertCost.

 operation signature: CampaignBudget::modifyCost(aValue : Money)

 logic description (pre- and post-conditions):

      pre: aValue.oclIsTypeOf(Money)

      post:totalAdvertCost = totalAdvertCost@pre + aValue
         totalAdvertCost =
         Campaign.CampaignSchedule.AdvertCollection.getCost()->sum

Other operations called: None

Events transmitted to other objects: None

Attributes set: totalAdvertCost

Response to exceptions: raises InvalidMoneyException

Non-functional requirements: enables improved response to Check Campaign Budget.

Operation specification: **adjustCost**

Operation intent: Synchronizing operation with CampaignBudget.

 operation signature: Advert::adjustCost( )

 logic description (pre- and post-conditions):

      pre: self.estimatedCost <> null

      post: if (self.actualCost.oclIsTypeOf(Money))
         then (result = advertCollection.Campaign.CampaignBudget
         .modifyCost(self.actualCost))
         else (result = advertCollection.Campaign.CampaignBudget
         .modifyCost(self.estimatedCost-self.estimatedCost@pre))

Other operations called: CampaignBudget::modifyCost(aValue : Money)

Events transmitted to other objects: None

Attributes set: None

Response to exceptions: None

Non-functional requirements: enables improved response to Check Campaign Budget.

Figure A3.5 *Selected class, operation and attribute specifications. (Continued.)*

A3.3 Interaction Diagrams

Figure A3.6 shows a revised sequence diagram for the use case Check Campaign Budget. This uses the probe notation of Jacobson et al. (1992) and Allen and Frost (1998) to show a uses association with use cases Find Client and Find Campaign. The new CampaignBudget class is shown in the diagram as this now has the responsibility for

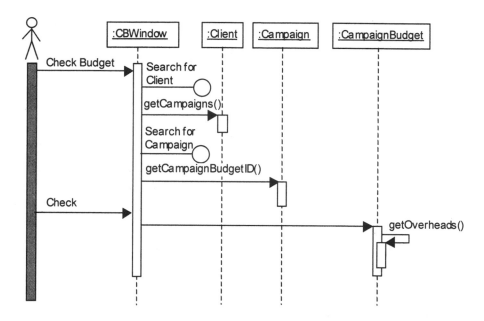

Figure A3.6 *Sequence diagram for* Check Campaign Budget.

financial aspects of the Campaign. The totalAdvertCost attribute is used rather than obtaining the cost of every advert, which was the design decision made to improve performance. The same use case is shown in the collaboration diagram of Figure A3.7. This shows the collection classes that are involved in the FindClient and FindCampaign use cases as part of this use case.

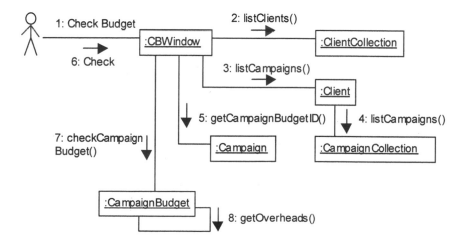

Figure A3.7 *Collaboration diagram for* Check Campaign Budget.

A3.4 Prototype Dialogue Design

The prototype dialogue design for the use case Check Campaign Budget from Chapter 16 is shown in Figure A3.8. The change of the implementation to include the new classes has a minimal effect on the interface classes.

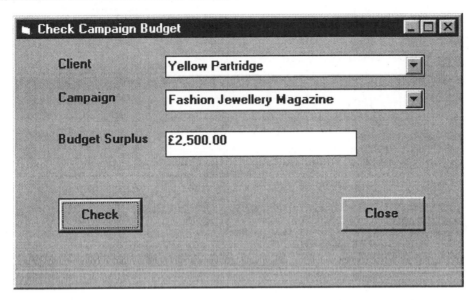

Figure A3.8 *Prototype interface design for* Check Campaign Budget.

A3.6 Statecharts for Interface

The behaviour of the user interface can be modelled with statechart diagrams. These are shown in Figure A3.9.

Figure A3.9 *Statecharts for user interface of* Check Campaign Budget. *(Continues on next page.)*

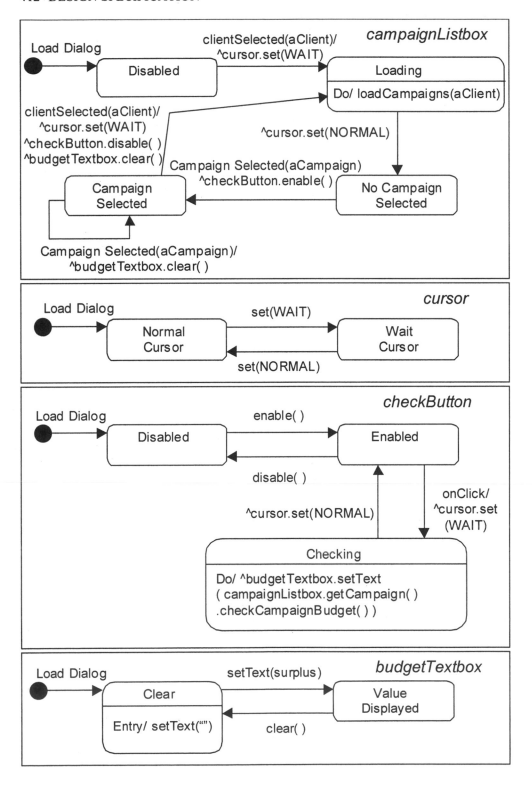

Figure A3.9 *Statecharts for user interface of* Check Campaign Budget. *(Continued.)*

18

Implementation

Objectives
In this chapter you will learn:
- how to draw component diagrams;
- how to draw deployment diagrams;
- the tasks that must be carried out in testing a new system;
- how to plan for data to be converted from an existing system;
- the alternative ways of introducing a new system into an organization.

18.1 Introduction

Implementation might be considered outside the scope of analysis and design. However, in projects that use rapid application development techniques, the distinction between different roles tends to break down (see Chapter 21). Analysts in particular may have a role during implementation in dealing with system testing, data conversion and user training. In this chapter we address some of the issues concerned with the implementation of systems and the role of analysts and designers in this process. We focus on the various UML diagrams that can be used to plan and document the implementation of the software. We also cover more general aspects of system implementation that are independent of the development approach that is taken.

A range of different software packages are required for implementation, for example, languages and database management systems. It is important to keep track of the relationship between different elements of the system such as source code files, object code files and software libraries. It is also important to maintain standards in the software development process: classes, objects and variables should be named in ways that make their purpose clear to others and make it possible to trace from analysis through design to code; programs should be self-documenting and well-structured. (This is all covered in Section 18.2.)

UML provides two diagrams that can be used to document the implementation of a system. Component diagrams are used to document dependencies between the different elements of the system (Section 18.3). They can also be combined with deployment diagrams (Section 18.4) to show how the software components relate to the physical architecture of the system. For a large system these diagrams may be an unwieldy way of documenting the implementation, and it may be simpler to use tables of information using a spreadsheet.

The testing of a new system is an essential part of implementation and includes testing of individual components, sub-systems and the complete system (Section 18.5). A major task when a new system is introduced is to take data from an existing system or systems and transfer it into the new system (Section 18.6). The existing data may be held on paper or in a computerized information system that is being replaced. Temporary staff may have to be employed in order to carry out the data conversion and may be required during the changeover to the new system. They and the existing staff will require training in how to use the system, and user documentation will have to be produced (Section 18.7). Finally, there are four different strategies for the introduction of a new system into an organization. These different approaches are appropriate to different circumstances, and each has its own advantages and disadvantages (Section 18.8).

18.2 Software Implementation

18.2.1 Software tools

The implementation of a system will require a range of tools. Ensuring that these are available in compatible versions and with sufficient licences for the number of developers who will be using them is part of the project management role. Many of these tools have been designed and developed to make the work of the system developer easier. In this section we describe each in turn.

CASE tools

Computer-aided software engineering tools allow the analysts and designers to produce the diagrams that make up their models of the system. CASE tools were discussed in detail in Section 3.6. Some of the diagrams in this book have been produced using a CASE tool. SELECT Enterprise and Rational Rose are two CASE tools that support UML notation. The data dictionary for the project should also be maintained using the CASE tool to link the textual and structured descriptions of every class, attribute, operation, state and so on to its diagrammatic representation.

To ensure that the implementation accurately reflects the design diagrams, it may be possible to generate code in one or more programming languages from the documentation in the CASE tool. SELECT Enterprise, for example, has code generators for languages such as Visual Basic, C++ and Java. It also supports the generation of the SQL statements to create relational database tables to implement the storage class diagram, and the generation of the CORBA IDL for a distributed system. The SQL of Figure 17.13 was generated in this way.

Some CASE tools also provide support for reverse engineering from code that has already been written into design models. When this is combined with code generation it is known as round-trip engineering.

Compilers, interpreters and run-time support

Whatever the language being used, some kind of compiler or interpreter will be required to translate the source code into executable code. C++ must be compiled into object code that can be run on the target machine. Smalltalk is interpreted, each command is translated as the program executes. Java is compiled into an intermediate bytecode format and requires a run-time program to enable it to execute. For applets, this run-time program is provided in the web browser, otherwise it is provided by the program called simply java or java.exe.

Visual editors

Graphical user interfaces can be extremely difficult to program manually. Since the advent of Visual Basic, visual development environments have been produced for a wide range of languages. These enable the programmer to develop a user interface by dragging and dropping visual components onto forms and setting the parameters that control their appearance in a properties window. All the user interface examples in Chapters 15 and 16 were produced in this way.

Integrated development environment

Large projects involve many files containing source code and other information such as resource files in C++. Keeping track of all these files and the dependencies between them, and recompiling all those that have changed as a project is being built is a task best performed by software designed for that purpose. Integrated development environments (IDEs) incorporate a multi-window editor, mechanisms for managing the files that make up a project, links to the compiler so that code can be compiled from within the IDE, and a debugger to help the programmer step through the code to find errors. An IDE may also include a visual editor to help build the user interface and a version control system to keep track of different versions of the software.

Class browsers

In an object-oriented system, a browser provides a visual way of navigating the class hierarchy of the application and the supporting classes to find their attributes and operations. Smalltalk-80 was the first language to provide this kind of browsing capability. Some IDEs now provide it. The Java Application Programming Interface (API) is documented in HTML and can be browsed with a web browser.

Component managers

Chapter 19 discusses how software reuse can be achieved through the development of reusable components. Component managers provide the user with the ability to search for suitable components, to browse them and to maintain different versions of components.

DBMS

A large-scale database management system will consist of a considerable amount of software. If it supports a client–server mode of operation, there will be separate client and server components as well as all the tools discussed in Section 17.2.6. To use ODBC or JDBC will require ODBC software installed on the client. For any database, special class libraries or Java packages may be required on the client either during compilation or at run-time or both. ObjectStore PSE Pro includes a post-processor that is used to process Java class files to make them persistent.

CORBA

An ORB is required in order to use CORBA. It will include the IDL compiler that takes interface definitions in IDL and produces the interface, stub and skeleton files necessary to use CORBA. The pure-Java Caffeine software that runs with Visigenic's Visibroker™ for Java ORB will generate the necessary classes directly from Java code, even from compiled class files.

Testing tools

Automated testing tools are available for some environments. What is more likely is that programmers will develop their own tools to provide harnesses within which to test classes and sub-systems. Section 18.3 covers testing in more detail.

Conversion tools

In most cases data for the new system will have to be transferred from an existing system. Where once the existing system was usually a manual system, most projects nowadays replace an existing computerized system, and data will have to be extracted from files or a database in the existing system and reformatted so that it can be used to set up the database for the new system. Packages like Data Junction® provide automated tools to extract data from a wide range of systems and format it for a new system.

Documentation generators

In the same way that code can be generated from the diagrams and documents in a CASE tool, it may be possible to generate technical and user documentation. In Windows there are packages such as Documentation Studio™ that can be used to produce files in Windows Help format. Java includes a program called javadoc that processes Java source files and builds HTML documentation in the style of the API documentation from special comments with embedded tags in the source code.

18.2.2 Coding and documentation standards

Even one person developing software on his or her own is likely to find at some point that they cannot remember the purpose of a class, an attribute or an operation in a program. On any project in which people collaborate to develop software, agreed standards for the naming of classes, attributes, operations and other elements of the system are essential if the project is not to descend into chaos.

Naming standards should have been agreed before the analysis began. In this book we have tried to conform to a typical object-oriented standard.

- Classes are named with an initial capital letter. Words are concatenated together when the class name is longer than one word. Capital letters within the name show where these words have been joined together. For example `SalesOrderProxy`.
- Attributes are named with an initial lower case letter. The same approach is taken as for classes by concatenating words together. For example `customerOrderRef`.
- Operations are named in the same way as attributes. For example `getOrderTotal()`.

There are other standards. In C++ one convention is to use *Hungarian* notation: all member variable (attribute) names are prefixed by an abbreviation that indicates the type of the member variable, for example b for a Boolean, i for an integer, f for a float and hWnd for a handle to a window object. This can be particularly useful in languages that are not strongly typed, like Smalltalk, as it helps to enforce the consistent use of the same variable for the same purpose by different developers.

Consistent naming standards also make it easier to trace requirements through from analysis through design to implementation. This is particularly important for class, attribute and operation names.

Not everything in a program can be deduced by reading the names of classes, attributes and operations. Beveridge (1996) gives five reasons for documenting code in a book on Java programming.

- Think of the next person. Someone else may be maintaining the code you have written.
- Your code can be an educational tool. Good code can help others, but without comments complicated code can be difficult to understand.
- No language is self-documenting. However good your naming conventions, you can always provide extra help to someone reading your code.
- You can comply with the Java coding standards. Your documentation will be in the same hypertext format as the Java API documentation.
- You can automate its production. The javadoc program is discussed below. It generates HTML from your comments.

Standards should be enforced for the way that comments are added to a program. This should include a block at the start of each class source file (or header file in C++) that describes the purpose of the class and includes details of its author and the date on which it was written. The amendment history of the source file can be included in this block. Every operation should begin with a comment that describes its purpose. Any obscure aspect of the code should be documented with a comment. If you are developing in Java, you can use the javadoc conventions to generate HTML documentation for your classes. You can embed HTML tags and javadoc tags in the comments, and javadoc will also use its own special tags to add information about the author and version to the HTML. The javadoc tags include:

@see classname—'See also' hypertext link to the specified class;

@version text— 'Version' entry;

@author name— 'Author' entry.

As well as technical documentation, there will be a need for user documentation that will be required for training users and for them to refer to once the system is in use. Standards need to be agreed with the users for the format of this documentation, which may be produced by analysts or by specialist technical authors.

18.3 Component Diagrams

In a large project there will be many files that make up the system. These files will have dependencies on one another. The nature of these dependencies will depend on the language or languages used for the development and may exist at compile-time, at link-time or at run-time. Component diagrams are one of the two types of implementation diagram in UML. Component diagrams show these dependencies between software components in the system. Stereotypes can be used to show dependencies that are specific to particular languages. Figure 18.1 shows a component diagram that represents the dependency of a C++ source code file on the associated header file, the dependency of the object file on both and the dependency of an executable on the object file. Stereotypes can be used to show

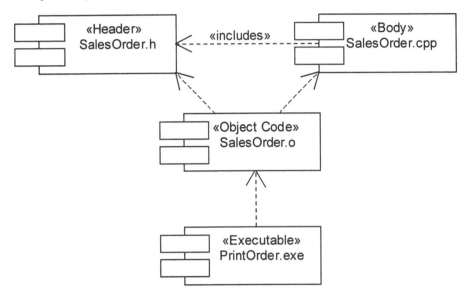

Figure 18.1 *Component diagram showing dependencies in C++.*

the types of different components.

An alternative representation of part of the diagram above is to use the UML interface notation to show the specification of a class (the header file in C++) as an interface and the

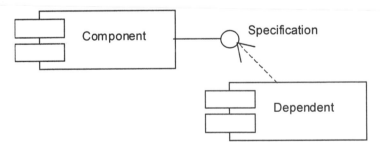

Figure 18.2 *Dependency of a component on the interface of another component.*

body as the component, as in Figure 18.2.

This notation can be used in Java to show the dependency of classes on the interfaces of other classes. This is particularly appropriate for distributed systems using CORBA in which applications running on a client are dependent on the interfaces of classes that are actually implemented on other machines on the network. Component diagrams need not be used at this low level, but can be used to show dependencies between large-scale components within a system as in Figure 18.3.

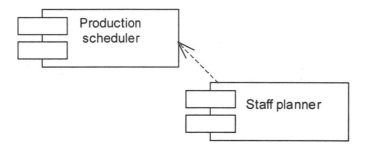

Figure 18.3 *Example dependency between higher level components.*

In Java, component diagrams can be used to show the dependency of classes on packages that contain the classes that they import. This is particularly important in a language such as Java where the availability of packages of classes at run-time is critical to the running of a program.

Active objects, typically processes running on a separate thread, can be shown in a component diagram. An example of this is shown in Figure 18.4.

Different authors use component diagrams in different ways. Fowler (1997) suggests that components correspond exactly to packages so that their use in deployment diagrams shows where each package is running on the system. Muller (1997) suggests that packages contain components as well as other packages. We would suggest the following distinction between them.

- Components in a component diagram should be the physical components of a system.
- During analysis and the early stages of design, package diagrams can be used to show the logical grouping of classes into packages.
- During implementation, package diagrams can be used to show the grouping of physical components into sub-systems.
- As we shall see in the next section, component diagrams can be combined with deployment diagrams to show the physical location of components of the system. The classes in one logical package may be distributed across physical locations in a physical system, and the component diagram and deployment diagram can be used to show this.

If you use component diagrams, it is advisable to keep separate sets of diagrams to show compile-time and run-time dependencies. However, this is likely to result in a large number of diagrams, and we suggest an alternative method of managing this information at the end of the next section.

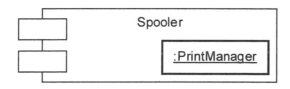

Figure 18.4 *Active object within a component.*

18.4 Deployment Diagrams

The second type of implementation diagram provided by UML is the deployment diagram. Deployment diagrams are used to show the configuration of run-time processing elements and the software components and processes that are located on them. They are made up of nodes and communication associations. Nodes are typically used to show computers and the communication associations show the network and protocols that are used to communicate between nodes. Nodes can be used to show other processing resources such as people or mechanical resources. Nodes are drawn as 3D views of cubes or rectangular prisms, and the simplest deployment diagrams show just the nodes connected by communication associations as in Figure 18.5.

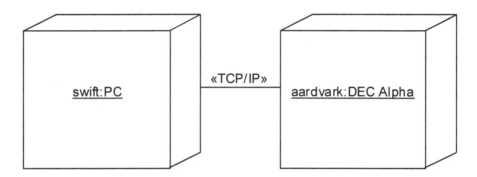

Figure 18.5 *Simple deployment diagram.*

Deployment diagrams can show either types of machine or particular instances as in Figure 18.5 where swift is the name of a PC. Deployment diagrams can be shown with components and active objects within the nodes to indicate their location in the run-time environment. Figure 18.6 shows the location of the Sales database on the server and some components on client PCs.

If you try to show all the components of a system in deployment diagrams they are likely to become very large or difficult to read. They can serve the purpose of communicating information about the location of key components to other members of the team or to users.

Figure 18.6 *Deploment diagram with components on PC clients and an active object on the server.*

Indeed, most computer professionals will have drawn an informal diagram like this at some time in their working lives to show where different parts of a system are to be located. Deployment diagrams show the physical architecture of the system.

If the aim of drawing component diagrams and deployment diagrams is to provide a complete specification of the dependencies between components at compile-time and run-time and the location of all software components in the implemented system, then this may be one of those cases where a picture is not worth a thousand words. Having implemented a system in Java that uses Java's class files, classes from a visual editor, JDBC drivers, CORBA and ObjectStore Persistent Storage Engine on PCs, and Oracle on a workstation, we know that keeping track of all these dependencies and documenting which components have to be on which machines is not a trivial task.

For most systems, this information may be easier to maintain in a tabular format, and a spreadsheet may be the best way of doing this.

Component diagrams can be replaced by a table that shows a list of all the software components down the rows and the same list across the top of the columns. It may be best to keep up to three tables (depending on the language used) for compile-time, link-time and run-time dependencies. For each case where a component is dependent on another, place a mark where the row of the dependent component intersects with the column of the component on which it is dependent. A simple example of this is shown in Figure 18.7.

In the same way, deployment diagrams can be replaced by a table that lists components down the rows and either types of machines or particular instances across the top of the columns. A mark is entered in the row–column intersection for every component that has to be on a particular machine or type of machine. If the exact location of components in a directory structure is important, then that location can be entered into the table. This is shown in Figure 18.8. Later this will form the basis of the information required for installing software onto users' machines for testing and eventual deployment.

Campaign database—compilation time dependencies				
	JDBC sun.jdbc.*	Campaign .java	Campaign Broker.java	Campaign Proxy.java
JDBC sun.jdbc.*				
Campaign. java				
Campaign Broker.java	✓	✓		
Campaign Proxy.java		✓	✓	

Figure 18.7 *Excerpt from example table to replace component diagram.*

Campaign database—run-time locations		
	Client PC	Database server
JDBCsun.jdbc.*	c:\jdbc	
Campaign.class	c:\agate\campaign	
CampaignBroker.class	c:\agate\campaign	
CampaignProxy.class	c:\agate\campaign	
SQL*Net		✓
OCI Listener		✓

Figure 18.8 *Excerpt from example table to replace deployment diagram.*

18.5 Software Testing

18.5.1 Who carries out the testing?

One view of testing is that it is too important to be left to the programmers who have developed the software for the system. This is not meant as a criticism of programmers but reflects the fact that it is important that testing is carried out by someone whose assessment of the software will be objective and impartial. It is often difficult for programmers to see the faults in the program code that they have written.

Some organizations employ specialist software testers. The following paragraph is an excerpt from an advertisement in the British computer press for a post as a tester.

> A leading financial institution has an opportunity for a systems tester to work on a business critical project. Testing throughout the project life cycle, you will liaise closely with developers and team leaders to implement test cases and organize automated testing scripts. All testing is organized within a fully automated environment. With a background in testing you will have strong business acumen...

However, not all organizations can afford the luxury of specialist testers. Most often the analysts who carried out the initial requirements analysis will be involved in testing the system as it is developed. The analysts will have an understanding of the business requirements for the system and will be able to measure the performance of the system against functional and non-functional requirements.

The systems analysts will use their knowledge of the system to draw up a test plan. This will specify what is to be tested, how it is to be tested, the criteria by which it is possible to decide whether a particular test has been passed or failed, and the order in which tests are to take place. Based on their knowledge of the requirements, the analysts will also draw up sets of test data values that are to be used.

The other key players in the process of testing new software are the eventual users of the system or their representatives. Users may be involved in testing the system against its specification, and will almost certainly take part in final user acceptance tests before the system is signed off and accepted by the clients.

18.5.2 What is tested?

In testing any component of the system, the aim is to find out if its requirements have been met. One kind of testing seeks to answer the following questions.

Does it do what it's meant to do?

Does it do it as fast as it's meant to do it?

This is equivalent to asking 'Never mind how it works, what does it produce?' and is known as *black box* testing because the software is treated as a black box. Test data is put into it and it produces some output, but the testing does not investigate how the processing is carried out. Mainly, this tests the analyst's understanding of the user's requirements. It is also necessary to check how well the software has been designed internally. This second kind of testing seeks to answer the following question.

Is it not just a solution to the problem, but a *good* solution?

This is equivalent to asking 'Never mind what it's for, how well does it work?' and is known as *white box* testing because it tests the internal workings of the software and whether the software works as specified. This tests the programmer's skill and understanding of the analyst's specification. In a project where reusable components are bought in, it may not be possible to apply white box testing to these components, as they may be provided as compiled object code. However, some suppliers will supply source code as well as compiled code, and there is a growing movement for suppliers to make source code available. As an aside, some organizations require, as part of their software contracts, that source code is placed in *escrow*. This means that a copy of the source code is lodged with a third party, usually a lawyer or a trade association, so that it is available to the client if the software company goes out of business. By this means the client ensures that they will be able to maintain and enhance the software even if its original developers are no longer able to.

Ideally, testers will use both white box and black box testing methods together to ensure:
- completeness,
- correctness,
- reliability, and
- maintainability.

However, the aim of any kind of testing is always to try to get the software to fail—to find errors—rather than to confirm that the software is correct. For this reason the test data should be designed to test the software at its limits, not merely to show that it copes acceptably with routine data.

Testing can take place at as many as five levels:
- unit testing,
- integration testing,
- sub-system testing,
- system testing, and
- acceptance testing.

In an object-oriented system, the units are likely to be individual classes. Testing of classes should include an initial *desk check*, in which the tester manually walks through the source code of the class before compilation. The class should then be compiled, and the compilation should be clean with no errors or warnings. To test the running of a class the tester will require some kind of test program (the term *harness* is often used) that will create one or more instances of a class, populate them with data and invoke both instance methods and class methods. If pre-conditions and post-conditions have been specified for operations, as suggested in Chapter 9, then the methods that have been implemented will be tested to ensure that they comply with the pre-conditions and that the post-conditions are met when they have completed. Statechart diagrams can be used to check that classes are conforming to the behaviour in their specification.

It may be difficult to test classes in isolation. For the reasons that are discussed in Chapter 19 on reuse, most classes are coupled in some way to other classes in the system. Unit testing merges into integration testing when groups of classes are tested together. The obvious test unit at this point is the use case. The interaction between classes can be tested against the specification of the sequence diagrams and collaboration diagrams. User interface classes and data management classes will also have to be tested in conjunction with the classes in the application logic layer. If scenario-based design has been used (see Chapter 15), then the scenarios can form the basis for testing scenarios in which a use case can be tested against a typical business situation.

Use cases that share the same persistent data should be tested together. This is one form of sub-system testing in which the sub-systems are built around different business functions that make use of the same stored data.

Testing is sometimes described as taking place at three levels.

Level 1
- Tests individual modules (e.g. classes).
- Then tests whole programs (e.g. use cases).
- Then tests whole suites of programs (e.g. the Agate application).

Level 2
- Also known as Alpha testing or verification.
- Executes programs in a simulated environment.
- Particularly tests inputs that are:
  - o negative values when positive ones are expected (and vice versa),
  - o out of range or close to range limits,
  - o invalid combinations.

Level 3
- Also known as Beta testing or validation.
- Tests programs in live user environment:
  - o for response and execution times,
  - o with large volumes of data,
  - o for recovery from error or failure.

A final stage of testing is *user acceptance testing* during which the system is evaluated against the original requirements before the client signs the project off. Documentation produced during analysis will be used to check the finished product.

Test no.	Test description	Test data	Expected result
234	Create a new Campaign.		Campaign Estimated Cost is set to £0.00
235	Add Advert 1 to Campaign.	Advert Estimated Cost = £500.00	Campaign Estimated Cost is set to £500.00
236	Add Advert 2 to Campaign.	Advert Estimated Cost = -£500.00	Negative value rejected. No change to Campaign Estimated Cost.
237	Add Advert 3 to Campaign.	Advert Estimated Cost = £300.00	Campaign Estimated Cost is set to £800.00
238	Set Advert 1 Completed.	Advert Actual Cost = £400.00	Campaign Estimated Cost is set to £700.00. Actual Cost is set to £400.00.

Figure 18.9 *Excerpt from test plan for Agate.*

18.5.3 Test documentation

Thorough testing requires careful documentation of what is planned and what is achieved. This includes the expected outcomes for each test, the actual outcomes, and for any test that is failed, details of the retesting. Figure 18.9 shows part of a test plan for the Agate case study. It shows details of each test and its expected outcomes. The results of the actual tests will be documented in a separate, but similar format. Many organizations have standard forms for these documents or may use spreadsheets to keep this information.

18.6 Data Conversion

When a new system is introduced, data from existing systems will have to be entered into the new system. The organization may have a mixture of existing manual and computerized systems that will be replaced by the new system. The data from these systems must be collated and converted into the necessary format for the new system. The timing of this will depend on the implementation strategy that is used (see next section), but it is likely to be a costly task, involving the use of staff time, the employment of temporary staff or the use of software to convert data from existing computer systems. These costs should have been identified in any cost benefit analysis that was carried out at the inception of the project.

If data is being collated from existing manual systems, it may be necessary to gather it from different sources. Data may be stored in different files, on index cards, in published documents, such as catalogues, or in other paper-based systems. If this data is going to be entered manually into the new system, by users keying it in, then the designers should draw up paper forms that can be used to collate the information so that it is all in one place when it is keyed in. Some data will only ever be entered when the system is started up, for example codes that are used in the system and will not be altered. Special data entry windows will be required for this kind of one-off activity.

Data from existing computer systems will have to be extracted from existing files and databases and reformatted to be usable with the new system. This provides an opportunity to clean up the data: removing out-of-date records and tidying up the values that are stored. Address and telephone number attributes of existing systems are likely to have been abused or misused by users. The work of converting the data may be done by using special programs written by the developers of the system, by employing consultants who specialize in this kind of work or by using commercial software that is capable of reading and writing data in a variety of formats, or even, as is the case with packages such as Data Junction, capable of working out the format of an unknown file.

The tasks involved in data conversion can be summarized as follows.

- Creating and validating the new files, tables or database.
- Checking for and correcting any format errors.
- Preparing the existing data for conversion:
  o verifying the existing data for correctness,
  o collating data in special forms for input,
  o obtaining specially written programs to convert and enter the data.
- Importing or inputting the data.
- Verifying the data after it has been imported or input.

All the converted data may have to be ready for entry into the new system to meet a tight deadline, or it may be possible to enter it over a period of time. It is best to convert relatively static data such as product information and customer details first and leave dynamically changing files of information such as orders or other business transactions until last. The implementation strategy will determine the timescale for conversion.

18.7 User Documentation and Training

18.7.1 User manuals

As well as preparing the technical documentation for the system, analysts will be involved in producing manuals for end-users. The technical documentation will be required by the system manager and other staff responsible for running the system, and by staff who have to maintain the system. Ordinary users of the system, who will be using it to carry out their daily work tasks, require a different kind of documentation.

Users will require two kinds of manual. During training they will need training materials that are organized around the tasks that they have to carry out with the new system. On-line computer-based training materials can be developed so that users learn the tasks in a staged way. These may be in the form of self-study tutorials that users can work through independently of any formal training that is provided.

The users will also need a reference manual that they can refer to while they are using the system. The reference manual should be a complete description of the system in non-technical language. Many software companies employ technical authors to write manuals in language that users can understand. The manual should be organized for ease of use. This involves the author understanding how the user will carry out their tasks and the kind of problem that they will face. The manual should be organized around the users' tasks, and should be supplemented with a comprehensive index based on the terms that the users will be familiar with rather than the technical terms used by the system developers. Particular attention should be paid to exceptional circumstances and not just routine tasks.

The reference manual may be replicated in the on-line help so that the users can refer to it while they are using the system. However, it should also be available as a paper manual that the users can refer to if there is a problem with the system.

18.7.2 User training

Temporary staff and existing staff will have to be trained in the tasks that they will carry out on the new system. Analysts are likely to be involved in the design of the training programme, the development of training materials, the planning of the training sessions and the delivery of the training itself.

Training programmes should be designed with clear learning objectives for the trainees. They will be using the system, and it is important that the training is practical and geared to the tasks that they will be performing. If it is too theoretical or technical, they will not find it useful. Training should be delivered 'just in time'—when the users need it—as they will forget much of what they are told within a short space of time, so training delivered a few weeks before it is required is likely to be wasted. On-line computer-based training using video and audio materials that users can refer to when they need it is likely to be of most use. If formal training sessions are used, then trainees should be given learning tasks to take away and carry out in their workplace. This implies that they will be allocated adequate time for training—skimping on training in order to save money is likely to be counter-productive. Staff will not get the best out of the system and are likely to become frustrated if they do not understand how to work the system.

18.8 Implementation Strategies

There are four main strategies for switching over to the new system:
- Direct changeover
- Parallel running,
- Phased changeover,
- Pilot project.

Figure 18.10 shows three of these changeover strategies in diagram form. Each of them has its advantages and disadvantages.

Direct changeover means that on an agreed date users stop using the old system and start using the new system. Direct changeover is usually timed to happen over a week-end to allow some time for data conversion and implementation of the new system. This does not mean that everything happens in a couple of days, as preparatory work will have been carried out in advance. The advantages and disadvantages of this approach are:
+ the new system will bring immediate business benefits to the organization, so should start paying for itself straightaway;
+ it forces users to start working with the new system, so they will not be able to undermine it by using the old system;
+ it is simple to plan;
− there is no fallback if problems occur with the new system;

– the plan must work without difficulties for it to be a success.

Direct changeover is suitable for small-scale systems and other systems where there is a low risk of failure such as the implementation of established package software.

Parallel running allows the existing system to continue to run alongside the new system. The advantages and disadvantages of this approach are:

+ there is a fallback if there are problems with the new system;
+ the outputs of the old and new systems can be compared—so testing can continue;
– there is a high cost as the client must pay for two systems during the overlap period, and this includes the staffing necessary to maintain information in the old system as well as the new;
– there is a cost associated with comparing the outputs of the two systems;
– users may not be committed to the new system as it is easier to stick with the familiar system.

Parallel running should be used in situations where there is a high level of risk associated with the project and the system is central to the business operations of the organization.

In a *phased changeover*, the system is introduced in stages. The nature of the stages depends on the sub-systems within the software, but introduction into one department at a time may be appropriate. The advantages and disadvantages are:

+ attention can be paid to each individual sub-system as it is introduced;
+ if the right sub-systems can be chosen for the first stages then a fast return on investment can be obtained from those sub-systems;
+ thorough testing of each stage can be carried out as it is introduced;
– disaffection and rumour can spread through the organization ahead of the implementation if there are problems with the early phases;
– there can be a long wait before the business benefits of later stages are achieved.

Phased changeover is suitable for large systems in which the sub-systems are not heavily dependent on one another. The different phases can be introduced on a geographical basis rather than by department.

A variation on phased changeover is the use of a *pilot project* approach. This involves trialling the complete system in one department or on one site. The decision on extending the system to the rest of the organization depends on the success of the pilot scheme. The

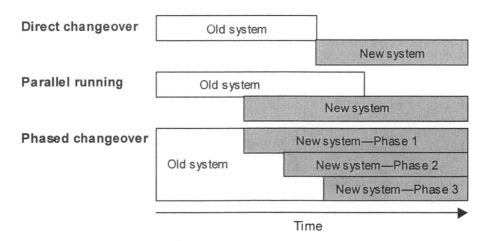

Figure 18.10 *Changeover strategies.*

pilot project can be used as a learning experience, and the system can be modified on the basis of the practical experience in the pilot project. As such, pilots are suitable for smaller systems, or packaged software, as it is unlikely that a large-scale system will be developed and then introduced in a way that makes its full-scale implementation dependent on the success of a pilot.

18.9 Summary

The implementation of a new system involves a large number of different types of software package that are used to produce and to support the finished system. Component diagrams and deployment diagrams are the two UML implementation diagrams that can be used to document the software components and their location on different machines in the system. For large, complex installations, these diagrams may become unwieldy, and a table format in a spreadsheet may be easier to maintain.

Analysts and designers have a role during the implementation stage in maintaining system and user documentation and in providing user training. They also plan and carry out testing, plan for data conversion from existing systems and assist the project management in planning the appropriate implementation strategy for the system.

Review Questions

18.1   List the different categories of software that may be used in developing a system.
18.2   What packages have you used and which categories do they fall into?
18.3   What is the difference between a package diagram and a component diagram?
18.4   Draw a component diagram to show the run-time dependency between a Java class file, the java.exe run-time program and the Java classes in a zip file.
18.5   Draw a deployment diagram to show how a web browser and web server are located on different machines and the communication protocol they use.
18.6   List five tests that you would carry out on the FoodCo use case `Start Line Run`.

Case Study Work, Exercises and Projects

18.A   Draw a component diagram for a program that you have written.
18.B   Find out about the Library system in your school, college, university or local library. Draw a deployment diagram to show the physical hardware architecture.
18.C   The FoodCo system will have a database on a central server and PCs in different offices that access that server. Draw a deployment diagram to show this architecture.
18.D   There are commercial packages to automate the installation of software such as InstallShield™ for Windows and RPM for Linux. Investigate how these packages maintain information about what components must be installed on a machine.
18.E   Read the user manual or on-line help for a software package that you use. What examples can you find of good practice and bad practice in what has been written? How would you improve it?

19

Maintenance and Reuse

Objectives
In this chapter you will learn:
- the tasks that are undertaken as part of system maintenance and review;
- how to plan systems development to achieve reuse of components;
- the different techniques for obtaining and managing reuse.

19.1 Introduction

In the system life cycle, implementation is followed by maintenance, when the new system has any remaining bugs removed and enhancements made to it. Using object-oriented technology does not make the problems of removing bugs and enhancing systems go away, but it does add a further stage to the life cycle—the reuse stage. In this chapter we look at how object-oriented software can be reused. In particular we look at the idea of *componentware*—software packaged into components that can be reused as part of other systems.

Section 19.2 addresses the tasks of review and maintenance that take place after a new system has been implemented. In Section 19.3 we discuss approaches to developing component-based software: we raise issues concerning planning for reuse during the project and consider how to package software to make it reusable. We provide an example of industrial strength componentware and present an example from the case studies used in this book which demonstrates how software can be designed and packaged for reuse (Section 19.4). Chapter 21 addresses the project management issues associated with planning for reuse.

19.2 Review and Maintenance

19.2.1 The next steps

The work of analysts, designers and programmers does not stop after a system has been implemented. There is a continuing requirement for staff to work on the new system. Firstly, it is important that the organization reviews both the 'finished' product and the process which was undertaken to achieve it. This may be for contractual reasons, in order to check that the product meets requirements. However, there is a growing recognition of the need for organizations to learn from experience and to record and manage the organizational knowledge which results from this learning. If there were any problems during the lifetime of the project, then these should be reviewed and conclusions drawn about how they might be avoided in the future. The amount of time spent on different tasks during the project can be used as the basis for metrics to estimate the amount of time that will be required for similar tasks in future projects. Secondly, it is unlikely that the system will be working perfectly according to the users' requirements, and further work will have to be done. Thirdly, in an object-oriented project, the design should be reviewed to identify candidate components for future reuse, although as we suggest later in this chapter, planning for reuse should begin in the early stages of a project. Identifying reusable software components is not an activity to be left until the completion of the project.

19.2.2 The review process and evaluation report

The review process will normally be carried out by the systems analysts who have been involved in the project from the start, although it is possible to involve outside consultants in the process for an impartial view. They will normally be supported by representatives of users and user management. The various stakeholders who have invested time, money and commitment in the project will all have an interest in the content of the evaluation report. The report can be very detailed or can provide an overview evaluation—like everything else in the project, there will be a cost associated with producing it. The report's authors should consider the following areas.

Cost benefit analysis. The evaluation should refer back to criteria that were set for the project at its inception. It may not be possible to determine whether all the benefits projected in the cost benefit analysis have been achieved, but most of the costs of development, installation, data conversion and training will have been incurred and can be compared with the projections.

Functional requirements. It is important to check that the functional requirements of the system have been met. Clearly, this is something that should have been taking place throughout the lifetime of the project, but a summary can now be produced. Any actions that were taken to reduce the functional requirements, perhaps to keep the project within budget or on schedule, should be documented for future action under the heading of maintenance. Major bugs should similarly be documented if they have emerged since the implementation of the new system.

Non-functional requirements. The system should be reviewed to ensure that it meets the targets for non-functional requirements that were documented during the requirements analysis stage. It is now possible to assess whether quantitative objectives for learnability, throughput, response times or reduction of errors have been achieved.

User satisfaction. Both quantitative and qualitative evaluations of the users' satisfaction with the new system can be undertaken, using questionnaires or interviews or both. The results should be treated carefully, as users can pick on quite minor problems or be influenced in their views of the finished product by experiences during the project.

Problems and issues. As stated above, this is an important part of the evaluation process. Problems that occurred during the project should be recorded. These problems may have been technical or political, and it is important to handle the political issues with tact. Including a section criticizing unco-operative users or obstructive user management will mean that some readers take no notice of the rest of the report. Solutions to problems should also be included in the report, as should an indication of who should be learning from this part of the process.

Quantitative data for future planning. The evaluation report provides a place in which to record information about the amount of time spent on different tasks in the project, and this information can be used as the basis for drawing up future project plans. The quantitative data should be viewed in the light of the problems and issues that arose during the project, as the amount of time spent on a difficult task that was being tackled for the first time will not necessarily be an accurate predictor of how much time will be required for the same task in the future.

Candidate components for reuse. If these have not already been identified during the project itself, then they should be identified at this stage. There will be different issues to be addressed, depending on whether the project has been carried out by in-house development staff or external consultants. For in-house projects, the reuse of software components should be viewed as a process of recouping some of the investment made and being able to apply those reusable elements of the system in future projects. For projects undertaken by external consultants, it may highlight legal issues about who owns the finished software that should have been addressed in the contract at the start of the project.

Future developments. Any requirements for enhancements to the system or for bugs to be fixed should be documented. If possible, a cost should be associated with each item. Technical innovations that are likely to become mature technologies in the near future and that could be incorporated into the system in an upgrade should also be identified.

Actions. The report should include a summary list of any actions that need to be undertaken as a result of carrying out the review process, with an indication of who is responsible for carrying out each such action.

The end of a project provides a good opportunity for managers of the development staff to review performance on the project as part of a separate review process, the results of which are not necessarily in the public domain. This may feed into staff appraisal procedures and can affect the payment of bonuses, promotion prospects and staff development issues such as the choice of the next project for a member of staff to work on or their need for staff development.

19.2.3 Maintenance activities

Very few systems are completely finished at the time that they are delivered and implemented, and there is a continuing role for staff in ensuring that the system meets the users' requirements. As well as maintenance of the system, there will be a need for support of users: providing initial and on-going training, particularly for new staff; improving documentation; solving simple problems; implementing simple reports that can be achieved using SQL or OQL without the need for changes to the system software; documenting bugs that are reported; and recording requests for enhancements that will be dealt with by maintenance staff. In large organizations these tasks are often handled by a helpdesk that supports all the organization's systems, and it may be appropriate for a member of the development team to join the helpdesk staff either temporarily or permanently. Whether or not this happens, helpdesk or support staff will need to be provided with training so that they can support the new system.

Maintenance involves more significant amendments to a system once it is up and running. Maintenance may be required for a number of reasons.

- There will almost certainly be bugs in the software that will require fixing. This is despite the fact that it is claimed that object-oriented development techniques reduce the likelihood of bugs. The use of object-oriented encapsulation should mean that it is easier to fix bugs without creating knock-on problems in the rest of the system. It is sometimes suggested that bug-fixing involves spending as much time fixing bugs that were introduced by the previous round of maintenance as it does in fixing bugs in the original system.
- Users request enhancements to systems virtually from day one after implementation. Some of these will be relatively simple, such as additional reports, and may be dealt with by support staff, while others will involve significant changes to the software and will require the involvement of a maintenance team. Often these user requests will reflect the fact that until the system is running and users have a chance to see what it can do, it is difficult for them to have a clear idea of what their requirements are.
- In some cases, changes in the way that the business operates or in its environment, for example new legislation, will result in the need for changes to the system.
- Similarly, changes in the technology that is available to implement a system may result in the need for changes in that system.
- Disasters such as fires that result in catastrophic system failure or loss of data may result in the need for maintenance staff to be involved in restoring the system from data back-ups. Procedures for handling disastrous system failure should be put in place before disasters take place.

In each of these cases, it is necessary to document the changes that are required. In the same way as it is necessary to have a system in place during a project for handling users' requests for changes to the requirements (a change control system), it is necessary to have a system for documenting requests for changes and the response of the maintenance team. This should include the following elements.

Bug reports. These forms should describe the bug in as much detail as possible. In particular, it is necessary to document the circumstances in which the bug occurs so that the maintenance team can try to replicate it in order to work out the cause.

Requests for enhancements. These should describe the new requirement in a similar amount of detail. Users should rate enhancements on a scale of priorities so that the maintenance team can decide how important they are.

Feedback to users. There should be a mechanism for the maintenance team to feed back to users on bug reports and requests for enhancements. Assuming that bugs affect the agreed functionality of the system, users will expect them to be fixed as part of the original contract or under an agreed maintenance contract. The maintenance team should provide an indication of how soon each bug will be fixed. Enhancements are a different matter. Depending on the contractual situation, enhancements may be carried out under a maintenance contract or they may be subject to some kind of costing procedure. Significant enhancements may cost large amounts of money to implement. They will require the same kind of assessment as the original requirements. They should not be left to maintenance programmers to implement as they see fit, but should involve analysts and designers to ensure that the changes fit into the existing system and do not have repercussions on performance or result in changes to sub-systems that affect others. This process itself may incur significant costs just in order to work out how much an enhancement will cost to implement. Significant enhancements should therefore be regarded as mini projects in their own right, and involve the professional skills of project managers, analysts and designers.

Implementation plans. The maintenance team will decide how best to implement changes to the system, and this should be carried out in a planned way. For example, significant additions to a class that affect what persistent data is stored in the database will require changes to the database structure, and may also require all existing instances of that class to be processed in order to put a value into the new attribute. This will probably have to take place when the system is not being used, for example over a weekend. Enhancements to the system may fall into one of four categories: those to be made at no cost; those that will be made at a cost to be agreed with the client; those that will be held over until a major upgrade to the software is made and that will be part of a future version; and those that cannot or will not be made in the foreseeable future.

Technical and user documentation. Amendments to a system must be documented in exactly the same way as the original system. Diagrams and data dictionary entries must be updated to reflect the changes to the system. If this is not done, then there will be a growing divergence between the system and its technical documentation; this will make future amendments all the more difficult, as the documentation that maintenance analysts consult will not describe the actual system. Clearly, user documentation, training and help manuals as well as on-line help must all be updated.

In large organizations with many systems, staff in the information systems department may spend more time on maintenance of existing systems than they do on development of new systems. There is a growing movement for organizations to *outsource* their maintenance. This means handing over the responsibility for maintenance of a system to an external software development company under a contractual agreement that may also involve the provision of support. Some companies now specialize entirely in maintaining other people's software.

The other impact of the growth of the maintenance task is that the time will come when a system is such a burden in terms of maintenance that a decision must be made about its replacement. The review process that takes place at this stage may lead to the inception of a new project and to the systems development life cycle starting again.

19.3 Reuse of Components

19.3.1 Achieving reuse

In Chapter 3 reusability was discussed as one of the reasons for adopting object-oriented development techniques and programming languages, and in Chapter 11 we highlighted reusability as one of the characteristics of a good object-oriented design. Object-oriented systems have not achieved the level of reuse that was expected of them. There are two reasons for this.

- The first reason concerns the level of coupling between different classes in an object-oriented design. When we come to design classes for different systems, it may be possible to identify similar classes that could be developed in a way that makes them of use in more than one system, but more often than not, the implementations of these classes will include attributes that tie them in to the particular application of which they are a part.

- The second reason concerns the lack of standards for reusable object classes. This has changed recently with developments in the technology of repositories in which to store objects and with the introduction of standards such as the Object Management Group's CORBA Version 2.0 (OMG, 1995) and Microsoft's COM (Component Object Model) and DCOM (Distributed Component Object Model).

If we consider the first of these two reasons, there is a need for a Client or Customer class both in the Agate system and in the FoodCo system. During analysis, these two classes may look very similar, but as we move into design, the associations between these classes and others in their system will be resolved into specific attributes. The Agate Client class will have attributes to link it to Campaigns while the FoodCo Customer class will be linked to SalesOrders. If the development of both systems is being carried out by the same software company, then it requires a novel style of project management and organization to recognize this commonality in two different projects. If the commonality is recognized, then either a Client class can be obtained from elsewhere and subclassed differently for each project, or a Client class can be written that is domain-neutral and then subclassed for each different project. This inheritance-based approach also helps to solve a problem that is common with software that is tailored to the needs of different customers of the software house: it clearly separates those parts of the class that are common to all users from those that have been tailored to specific needs. This helps with the installation of upgrades and prevents the changes made for one customer being implemented for all customers.

However, it is in the nature of system development projects that reuse is most often addressed after the end of a project, at the same time as maintenance is taking place, when developers look at the classes in a project to see whether it is possible to reuse some of them in a subsequent project. Unless action has been taken to plan for reuse early in the project life cycle, it is likely that these classes will suffer from the problem of coupling to other classes mentioned above. Allen and Frost (1998) argue that the class is the wrong level of granularity at which to apply reuse. They argue that reuse should take place at the level of components rather than classes. They define a component as follows.

> A component is an executable unit of code that provides physical black-box encapsulation of related services. Its services can only be accessed through a consistent, published interface that includes an interaction standard. A component must be capable of being connected to other components (through a communications interface) to form a larger group.
>
> (Allen and Frost, 1998, p.4)

CORBA's Interface Definition Language (IDL) provides the kind of standard for the interface to components, while the mechanisms that CORBA provides for communication between distributed objects provide the communications interface. Microsoft's COM and DCOM provide similar standards and mechanisms.

Meyer (1997) also makes the case for reuse of what he calls *abstracted modules*, and states that the objective of his book is to devise the form of such abstracted modules and to explore their properties.

A number of programming languages and development environments provide mechanisms by which developers can package software into components. Table 19.1 lists some of these. This table shows that the search for ways of promoting reuse through some kind of modular architecture is not new in the software development industry. Reuse has been an objective that has driven the design of programming languages and has informed the development of programming styles. However, the potential for developing reusable components has been increased recently by three factors.

- The development of CORBA as a standard for interoperability of components written in different languages and running on different platforms.

- The promotion of Java as an object-oriented language with relatively straightforward mechanisms for producing software in packages to deliver different services.
- The growth of the Internet and the World Wide Web, which has made it possible for people to make their software components available to a wide marketplace of potential reusers.

Language or development environment	Mechanism for component reuse
Borland Delphi	Object Pascal units compiled into .dll files—Dynamic Link Libraries
Microsoft Visual Basic	.vbx files—Visual Basic Extensions .ocx files
Microsoft Windows	.ole files—Object Linking and Embedding DDE—Dynamic Data Exchange .dll files—Dynamic Link Libraries COM—Common Object Model DCOM—Distributed Common Object Model
CORBA	.idl files—Interface Definition Language IOP—Inter-ORB Protocol
Java	.jar files—Java Archive packages JavaBeans ™

Figure 19.1 *Languages and development environments with mechanisms for reuse.*

It is the platform independence of Java and CORBA that makes them different from the other languages and environments shown in Figure 19.1.

However, the existence of CORBA, Java and the WWW does not guarantee that effective reuse of components will take place. A strategy needs to be put in place within the organization to ensure that reuse is built into the systems development life cycle.

19.3.2 Planning a strategy for reuse

Allen and Frost (1998) describe the SELECT Perspective approach to the development of reusable components. At the level of practical techniques, this includes guidelines for the modelling of business-oriented components and for wrapping legacy software in component wrappers. They distinguish between reuse at the level of component packages, which consist of executable components grouped together, and service packages, which are abstractions of components that group together business services. The focus of this approach is to identify the services that belong together and the classes that implement them. Service classes in a single package should have a high level of internal interdependency and minimal coupling to classes in other packages.

In order to develop reusable components while achieving the development of a system to meet users' needs, the Perspective approach breaks the development process into two parts: the solution process and the component process. These two parts run in parallel and feed off each other.

- The solution process focuses on specific business needs and delivering services to meet the users' requirements. Its products have immediately definable business value. During the solution process, developers will draw on the component process in their search for reusable components that can be applied to the project.
- The component process focuses on developing reusable components in packages which group together families of classes to deliver generic business services. During the component process, the developers produce components that can be reused in the solution process. The component process also searches out opportunities to reuse services from existing legacy systems and legacy databases and from other packages of components.

Allen and Frost use the analogy of sowing and harvesting reusable services: the component process sows reuse and the solution process harvests services. Figure 19.2 shows the relationship between the two in diagrammatic form.

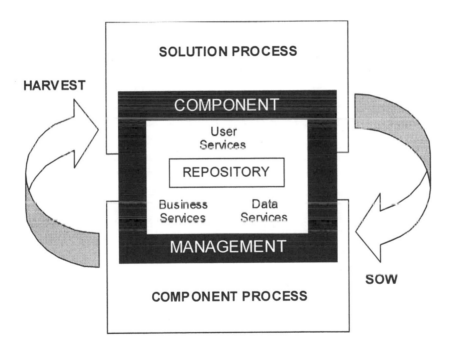

Figure 19.2 *The SELECT Perspective service-based process (adapted from Allen and Frost, 1998).*

Central to Figure 19.2 is a *repository*. Software support is needed for effective component reuse to take place. This support takes the form of repository-based component management software. Components are placed in the repository as a means of publishing them and making them available to other users. The repository is made up of catalogues and the catalogues contain details of components, their specifications and their interfaces. Component management software tools provide the functionality for adding components to the repository and for browsing and searching for components. Component management software may be integrated with CASE tools to allow the storage of analysis and design models as well as source code and executables.

19.3.3 Commercially available componentware

Until recently, most commercially available components took the form of utilities or graphical user interface components. The best example of this is the wide variety of controls that are available for use with Microsoft Visual Basic. Originally these were supplied as add-ins in the form of .vbx files which could be included in the Visual Basic toolbar in the same way as the built-in controls, or in the form of OLE (Object Linking and Embedding) objects which could allowed the functionality of other software packages such as word-processors to be embedded in applications. With the introduction of ActiveX®, based on the Microsoft COM architecture, these add-ins are now available as .ocx files. If you look through the catalogue of a good software supplier that sells development tools, you will find pages of ActiveX controls that can be used in your applications and that provide the developer with the possibility of building a wide range of different functions into their software without having to reinvent the wheel. Examples include:

- serial communications,
- computer-aided design,
- project management including Gantt charts,
- spreadsheets,
- scientific charts,
- barcode reading and printing.

The use of standardized mechanisms to access the functionality of these controls has meant that other software companies can also write interfaces to them, and it is just as possible to use ActiveX controls in Borland Delphi or in Java (using Microsoft Visual J++®) as it is in Visual Basic or Visual C++.

For applications written in Java, however, there is another mechanism that can be used: the JavaBean. JavaBeans (or just Beans) are components written in Java that typically combine a graphical element with the functionality to support it. Beans support the Component–Container model in the Java Abstract Windowing Toolkit (AWT), which means that Beans that encapsulate access to particular domain objects can be added into applications in the same way as other visual components can be added to the applications.

Most of these add-in controls provide generic capabilities rather than reusable components for particular types of business operations. Suppliers are now producing reusable components for business functions. Most notable among these developments is IBM's San Francisco project.

The San Francisco project provides distributed, server-based components for different types of business processes. San Francisco uses a layered architecture (shown in Figure 19.3). The Foundation layer provides a programming model to support distributed transactions, and uses a set of distributed object services and utilities written entirely in Java. It also provides an optional GUI framework written using JavaBeans. The Common Business Objects layer implements general purpose business objects together with the facts and rules required for any business application. This includes business objects such as company, business partner, address and calendar. Four components are provided in the Core Business Processes layer at present. These are:

- general ledger,
- accounts receivable and accounts payable,
- warehouse management, and
- order management.

These have been built using design patterns and provide support for electronic commerce.

Figure 19.3 *Layered architecture of the San Francisco project (adapted from IBM, 1998).*

It is IBM's intention to extend the range of components in the Core Business Processes layer. Other developers are likely to follow suit with the development of frameworks for delivering reusable and customizable components that can be used in software projects.

19.4 Case Study Example

A common feature of many applications is the need to control the access of members of staff to the different programs that make up a system. A non-functional requirement both for Agate and for FoodCo is to restrict access of staff to the use cases that they are permitted to use. This requirement can be based on the use case diagrams that have been produced during the development of the project, and can be summarized as follows.

> *Each program in the system will be represented by a use case in the use case diagram. One or more actors will be associated with each use case, and each actor may be associated with more than one use case. A member of staff will fill the role of one or more actors, and each actor will be represented by one or more members of staff. Each actor will be authorized to use specific use cases (programs). A member of staff may only use those use cases (programs) for which one of the actor roles they fill has been authorized.*

This requirement can be modelled in a class diagram in the same way as the business classes that meet the functional requirements of the system. Figure 19.4 shows the class diagram for this requirement. (A Password attribute has been added to three classes to reflect the fact that passwords may be associated with individual staff members, the actors or the use cases.) There will be a need to maintain persistent data to support this security system in exactly the same way that there is a need for data for the main application, and as such there will be a need to define the use cases that are needed to maintain this data. Figure 19.5 shows the use case diagram for the security system. Each of these use cases

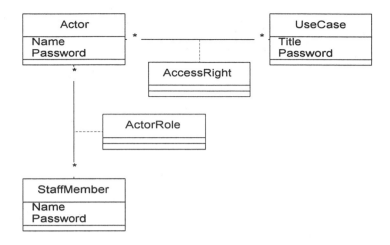

Figure 19.4 *Initial class diagram for security requirement.*

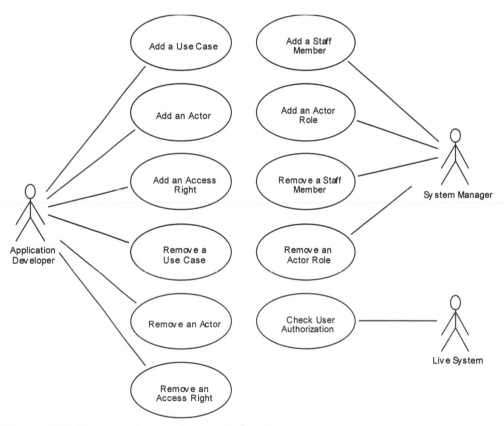

Figure 19.5 *Use case diagram for security requirement.*

will have a sequence diagram to model the interaction that takes place between objects, and Figure 19.6 shows a sample sequence diagram for the use case Add an Actor Role. The security requirement is not part of the business requirements of the domain, and we want to reuse the security software in other applications, so it makes sense to separate these

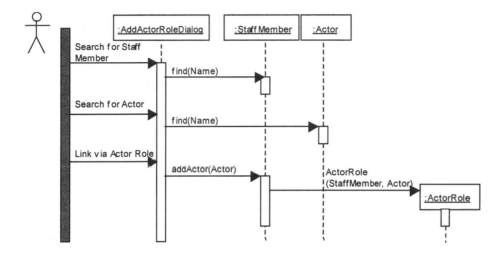

Figure 19.6 *Sequence diagram for* Add an Actor Role.

classes from the rest of the software and put them in a package of their own. The security classes will require their own user interface classes which will be a separate package within the overall security package. They will have dependencies on other packages that provide these services, such as the Java AWT. The application logic classes that control the use cases for the main application will have a dependency on the security classes. Figure 19.7 shows these package dependencies.

We might choose to design the system so that when a user starts running the application, they are prompted for their name or user ID and a password. Alternatively, if they are required to log into a network anyway, the software could obtain the user's ID from the network operating system. Each time a user starts a new use case in the main application, the application will need to check with the security classes whether that user is authorized to use that particular use case.

Clearly, the security package itself should not be accessible to unauthorized users, so

Figure 19.7 *Package diagram showing security classes and dependencies.*

there will be a requirement for the security package to use its own services to restrict access to the developers of the system, who will set up the associations between use cases and actors, and to the system manager who will authorize staff within the company to use particular use cases by linking them to specific actor roles.

One implication of introducing this security package is that the programmers writing the code for the main application will have to understand the security package as well as the application itself. One way of making this easier for them is to provide a single object as an application programming interface (API) to which all messages concerning the security package can be sent. There is a design pattern that we can use to do this, called the Façade pattern. (See Chapter 14 for more on design patterns.) The Façade pattern is a Gang of Four structural pattern, and is described in Gamma et al. (1995) in the following terms.

Intent

Provide a unified interface to a set of interfaces in a subsystem. Façade defines a higher-level interface that makes the subsystem easier to use.

....

Applicability

Use the Façade pattern when

- you want to provide a simple interface to a complex subsystem ...

- there are many dependencies between clients and the implementation classes of an abstraction. Introduce a façade to decouple the subsystem from clients and other subsystems, thereby promoting subsystem independence and portability.

- you want to layer your subsystems ...

(Gamma et al., 1995, pp. 185–186)

The structure of the Façade pattern is shown in Figure 19.8. We can use this structure by adding a single class, which we shall call `SecurityManager` which provides the API to the functionality in the security package. This is shown in the class diagram in Figure 19.9.

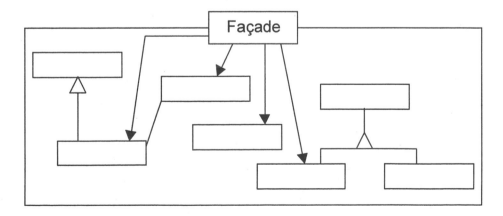

Figure 19.8 *Structure of Façade pattern (adapted from Gamma et al., 1995, p.187).*

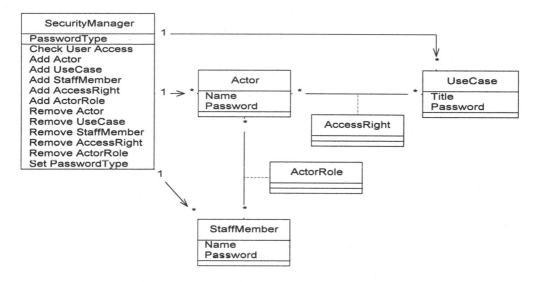

Figure 19.9 *Class diagram showing* `SecurityManager` *façade class.*

In order to make the security package as reusable as possible, it either needs to make use of whatever data storage mechanisms are used in the application with which it is supplied, or it needs to have its own mechanism for persistent storage. The simplest approach is to provide the security package with its own persistence mechanism. Given the relatively small volumes of data that are involved, it is possible for the persistence to be provided by using a system of files. (Alternatively, the persistence services could be provided by a persistence framework such as ObjectStore PSE or by CORBA.) If the data is to be stored in simple files, then it should be encrypted before storage. An encryption package could be added.

The security package could be implemented in Java, compiled and stored in a Java Archive file (.jar). This can be shown in a component diagram, as in Figure 19.10, which shows the dependency of the Agate application logic classes (also stored in a .jar file) on the Security Manager interface that the security package presents. A similar diagram could be drawn for FoodCo's project, and this component can be used in both the Agate and FoodCo projects and in other future projects.

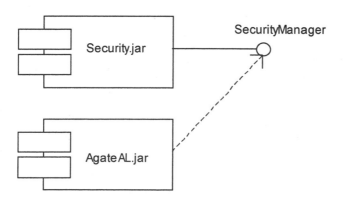

Figure 19.10 *Component diagram showing dependency on the* `Security` *package.*

19.5 Summary

After a new system has been implemented, it is customary and advisable to carry out a post-implementation review. This will result in the production of an evaluation report to stakeholders that will measure the success of the project and identify issues and problems from which the organization should learn. Typically the evaluation report will include the following sections.

- Cost benefit analysis.
- Functional requirements.
- Non-functional requirements.
- User satisfaction.
- Problems and issues.
- Quantitative data for future planning.
- Opportunities for reuse.
- Future developments.
- Actions.

New systems rarely work exactly as expected, and maintenance must be carried out in order to ensure that the system is bug-free and meets users' requirements. Procedures must be put in place for the maintenance team (project manager, analysts, designers and programmers) to document the process and the changes that are made.

In object-oriented systems, the maintenance stage offers an opportunity to identify potential sub-systems for reuse. Reusable components include groups of classes that are closely linked and which provide coherent business services. The SELECT Perspective is a methodology that focuses on techniques and project planning to achieve reuse. For reuse to be effective, it should not be left until the maintenance stage but should be planned into the project from the outset. The introduction of CORBA, Java and the WWW make it easier for developers to produce and distribute reusable software components. Commercially available components include the products of IBM's San Francisco project which provides business components for general ledger, for accounts receivable and payable and for warehouse and order management. The Façade pattern provides a means for hiding the complexity of components from application programmers by means of a class that handles the API to the classes in the component.

Review Questions

19.1   List the sections that you would include in a post-implementation evaluation report and explain the content of each section.

19.2   What is the difference between maintenance work carried out to fix bugs and work carried out to add requested enhancements to a system?

19.3   Why should decisions about enhancements not be left to maintenance programmers?

19.4   What tasks do maintenance staff undertake?

19.5   Give a definition of a 'component'.

19.6   Name three mechanisms for creating reusable components in different programming languages.

Case Study Work, Exercises and Projects

19.A What would you include in the forms for users to fill out for bug reports and enhancement requests? Produce a draft design for each of these two forms.

19.B Describe the mechanisms that are available in an object-oriented language with which you are familiar for creating reusable components.

19.C The data encryption package to be added to the security package is required to provide services to encrypt different data types, such as the `Integer`, `String` and `Date` classes. Draw a class diagram to show new classes `CryptInteger`, `CryptString` and `CryptDate`. Encryption and Decryption services are to be provided by a single Façade class called `EncryptionManager`. Include this in your class diagram.

19.D Redraw the package diagram in Figure 19.7 to include the encryption package and add any new dependencies necessary.

19.E Draw a deployment diagram for FoodCo showing the security package residing on a server and the application logic for the system on a separate server. (You may want to refer back to Chapter 18.)

19.F Complete the design of the security package and implement it in an object-oriented language such as Java, Smalltalk or C++. (This is suitable for a coursework assignment or small project.)

Further Reading

Allen and Frost (1998) provide an argument for component-based development, a worked example of a case study based on the SELECT Perspective, their methodology, and guidelines on how to plan for reuse. For a comparison of the mechanisms involved in CORBA and DCOM (and some other approaches to distributed systems), Orfali and Harkey (1998) provide a clear and readable coverage which includes detailed instructions on how to implement systems which use these techniques.

To find the wide range of software add-ins that are available as ActiveX controls, look at the web-site of a supplier such as System Science: www.systemscience.co.uk (in the UK). Look at IBM's web-site at www.ibm.com/java/sanfrancisco/ for information about the IBM San Francisco project. Details of The SELECT Perspective, SELECT Component Manager® and SELECT Software Tools' other products are available at www.selectst.com.

20

Systems Development Methodologies

Objectives
In this chapter you will learn:
- what a systems development methodology is;
- why methodologies are used;
- the need for different methodologies;
- the main features of one methodology.

20.1 Introduction

Until this point in the book, our main focus has been on describing the techniques used during the analysis and design of an object-oriented information system. Underpinning this is an implicit assumption that projects move from one task to another, sometimes in sequence, sometimes looping back to repeat a task if it requires further work, sometimes choosing between alternative ways of proceeding, but always with a particular purpose in mind, and always under someone's control. In practice, the planning, organization and control of any large-scale project is itself a significant problem, and developers will need guidance on a wide range of questions that arise when the techniques are applied in a real-life situation. The use of a set of modelling or documentation standards (such as that provided by UML) has an important part to play, but is not on its own enough. There is a need for a higher level view of the system development process, which promotes ways of organizing and managing the many different activities involved. This need is partly met by systems development methodologies. However, while the use of an appropriate methodology can help to guide the system development process and thus reduce the impact of many of the problems introduced in Chapter 2, this too cannot guarantee success.

In this chapter we describe what is meant by 'methodology', differentiating this from 'method' (Section 20.2). We discuss some reasons why methodologies are widely used in IS development, and also why they continue to evolve today (Section 20.3). Some leading object-oriented methodologies are briefly described (Section 20.4), and one, Objectory, is given a more detailed treatment (Section 20.5). We discuss some relative strengths and weaknesses of several different methodologies. These factors are then considered in light of the need to select a methodology that is suited both to the organizational context and to the kind of projects to be carried out (Section 20.6). Finally, we touch on some of the competing intellectual traditions that gave rise to the many methodologies in use today (Section 20.7), in particular the distinction between the so-called soft and hard views of systems development.

20.2 'Method' and 'Methodology'

The techniques of system development must be organized into an appropriate developmental life cycle if they are to work together. For example, once an analyst has completed a first draft class model, should he or she next construct ISDs for the main use cases, or begin to write operation specifications? Both of these need to be completed at some point, but how is the analyst to know which is more appropriate at this specific point in the project? The class model itself contains nothing that helps to make this decision.

The organization of tasks is not contained within the techniques themselves, and must be described at a higher level of abstraction. The *method* of a project is the term given to the particular way that the tasks in that project are organized. Sometimes this is called the *process* of software development, although process can also have a more all-embracing meaning, that includes what the tasks are, how they are carried out and how they are organized.

The words 'method' and 'methodology' are used interchangeably by many authors, but their meanings actually differ in a significant way. In order to plan and organize for the next project, project managers must be able to think at a still higher level. It is at this level that the term 'methodology' applies. A method is a step-by-step description of the steps involved in doing a job. Since no two projects are exactly alike, any method is specific to one project. A methodology is a set of general principles that guide a practitioner or manager to the choice of the particular method suited to a specific task or project. Or, to put it in familiar object-oriented terms, a methodology is a type while a method is its instantiation on one project. Figure 20.1 summarizes the different levels of abstraction involved.

Increasing level of abstraction	Example of application	Typical product
Task	Developing a first-cut class diagram for FoodCo.	A specific version of the FoodCo class diagram.
Technique	Description of how to carry out a technique, e.g. UML class modelling.	Any UML class diagram.
Method	Specific techniques used on a particular project (e.g. FoodCo use cases, class model, collaboration diagrams, etc.) that lead to a specific product.	FoodCo's product costing system.
Methodology	General selection and sequence of techniques capable of producing a range of software products.	A range of object-oriented business applications.

Figure 20.1 *Increasing level of abstraction in defining a project.*

An analogy might be the difference between the method followed by an amateur cook, who prepares a dish by following the recipe in a book, and a master chef, who has a deep understanding of ingredients, flavours, cookery tools and techniques. The amateur may use the recipe in a prescriptive way, while by contrast, a master chef may use the recipe as a guide, but will also vary the steps and ingredients in a creative way. This approach of selecting and adapting existing methods could reasonably be called a methodology. The difference lies in not always being bound slavishly to follow specific, pre-defined methods.

A methodology in the domain of IS must cover a number of aspects of the project, although coverage varies from one to another. Avison and Fitzgerald (1988) describe a methodology as a collection of many components. Typically, each methodology has procedures, techniques, tools and documentation aids that are intended to help the system developer in his or her efforts to develop an information system. There is usually also some kind of underlying philosophy that captures a particular view of the meaning and purpose of information systems development.

Checkland (1997), in a conference address that discussed the potential contribution of the wider systems movement to information systems development, gave a more general definition that captures well the notion of methodology as a guide to method. In his view, a methodology is a set of principles that in any particular situation has to be reduced to a method uniquely suited to that particular situation.

To give some examples of these aspects:

- The UML class diagram is a technique, and so is operation specification.
- SELECT Enterprise is a tool.
- The activity represented by 'find classes by inspecting the use case descriptions' is an aspect of process. So is the advice that an analyst is usually the best person to write test plans.
- The advice that 'operation specifications should not be written until the class model is stable' is an aspect of structure, as it identifies a constraint on the sequence in which two steps should be performed.
- Analysis and design are (reasonably) distinct stages.
- The statement 'object-oriented development permits the construction of software which is robust and resilient to change' is an element of a systems development philosophy.

A package that contains enough information about each of these aspects of the overall development process is fit to be named a methodology. Many attempts have been made to capture the essence of methodology for software development, and the resulting methodologies are almost as varied as are the projects themselves. In practice, methodologies vary widely in philosophy, in completeness of definition or documentation, in coverage of the life cycle, and even in the type of application to which they are best suited.

Some authors even believe that the idea of a published methodology is misleading, if it is taken to mean that following a methodology leads inevitably to a good product. For example, Daniels (co-developer of the Syntropy methodology) argued in a conference presentation (1995) that an IS methodology is a means of *learning* the process of systems development, not a recipe for practising it. He compared a methodology to a ladder that, once used, can be thrown away (the metaphor is borrowed from Wittgenstein). We are safe to 'throw the ladder away', not because we do not need one, but because we now know how to build one of our own that is better suited to our organization and projects.

20.2.1 Logical views of a system

At a very abstract level of discussion, three complementary views of a real-world system must be understood in order to model it adequately for the purpose of conducting software development. These are:

- a data view, that describes the real-world system in terms of attributes and associations that must be stored within the software;
- a process view, that describes the operations that are (or need to be) carried out on that data;
- a temporal view, that captures the time sequence and time constraints on individual processes, and also the possible sequences of events that may impinge on the system.

In order to be useful, any IS methodology must include techniques for discovering, analysing and modelling the relevant content for each of these abstract views. These are packaged in different ways by the various modelling approaches. The history of information systems development represents an evolution of representations of the three system views, which have necessarily become much more sophisticated over time, as the systems being modelled have themselves grown larger and more complex.

20.3 Why Use a Methodology?

Over many decades, IS methodologies have been developed and introduced specifically to overcome those problems of software development projects that were perceived to be important at the time. However, to date, no methodology has been wholly successful in fulfilling its objectives, partly because computing is a highly dynamic field, and the nature of both projects and their problems is constantly changing. In a changing world, it is unlikely that yesterday's solution will ever completely solve today's problems.

Nevertheless, many advantages have been claimed for the use of a methodology, including the following.

- The use of a methodology helps to produce a better quality product, in terms of documentation standards, acceptability to the user, maintainability and consistency of software.
- A methodology can help to ensure that user requirements are met completely.
- Use of a methodology helps the project manager, by giving better control of project execution and a reduction in overall development costs.
- Methodologies promote communication between project participants, by defining essential participants and interactions, and by giving a structure to the whole process.
- Through the standardization of process and documentation, a methodology can even encourage the transmission of know-how throughout an organization.

If all these claims can be fulfilled in practice, the benefits would be clear. The evidence is mixed, partly because of the many dissimilarities between organizations, and also between the different types of project within any one organization. However, the picture in the UK is that approximately two thirds of businesses use some form of IS methodology. Even where this actually means little more in practice than an in-house set of standards for documentation and procedures (as it certainly is in at least some cases), this still shows a consistent level of faith in the utility of a methodology of some kind.

20.4 A Brief Historical Review

Viewed historically, the earliest approach that might be thought of as a step towards an IS methodology is the traditional life cycle (TLC). Other life cycle models such as the spiral life cycle and the OMG life cycle (described in Chapter 3) are further steps towards a methodological approach. However, as we saw in Section 20.2.1, a methodology is rather more than a life cycle.

20.4.1 Structured methodologies

Early structured methodologies, such as those authored by DeMarco (1979) and Gane and Sarson (1978), were introduced to overcome some of the problems encountered with the TLC. The central aim, suggested in the name, was to produce a 'structured' specification of the proposed software, with all functions, data storage, and interfaces between sub-systems clearly defined. To achieve this, it is necessary to define the techniques to be used and the deliverables at the end of each stage, as well as the stages of the life cycle. This all provides a more solid basis for project management, since the development process is more visible, and progress can be measured in terms of deliverables. A project manager using a structured methodology might be able to apply tests like this: 'If the data flow model and entity-relationship diagram have been approved, the analysis phase is complete'.

However, structured methodologies did not overcome all of the difficulties in systems development. For example, a manager still sometimes had no sound way of knowing whether a team was being optimistic in its progress reports. Anecdotes about project management in this period often take a rather ironic form: 'It's funny how a team can complete 90% of the data flow diagrams in 10% of the time, but then the other 10% of the work takes the other 90% of the time.' Secondly, most structured methodologies focus either on functional or data aspects of the system being modelled. For example, Yourdon's (1989) structured methodology is primarily process-oriented, while Information Engineering (Finkelstein, 1989) is primarily data-oriented. Whichever view holds sway, there is a danger of neglecting important aspects of the other. Even where both are given equal emphasis, there can be difficulty in co-ordinating the models of the two views. Some methodologies attempted to cater for all views in a balanced way. One example is SSADM (the acronym stands for Structured Systems Analysis and Design *Method*—a good example of the method/methodology confusion!), for many years a UK government standard for systems development and procurement (Skidmore et al., 1994). This has separate but carefully correlated models of processes, data and temporal sequence. While SSADM was undoubtedly very successful, any methodology that attempts this degree of co-ordination inevitably tends to grow large and unwieldy, partly due to the need for continual checking and cross-referencing between the separate models.

20.4.2 Object-oriented development

A more important difficulty of structured methodologies in general is that most remain tied to a waterfall life cycle. This runs directly counter to the natural, iterative style of development for object-oriented software, thus making it very difficult for a structured analysis and design approach to lead to an object-oriented implementation. In object-oriented development, as we saw in Chapter 3, an iterative cycle of development is both necessary and desirable. For this reason, throughout the 1990s, a great deal of research effort has resulted in several methodologies better suited to object-oriented software development. Some are briefly reviewed in the following sections.

The reader should note, however, that this gives only a flavour of the field, being neither comprehensive over the range, nor detailed in discussion of any one methodology. It would be pointless, in any case, to present here a detailed introduction to contemporary object-oriented methodologies. At the time of writing, three of the most significant (those of the 'Three Amigos', Booch, Jacobson and Rumbaugh, the co-founders of UML) appear to have been superseded by a methodology that merges the three into a single *Unified Software Development Process* (Jacobson et al., 1999).

OMT

The Object Modelling Technique (OMT) was developed by Rumbaugh and his co-authors at General Electric's Shenectady Research and Development Center (Rumbaugh et al., 1991). OMT was an important milestone on the road that led to the Unified Modelling Language. OMT's class diagram notation was incorporated into UML almost unaltered. But, unlike UML, OMT is much more than a language, or set of techniques and notations. It also provides a methodological framework of defined procedures, organized into phases and sub-phases, each with distinct deliverables, the whole supported by documentation and under-pinned by a philosophy.

In OMT, the development life cycle is divided into three main phases: Analysis, System Design and Object Design. Implementation is regarded as outside the scope of the methodology. (The book also gives some advice on implementing an object-oriented design in both object-oriented and non object-oriented environments, but given the early publication date much of this now appears rather dated. For example, Smalltalk, C++ and Eiffel are addressed but object-oriented databases are not covered, due to their relative immaturity at the time, while Java had not yet been developed.)

Within each OMT phase, a set of steps is defined, each with a procedure that uses one or more of the techniques. Collectively, the steps in each phase lead to one or more end-of-phase deliverables. Some steps in OMT's analysis phase are shown in Figure 20.2. The description of steps conveys the overall structure of the life cycle, the procedures used and a definition of the deliverables. The authors stress that, while the order of steps in OMT is not meant to be prescriptive, it nevertheless provides important guidance on how the different tasks should be combined, and also on the necessarily sequential relationship of some of the intermediate products of the overall process. However, there is a degree of flexibility to be exercised according to the judgement of the developers.

> Experienced developers are able to combine several steps or perform certain steps in parallel for portions of a project...
> Iteration of the steps is necessary at successively lower levels of abstraction, adding more detail to the model.
>
> (Rumbaugh et al., 1991)

An underpinning philosophy can be discerned in the following statement.

> The greater emphasis on the essential properties of an object forces the software developer to think more carefully and more deeply about what an object is, with the result that the system is usually cleaner, more general and more robust than it

would be if the emphasis were only on the use of data and operations.

(Rumbaugh et al., 1991)

Methodology step	Comment
1. Write or obtain an initial description of the problem (Problem Statement).	*A major task*
2. Build an Object Model: • Identify object classes. • Begin a data dictionary containing descriptions of classes, attributes and associations. • Add associations between classes. • Add attributes for objects and links. • Organize and simplify object classes using inheritance. • Test access paths using scenarios and iterate the above steps as necessary. • Group classes into modules, based on close coupling and related function.	*A major task broken down into several more detailed tasks*
⇒ **Object Model** = object model diagram + data dictionary.	*An end-of-phase product*

Figure 20.2 *Part of OMT's analysis phase (adapted from Rumbaugh et al, 1991).*

OMT is an example of what is sometimes called the 'second generation' of object-oriented methodologies. This derives from a rather ill-defined sequence that usually locates Shlaer and Mellor in the first generation, Booch, Coad and Yourdon, Objectory and OMT in the second, with Fusion and Syntropy representing the third generation.

Perhaps in due course OPEN—described in the next section—and the new methodology from the Three Amigos will come to be called the fourth generation. Since only ten years have elapsed this gives some indication of the typical lifetime of an information systems development methodology.

OPEN

Object-oriented Process, Environment and Notation (OPEN) is a methodology that has a 'tailorable life cycle model' (Graham et al., 1998). Its authors define a methodology as consisting of the distinct elements shown in Figure 20.3. As a package, OPEN consists of a number of components, described in the following paragraphs.

Activities. An activity is a collection of tasks seen from a project manager's perspective. Activities are similar to stages and phases in other methodologies. Activities have both pre- and post-conditions, and are carried out within timeboxes (see Chapter 21). Some examples of activities are: *project initiation, requirements engineering, analysis and model refinement, project planning* and *build*.

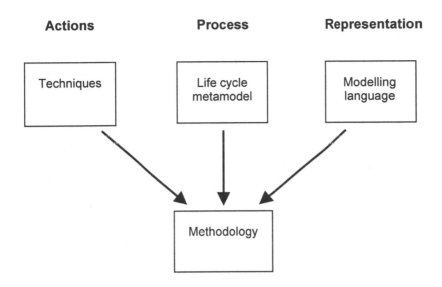

Figure 20.3 *The OPEN view of a methodology.*

Tasks. Within each activity, one or more tasks are carried out. These primarily represent the developer's view of the project, although management tasks are also included in a very comprehensive list (five pages of task names alone are given in *The OPEN Process Specification*).   Some examples of tasks are: (draw) *rich pictures* (adopted from Soft Systems Methodology—see Section 20.7), *analyse user requirements, choose project team, design UI* (user interface) and *perform class testing.*

Techniques. Each technique describes how to carry out one or more tasks. The list of techniques is also comprehensive, and includes many that are not original to OPEN. For example, the *rich pictures* task (mentioned above) involves use of the *rich picture* technique and the CRC technique is also used (see Chapter 8). Other examples of techniques are: *class internal design* and *object life cycle histories*. Many OPEN techniques use the COMN notation (see below).

Deliverables. Deliverables are the post-conditions for activities, and often also the pre-conditions for other activities. Some examples of deliverables are: *user requirements statement* (the main deliverable from the requirements engineering activity), *requirements specification* (this includes a physical prototype of the proposed system) and many diagrams such as *inheritance diagrams* and *deployment diagrams*.

Notation. For object modelling, OPEN has its own notation, called COMN (Common Object Modelling Notation), which includes a metamodel written in OML (Object Modelling Language) that is in many respects a rival to UML. We give no example of this here, but we note that its authors claim that COMN is in some respects more expressive and more precise than UML. Further discussion of this topic is beyond the scope of this book.

20.5 An Outline of Objectory

Objectory is, like OMT, a key contributor to the UML notation set, lending it the use case diagram and the interaction sequence diagram. Although Objectory may, also like OMT, now be regarded as rather dated, we present it here in some detail since we regard it as among the most complete and most useable of all the object-oriented methodologies that have been produced to date. Objectory is said to derive from three long-established traditions: object-oriented programming, conceptual modelling and block design, an approach to the modular design of electronic hardware components that originated in the 1970s at the Ericsson telecommunications company (Jacobson et al., 1992). The Objectory life cycle consists of three high-level processes (not to be confused with sequential stages), that produce five distinct models. The diagram in Figure 20.4 shows how these relate to each other. The models are described in more detail in the following sections.

The overall philosophy of Objectory identifies two important characteristics of the way that the models are logically related to each other. Firstly, the transition from one model to another is seamless—it is easy for the developer to see in advance how to get from elements in one model to elements in the next. This is regarded as important because it enables a successful process to be repeated. Secondly, all transformations from one model to another are traceable, meaning that it can be seen in retrospect how objects in one model have been derived from others in earlier models. This is important because it makes it possible to verify that the delivered software meets its specified requirements.

Figure 20.4 *Processes and models in Objectory, adapted from Jacobson et al, 1992.*

A further important aspect of Objectory's philosophy is that the emphasis of the development process is on the development, delivery and maintenance of a certain product. For this reason, the structure of the methodology is not a waterfall model, but instead is based on dividing the development work into processes that interact with each other to produce a software system. During the course of a project, all three main processes may exist concurrently for the entire life cycle.

20.5.1 Requirements model

The aim of this model is to capture the functional requirements for a new system. The requirements model consists mainly of a use case model. It also sometimes includes an interface model (in the form of prototype screens) and a domain object model (similar to the analysis class model described in Chapter 6), but these can be omitted at this stage unless it is felt they will help to extract user requirements.

The principal activities in producing a use case model are identifying actors and use cases, and describing the use cases. These are essentially the same processes as those described earlier in Chapter 5. A final refinement of the use case model involves the identification of abstract use cases, that consist of common elements abstracted out of the concrete use cases derived from user requirements.

Abstract use cases

In Objectory, the process of abstraction through generalization begins with use cases. Since these remain central to the development process throughout the entire life cycle, there is an advantage in generalizing the system's behaviour from the user's perspective before much object identification has taken place. Sometimes this is best done by identifying abstract actors. For example, if two actors, `salesStaff` and `salesManager` can both receive a receipt, we can add the abstract actor `receiptReceiver` to express this common behaviour.

20.5.2 Analysis model

The focus of the analysis model is on structuring the requirements model. The system is described in terms of three types of objects: entity objects, control objects and analysis objects. The notation is illustrated in Figure 20.5.

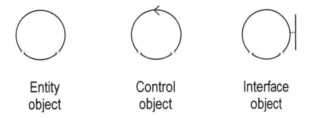

Entity object   Control object   Interface object

Figure 20.5 *Objectory notation for object types in the analysis model.*

These represent concepts that are already familiar from Section 11.4.3, where the various components or layers that make up the architecture of a system were identified, and also from Section 12.3.3, where the Model-View-Controller architecture was described.

Identify analysis objects

Objects are identified from the use case descriptions, and this process is normally carried out separately for each use case in turn. Attributes and operations are assigned as they become apparent from analysis of the use case. Communication associations are also identified at this stage, as these connect together objects that need to use each other's operations or access each other's attributes. Figure 20.6 gives an example of a diagram with all three types of object present (for simplicity, no attributes are shown).

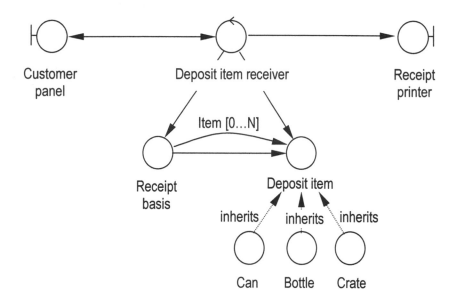

Figure 20.6 *Analysis object model for the use case* Return an item *(adapted from Jacobson et al., 1992).*

Each object is informally specified in terms of its role and its responsibilities. No formal specification of operations is attempted yet.

Identify sub-systems

In order to keep the model to a manageable size, it is partitioned into sub-systems at the analysis stage. One approach to this is to look for 'change units', or 'service packages'. These represent clusters of system functionality that are relatively free-standing, and each is delivered to the customer in its entirety or not at all.

Each sub-system should have strong functional coupling internally, but only weak coupling with other sub-systems. Often, this means that a sub-system is centred on a single control object together with any closely dependent interface and entity objects. The objects in Figure 20.6 would probably make a reasonable sub-system, although any decision would clearly depend on a wider view that includes objects from other related use cases. Dependencies between use cases should also be identified.

20.5.3 Construction

This phase (i.e. phase of development, not a project phase, since the latter may not map directly onto a particular development activity) aims to refine the analysis model by adding detail and formality. The analysis model must also be adapted to meet any constraints that apply to its implementation (discussed in Chapter 11). Two main models are produced: the design model and the implementation model. Both depend on an understanding of the implementation environment, so the construction phase begins with identifying and describing this in terms of its constraints, and reaching policy decisions on those issues that will require a design trade-off.

Develop block structure

In Objectory, a *block* is a component of the system that is intended to be implemented as a discrete software module. In other words it is equivalent to a design class in UML. So the first attempt at a design model is no more than a direct, one-to-one mapping from the analysis model, in which each analysis object initially becomes a block. This structure is preserved through all ensuing transformations, except where there are sound design reasons for changing it.

Whether or not the structure of the analysis model can be preserved, other structures will probably need to be added around it, for example to provide services related to the implementation environment. A common example is when an application will handle files in the operating system; here a new sub-system should be designed to act as a 'file manager', interfacing between the application's file handling and the operating system's file handling. Such a 'block' encapsulates the operating system and machine levels from the perspective of the application, and may also be a reusable component   or may already exist from an earlier project.

Other blocks may be added to handle other services that are not present in the analysis model. Occasionally blocks may be deleted, split or joined in order to optimize the design, e.g. to improve system performance by shortening access paths. Any of these actions should be considered with care, since, wherever a gain is made against one design criterion, there is certain to be a balancing cost against another.

Each use case is modelled in detail using interaction diagrams. Since these are very similar to sequence diagrams in UML (see Chapter 8), they are not covered here.

Design blocks

The design of an individual block begins with its interface. This can be extracted from message protocols on the interaction diagrams for use cases in which that block participates, and translated into the target programming language. Since the interface for the block can now be regarded as 'frozen', its internal details can now be designed. Behaviour can be described in any convenient notation, including some of those described in Chapter 9 of this book.

Following this, implementation of the design in software is a straightforward task, since all difficult decisions have already been made.

20.5.4 Testing

The testing phase of Objectory's life cycle is concerned with verifying the resulting code, and so is beyond the scope of this book.

20.5.5 Components

Up to this point, Objectory has been described as if it were a methodology that only applied to single projects. However, the importance of components is recognized to some extent by the presence of a fourth process, *Component Development*, which, in a sense, stands to one side of the normal project life cycle. Its relationship to the main development processes described in earlier sections is shown in Figure 20.7.

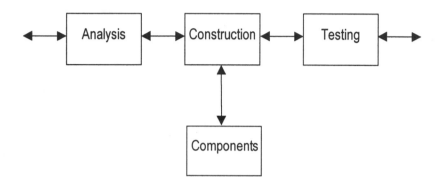

Figure 20.7 *The relationship between Objectory's component development activity and the main development cycle (adapted from Jacobson et al., 1992).*

This is one of the only full life cycle methodologies of its time (the early 1990s) to seriously address questions of component building for reuse. Even so, components are still seen primarily in terms of single objects, and, apart from a brief mention of *frameworks*— 'an entire skeleton of a design to build your application in' (Jacobson et al., 1992)— discussion of potential reuse is mostly at the class level. As we saw in Section 19.3.1, some more recent authors have argued that the class is often too low a level of granularity at which to find successful components.

20.6 Issues in Choosing a Methodology

The introduction of a methodology to an organization is not a trivial matter. There are many costs, some quite difficult to estimate. Staff must be trained in the techniques, structure and management of the new methodology, documentation must be purchased and software licences must be obtained for CASE tools that support the methodology. The indirect, hidden costs are often under-estimated. Productive time is lost during training, and for some time after the change there is also a reduction in productivity and an ongoing need for support from external consultants. This is true whether or not the organization already uses a methodology. Even with careful evaluation before a decision is made, followed by careful planning of the change, it is still wise to conduct a full-scale trial of a new methodology on a pilot project, which must also be chosen carefully. It would be unwise to risk the failure of a critical system, yet a pilot project must be sufficiently complex to put the new methodology to a thorough test.

The choice of the 'right' methodology is also fraught with difficulties, as there are now many hundreds to choose from, and these differ radically in their philosophies, their coverage of the life cycle and their suitability to particular application domains. Many factors that affect the appropriateness of a methodology, including type of project (large, small, routine or mission-critical), application domain (e.g. real-time, safety-critical, user-centred, highly interactive, distributed or batch-mode) and nature of the IS development organization.

An influential thinker on the management of software development is Humphrey (1989), whose 'process maturity model' suggests that organizations evolve through five successive levels of maturity. First comes an 'initial' level, where development activities are chaotic, and each developer uses ad hoc procedures that they have probably devised themselves. There are no common standards and no dissemination of good practice through the organization, so the success of any project depends solely on the skill and experience of the development team. Many organizations today are still at this level, where there is no point in introducing any methodology, since management have neither the skill nor the structures required to control it. Next comes the 'repeatable' level, where an organization has adopted some development standards and project management procedures. These allow successes to be repeated on later projects, and the organization can benefit from a methodology, since management procedures are capable of enforcing its application. However, while individual managers may repeat their successes, there is no clear understanding of which specific factors led to each success. It is unlikely that success can be generalized to different kinds of project or application, and the flexibility of the organization is still limited. A prescriptive methodology that defines all steps in some detail is more likely to be successful.

An organization at the 'defined' level has produced its own definition of the software process, and is able to standardize activities throughout the organization. A methodology can now be introduced more readily and is likely to produce greater benefits, since the organization already has a culture of working to defined procedures. But staff still adapt much more readily to a methodology that is in harmony with their current ways of working. The next step is typically to introduce a metrics programme (see Section 21.5), which, if successful, can lift the organization to the 'managed' level—but few organizations are yet at this level. Only a tiny handful have reached the final 'optimizing' level, where there is a capability for continuous improvement in all activities (corresponding to the general management approach called 'Total Quality Management').

20.7 Hard *versus* Soft Methodologies

This chapter would not be complete without some mention of the critical debate that turns on the distinction between *hard* and *soft* methodologies. The distinction has emerged principally from the broad systems movement, and, while there is no one precise definition of the difference, it is summarized in Figure 20.8.

Hard systems view	Soft systems view
The activity of IS development is all about building a technical system that is made only of software and hardware.	An IS also comprises the social context in which the technical system (software and hardware) will be used.
Human factors are chiefly important from the perspective of the software's usability and acceptability. Politics and group behaviour are only an issue for project managers.	A new IS impacts on interpersonal communication, social organization, working practices and much more, so human and social factors are paramount.
Organizations exist only to meet rational objectives, through the application of rational principles of business management. It is possible to be both rational and objective about the requirements for a new system.	Organizations are made up of individuals with distinct views and motivations, so any picture of requirements is subjective. It is not always possible even to reach a consensus. In practice, this means that the powerful decide, not the wise.
When requirements are uncertain or unclear, it is up to management to decide. Setting objectives is a principal role of management, and others should follow their lead.	If management has not accommodated the full range of views in the organization, encouraging managers to decide on the requirements may be completely counter-productive.

Figure 20.8 *Some underlying assumptions of hard and soft systems approaches.*

To summarize, 'hard' is usually taken to mean objective, quantifiable or based on rational scientific and engineering principles. In contrast, 'soft' means complex, ambiguous, subjective. The methodologies that we have discussed so far in this chapter all derive mainly from the hard tradition, although some influence of a soft approach can be discerned in Objectory's use case technique, since this aims at eliciting the practical, context-based requirements of individual users.

On the whole, those methodologies that might be characterized as principally soft in their orientation tend to focus more on making sure that the 'right' system is developed, than on how to actually develop the system. Their intellectual antecedents are diverse. For example, Mumford's Effective Technical and Human Implementation of Computer Systems (ETHICS) (Mumford, 1995) derives from the socio-technical school of social theory, while Checkland's influential Soft Systems Methodology (SSM) (Checkland, 1981; Checkland and Scholes, 1990) is based on some philosophical ideas about the nature of systems at a conceptual level. However, in spite of their very different origins, both provide ways of exploring and agreeing the characteristics of the organization as a system, before any attempt is made to define a specific information system that will help users meet their goals. This approach is very different from that taken by most 'hard' methodologies, which tend to assume that the purpose and nature of the organization can, to a large extent, be taken for granted.

One way of looking at this is to say that soft and hard methodologies cover different parts of the life cycle. In this view, a soft methodology is more useful in the earlier stages of the life cycle, particularly when there is uncertainty about the goals or strategy of the organization as a whole. A hard approach will be more appropriate once any initial uncertainties and ambiguities have been resolved (insofar as this is possible), since the emphasis then shifts to a specific project with relatively clear goals and boundaries.

This has led to the suggestion that, in certain situations, hard and soft methodologies can complement each other, and can be used together to help overcome some of the perennial difficulties in systems development. Flynn proposes a 'contingency framework' shown in

Figure 20.9, which aims at helping to select an appropriate methodology for a specific organizational context.

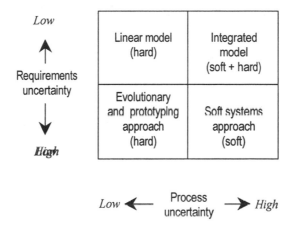

Figure 20.9 *The contingency framework can be used to help select an appropriate methodology for a given organizational context (adapted from Flynn, 1998).*

'Requirements uncertainty' is the extent to which requirements are unknown or subject to debate or disagreement, and also whether they are expected to change during development. For example, a new system intended primarily to automate an existing manual system may have relatively low requirements uncertainty. Agate's campaigns system might fall into this category. 'Process uncertainty' refers to the degree of doubt about the best way to build the proposed system. For example, it may be very difficult to decide on the best process if a new system will use untried technology, with unpredictable effects on the organization's employees. A project intended to introduce Electronic Commerce to an organisation with no experience of it might fall in this category.

A project is rated along both dimensions, and this helps to indicate an appropriate development approach. For example, where both the requirements and process are clear from the outset, a linear model of development is recommended, which in practice might either mean using a traditional structured methodology, or procuring a ready-made solution. At the other extreme, a soft approach is recommended, so that the character of the problem is clarified before any further action is taken.

In seeking to merge together a soft and a hard methodology, the development team is really trying to devise a unique method suited to the project. This implicitly recognizes the complementary nature of their strengths and weaknesses.

20.8 Summary

We began this chapter by considering two definitions of 'methodology', and went on to discuss how the concept of methodology differs from method. This is an important distinction, since the development approach is an important factor in the success or failure of a project. Many methodologies have been developed over the years, stemming from very different traditions, and each in some way attempting to counter a perceived shortcoming in other contemporary methodologies. The 1990s have been a prolific time for the spread of object-oriented methodologies, among which OMT and Objectory have been two of the most popular. The last decade has also seen a great deal of research into the feasibility of merging hard and soft methodologies, in the hope of being able to meet a wider range of demands, and improve the overall success rate of IS development projects.

Review Questions

20.1 What is the difference between 'methodology' and 'method'?
20.2 Distinguish between 'task' and 'technique', and give some examples of each.
20.3 What are the three logical views of an information system?
20.4 What are the four main high-level processes in Objectory?
20.5 What is meant by an 'abstract use case'?
20.6 Name the five levels of Humphrey's model of process maturity.
20.7 Distinguish between the hard systems view and the soft systems view.
20.8 Why might a methodology based on a hard systems approach be unsuccessful in a situation where the goals of the organization are unclear?
20.9 What general advantages are claimed for using a methodology?
20.10 What might be the disadvantages of using an inappropriate methodology?

Case Study Work, Exercises and Projects

Do some research in your library or on the Internet, and collect material that describes four or more different systems development methodologies. Try to make these as different from each other as possible, for example by choosing one that is structured (e.g. SSADM), one that is object-oriented (e.g. OMT), one based on a soft systems view (e.g. SSM) and one from another tradition (e.g. ETHICS). Then use the following questions as a basis for comparison.

20.A What techniques are used by each methodology? In particular, how do they represent the process view, the data view and the temporal view of the system?
20.B To what extent does each methodology cover the full project life cycle, from project selection through to implementation and maintenance?
20.C How far do you think each methodology can be adapted to suit differing projects or circumstances?
20.D Can you find a statement that gives an underlying philosophy for each methodology? If not, is it possible to identify the intellectual tradition from which the methodology has been derived?

Further Reading

For more information about any specific methodology, readers should consult a primary source, e.g. Jacobson et al. (1992) for Objectory, Rumbaugh et al. (1991) for OMT, Checkland and Scholes (1990) for Soft Systems Methodology, Mumford (1995) for ETHICS, Graham et al. (1998) for OPEN, and so on.

Flynn (1998) introduces the many traditions behind today's methodologies, and discusses why there is still disagreement about the 'right' way to do systems development. Jayaratna (1994) offers an alternative framework for the evaluation of methodologies. Both of these books should be read by anyone who has the responsibility for recommending a methodology for adoption by an organization.

The new publications from the Three Amigos are of particular interest from a UML perspective (Jacobson et al., 1999; Booch et al., 1999; Rumbaugh et al., 1999).

21

Managing Object-Oriented Projects

Objectives
In this chapter you will learn:
- how to manage iterative projects;
- how to use project planning tools;
- what metrics are available for object-oriented projects.

21.1 Introduction

Information systems development is a complex activity that requires careful management. We discussed the need for various models of the systems life cycle in Chapter 3 and the way in which they bring structure to a project. The discussion of the UML techniques in this book highlights the need to plan and manage the whole process. There are inter-dependencies between the artefacts of software development and their production has to be co-ordinated if the process is to be efficient. A large software development project may involve many developers, some with specialized skills. The requirements capture specialist is required early in project, the expert in OODBMS implementation is needed during design and construction, the installation and support teams become involved to some extent during design and construction and more fully when the information system is complete. The different activities may require different resources whose availability has to be planned. As suggested in Chapter 2, the timing of the installation may be critical for the success of the project. The management process is further complicated by the fact that the sequence of some activities may be significant. For example, testing of a system can only begin when at least some elements have been constructed, and testing a sub-system requires all elements of that sub-system to be complete. Resource allocation and project planning are considered in Section 21.2.

Iterative development approaches present particular challenges. One critical issue is how to control the number of iterations. This was discussed briefly in Chapter 3 and we examine the question in more detail in Sections 21.3, and also in 21.4 where we introduce the Dynamic Systems Development Method (DSDM).

A significant part of project planning is the determination of the resources required for each stage in the project. It is also important to establish the period for which each resource will be required. This can be based only upon an estimate of how long each task is likely to take. Estimates of task duration rely heavily upon experience but can also involve the measurement of various aspects of the proposed system. Many factors affect the length of time that it takes to complete a given activity and it is not possible to quantify them all. Some factors are measurable (at least to some extent), for example, system size and complexity. Less tangible aspects of the project such as staff expertise, team morale or the difficulty of using new technology are harder to measure. Section 21.5 introduces software metrics that can be used to estimate the influence that project characteristics have upon overall development time.

The patterns that were introduced in Chapter 14 are mainly useful in a design context. Other patterns have been developed that are applicable to the management of the development process. In Section 21.6 we briefly introduce the idea of development organization patterns. One of the difficulties of introducing object-oriented development to an organization is that, initially at least, some systems are based on an object-oriented approach while others are not. In Section 21.7 we outline some of the problems to which this gives rise.

The move towards object-orientation introduces a series of challenges for an organization. Staff must be trained and existing software systems must be integrated with the new approach. Any major change (a change in development method is only one example) requires careful planning in order to minimize the attendant risks (Section 21.8).

21.2 Resource Allocation and Planning

Given the complexity of project management there is a need for tools and techniques to support the process. Yourdon (1989) identifies three particular areas of the management of software development where modelling techniques can play a useful role:

- in the estimation of money, time and people required;
- in assisting the revision of these estimates as a project continues;
- in helping to track and manage the tasks and activities carried out by a team of software developers.

Many tools (e.g. CA SuperProject) have been developed to support the management of any type of project, not just those that are focused on software development.

21.2.1 Critical Path Analysis

The technique known as Critical Path Analysis (CPA) was developed in the late 1950s for use on major weapons development projects for the US Navy (Whitten, Bentley and Barlow, 1994). Originally known as Project (or Program) Evaluation and Review Technique (PERT)[1], it is also called Network Analysis and it has been widely used on many different types of project.

For the purposes of carrying out a critical path analysis, a project is viewed as a set of activities or tasks, each of which has an expected duration. Completion of an activity corresponds to a milestone or event for the project. Each milestone also represents the start of activities that are directly dependent on the completion of the predecessor(s). CPA is based on an analysis of sequential dependencies among the activities, and uses the expected duration for each task to derive an estimate of the overall duration of the project. In particular, it identifies any inter-task dependencies that are critical to the project duration—

---

[1] Strictly speaking PERT is more elaborate than CPA, using statistical measures in addition to critical path analysis, but the terms are generally used synomously. CPA charts are sometimes known as activity diagrams (Skidmore et al., 1994) but this name risks confusion with UML activity diagrams.

collectively these are known as the *critical path*. The preparation of a CPA chart involves the following steps.

List all project activities and milestones

A sample list for the development of the Agate campaign management system is shown in Figure 21.1. Each activity is labelled with a letter and has a short description. The third column in the table contains a milestone number that represents the completion of that activity.

Activity	Description	Milestone	Preceding activities	Expected duration	Staffing
A	Interview users	2	-	5	2
B	Prepare use cases	3	-	2	See A
C	Review use cases	4	A, B	2	3
D	Draft screen layouts	5	C	2	2
E	Review screens	6	D	2	2
F	Identify classes	7	C	2	3
G	CRC analysis	8	F	4	3
H	Prepare draft class diagram	9	F	5	3
I	Review class diagram	10	G, H	4	4

Figure 21.1 *Project activity table for Agate.*

Determine the dependencies among the activities

Some activities cannot start until another (sometimes more than one) has been completed. The predecessors of each activity are listed in column 4. For example, in Figure 21.1 the activity 'review use cases' must be completed before the activity 'identify classes' can begin.

Estimate the duration of each activity

There are several different approaches to this, due to the uncertainty involved in estimating task duration. One that is used widely is given by the following formula:

$$ED = \frac{MOT + (4 \times MLT) + MPT}{6}$$

where *ED* is the expected duration of a task, *MOT* is the most optimistic time, *MLT* is the most likely time and *MPT* is the most pessimistic time for its completion. The *ED* is thus a weighted average of the three estimates. Each *MOT* assumes that a task will not be delayed, even by likely events such as employee absence. The *MPT* assumes that most things that can go wrong will go wrong, and that completion of the activity will be delayed to the maximum plausible extent. Equipment will arrive late, technical problems will occur and some staff will be ill (it would be unrealistic to take this to the extreme, for example, all development staff will be struck down by an influenza epidemic—estimates should be realistic). The *ED*s are entered in the fifth column of the table. Note that in this example the staff requirements for activities A and B have been treated as one since the two activities are highly interdependent.

Draw the CPA chart

Two main styles of notation are used for CPA charts, known respectively as 'activity on the node' and 'activity on the arrow' diagrams. Both show the same information, but they look very different to each other. To avoid confusion we present only 'activity on the arrow' notation. In this style each milestone is represented by a circle divided into three compartments. One compartment is labelled with the milestone number, and the other two will hold the earliest start time (EST) and the latest start time (LST) (these terms are explained below) for all activities that begin at that milestone. Figure 21.2 illustrates the notation.

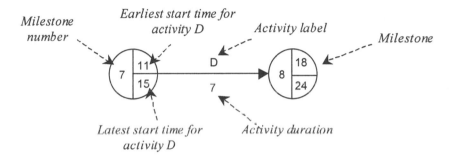

Figure 21.2 *CPA notation.*

The first draft of a CPA chart shows dependencies between activities and their expected durations, since this information is known or can be estimated before the diagram is drawn. The partially completed CPA chart in Figure 21.3 represents graphically the activity precedences listed in table form in Figure 21.1.

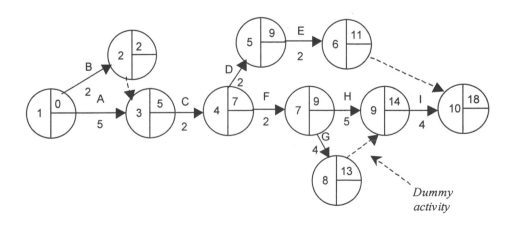

Figure 21.3 *Partial CPA chart for the Agate advertising sub-system project.*

Note the *dummy activity* between milestones 8 and 9 (there are others between 2 and 3 and between 6 and 10). The explanation for this is that activities H and G both depend on milestone 9 (the completion of activity F), and are also both predecessors to milestone 10 (where activity I begins). Since H and G may not necessarily finish at the same time, an extra milestone is needed for one of these events. In effect, milestone 8 represents the completion of G, regardless of whether or not H has finished (it is not significant which of the two activities is chosen to terminate at the extra milestone). A dummy activity (with ED of 0) then needs to be introduced in order to connect milestone 8 to milestone 9, thus preserving the sequence of dependencies.

The next step is to enter an earliest start time for each milestone. This is done by working through the diagram from the very first milestone to the very last (sometimes called a *forward pass*). It is a convention that the EST for the first milestone is set to 0 (elapsed time is usually measured in days, but any other unit of time can be used equally well). The EST for most other milestones is calculated simply by adding the EST of the immediately preceding activity to its duration. For example, activity C (the immediate predecessor for D) has an EST of 5 and an ED of 2, therefore the EST for activity D is 7. Where an activity has two or more predecessors, its EST is determined by the predecessor that has the latest completion time. For example, activity I is dependent upon the completion of both G and H, so its EST is set to the later of the two calculations. In general the EST for any milestone is set to the earliest time that *all* predecessor activities can be completed.

The latest start time for each milestone is entered by working back from the last milestone (sometimes called a *backward pass*). By convention the LST for the last milestone is set equal to its EST. Each preceding LST is then calculated as follows. The LST for most milestones equals the LST for its successor minus the ED of the intervening activity. For example, the LST for milestone 9 is 18 − 4 = 14 (the LST for milestone 10 minus the ED for activity I). Milestone 7 presents more of a problem as this has two successor activities, G and H. In such cases, two calculations are performed (or more, if there are more than two successors) and the earlier of the two answers is taken. For example, if the LST for milestone 7 were determined purely by activity G, this would give 14 − 4 = 10 (the LST for milestone 8 is 14). This means that activity G can afford to begin as late as time 10 without delaying any other activities. But a similar calculation for activity H gives 14 − 5 = 9. This means that if H begins later than time 9, its completion will be delayed beyond time 14, which in turn would delay milestones 9 and 10 and thus also activity I and the project completion. The LST for milestone 7 is therefore set to 9. In general, the LST for a milestone is set to the latest time that allows *every* activity that begins at that milestone to be completed by the LST for its succeeding milestone.

Identify critical path

Once all LSTs have been entered onto the diagram, the *slack time* (or *float*) for each activity can be calculated. This is the difference between an activity's EST and its LST, and it represents the time by which that particular activity can be delayed without affecting the overall duration of the project. The path through all milestones that have a slack time of 0 is called the *critical path*. This is indicated by a double bar across activity arrows that connect the milestones. These milestones, and their intervening activities, are critical to the completion of the project on time. Milestones that have an EST that is different from their LST are not critical in the sense that they have some scheduling flexibility. The completed diagram is shown in Figure 21.4.

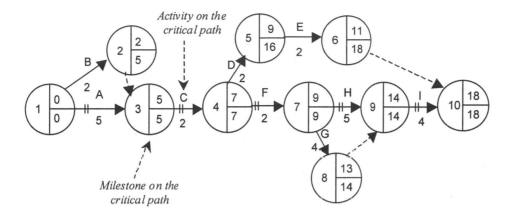

Figure 21.4 *The critical path for the Agate advertising sub-system project.*

A CPA chart is an effective tool for identifying those activities whose completion is critical to the completion of a project on time. If any activity that is on the critical path falls behind schedule then the project as a whole is behind schedule. However, while critical activities naturally receive the closest scrutiny, all project activities should be monitored. Delay even in a non-critical activity can, if it is sufficiently severe, alter the critical path.

CPA charts have limitations, chief among which is that they are not very useful for representing the extent of any overlap between activities. As a result, they are often used in conjunction with other techniques, particularly the Gantt chart (described in the following section).

Gantt charts

The Gantt chart (named for its inventor Henry Gantt) is a simple time-charting technique that uses horizontal bars to represent project activities. The horizontal axis represents time and is often labelled with dates or week numbers so that the completion of each activity can be monitored easily. Activities are listed vertically on the left of the chart, and the length of the bar for each activity corresponds to its ED.

A Gantt chart shows the overlap of activities clearly and this provides an effective way of considering alternative resource allocations. The Gantt chart can be drawn with either dashed lines or dashed boxes that show the slack time for non-critical activities.

A Gantt chart is also easy to convert into a stacked bar graph that can be used to show the way that the total resource allocation for a project changes over time. Figure 21.5 shows a Gantt chart and a staffing bar chart for the Agate advertising sub-system project. The staffing chart is derived as follows. The final column of Figure 21.1 gives a staff allocation for each activity. The Gantt chart is read vertically for each successive time interval to calculate the total number of staff required for all project activities combined. The result is shown as a vertical bar that indicates the total staffing required at that time.

Activities that are not on the critical path can have their start time adjusted. For example, activity E cannot begin until time 9 at the earliest, but it could start as late as time 16 without affecting project completion. A project manager can adjust the resource profile to accommodate staff availability, a process known as resource *smoothing*. For example, activities E, G and H can occur concurrently. H is on the critical path and cannot be moved, but the slack time for E allows it to begin at time 16 instead of time 9. The manager can minimize the overall resource requirement by rescheduling activity E to begin at time 14. This should be done with care, however, as it may move an activity onto the critical path. Figure 21.6 shows the smoothed resource profile.

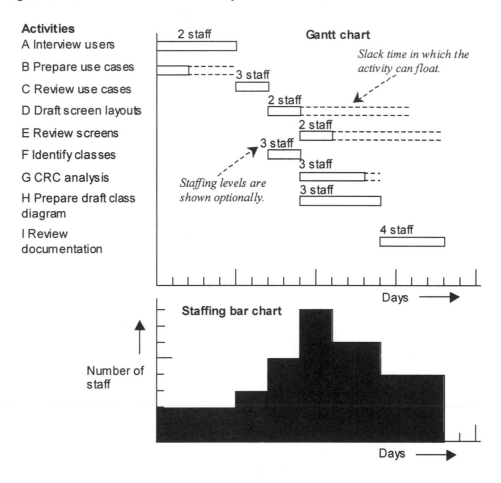

Figure 21.5 *Gantt Chart with staffing bar chart.*

The Gantt chart is a useful tool for resource monitoring. The progress of each activity can be shown independently and compared against planned progress. In Figure 21.6 the Gantt chart reflects the current state of a project. Activities A, B, C and F are complete. D has not been started and will shortly become critical. G is on schedule but H is behind schedule. Activities D and H need to be investigated by the project manager. Possible reasons for the delay include:

- an unexpected technical problem;
- staff absence;
- the complexity of the activity has been underestimated.

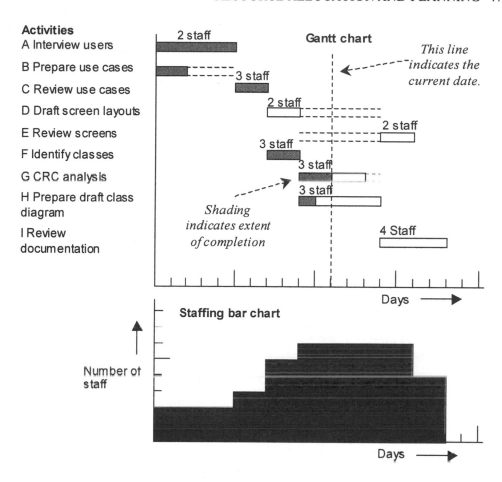

Figure 21.6 *Gantt chart showing activity completion and smoothed staff profile.*

The project manager must decide how best to get the project back on schedule. Perhaps the allocation of more staff to activity H would resolve the difficulty, but throwing staff at an activity can be counter-productive for several reasons. Firstly, if the additional staff are not familiar with the activity, they may require extensive briefing, thus reducing the time that is available to complete the work. Secondly, as team size increases so do the communication overheads. Thirdly, some activities are limited in the maximum number of staff that can be involved. For example, a car repair may require 20 person-hours of work to complete. This does not mean that 20 mechanics working together could do the work in one hour. It is more likely that only, say, five mechanics could work productively on one car at any one time. The best possible repair time would therefore be four hours.

When a critical path activity is behind schedule it may not be possible to regain the lost time. A project manager then has really only two options, and ideally the choice should be discussed with the client. Firstly, the project deadline can be moved to accommodate the delay. Secondly, the scope of the project or the quality of the product can be reduced to permit completion on time. The last approach requires an analysis of all uncompleted critical path activities in order to identify what can be omitted to reduce their ED's. The resultant changes may alter the critical path, and activities whose completion was not critical before may now become critical. Any attempt to reduce the scope of a project requires user involvement to ensure that only non-critical features are omitted, particularly during the first increment.

The process of planning a project is iterative. An alteration to the staff allocation for an activity will probably change its ED, which in turn may change the critical path. A project manager must operate within the constraints of staff and resource availability.

21.3 Managing Iteration

In Chapter 3 we considered prototyping as an iterative development activity and discussed the need to specify objectives for the prototyping activity so that they can provide criteria to control the number of iterations (Figures 3.6 shows an example of an iterative life cycle). At the end of an iteration the product, say a prototype, is evaluated against pre-defined objectives. In general terms this seems straightforward but in practice it can be difficult to determine whether the objectives of the iterative activity have been achieved.

Let us suppose that the interface for the Agate campaign management system is to be developed by prototyping, with the explicit objective of producing an interface with which the campaign staff are happy. However, although this objective may be worthwhile it is of little use for the management of the activity. Imagine that the users are never completely happy at the end of any iteration: they will continue to suggest further improvements without end. As the process continues, the nature of the modifications that are suggested at each iteration will change. Over time the improvements will become cosmetic and ultimately peripheral to the utility of the system. It would be sensible to end the iterative process before this point is reached. A more suitable objective for the exercise might be phrased as follows:

> *Continue the iterations until fewer than five cosmetic changes are requested on a single iteration.*

It is still not clear how many iterations will be needed to satisfy this criterion. If the project has unlimited time and an unlimited budget this may not be a problem but this is unlikely to be the case. Additional criteria can be added to tighten up the objectives, such as the following.

> *The prototyping phase must be completed before the end of October and must not exceed 50 developer-hours.*

21.4 Dynamic Systems Development Method

The Dynamic Systems Development Method (DSDM) is a management and control framework for rapid application development (RAD). The distinction between RAD and prototyping is sometimes blurred. A RAD approach aims to build a working system rapidly while a prototyping approach also builds rapidly, but usually only produces a partially complete system, typically to confirm some aspect of the requirement. Because both approaches aim to build software quickly, similar development environments are used and one approach to prototyping continues the development of a prototype incrementally until it becomes a working system. In effect this is a RAD development approach.

The traditional waterfall approach to systems development has deficiencies, particularly the time taken to deliver a working system and the inflexibility of the approach to requirements change. Iterative approaches to development can also be problematic. As suggested above it is sometimes difficult to cease the iterations when they become unproductive. In the early 1990s RAD became much more popular and was viewed as a way of matching systems development to the fast changing needs of business. However, there were until recently no commonly accepted structures for either the use or the management of RAD. The DSDM consortium was formed in 1994 to produce an industry standard definition of the RAD process and DSDM was subsequently defined. The DSDM framework defines structure and controls to be used in a RAD project but does not specify a development methodology. DSDM may be used with either an object-oriented or a structured methodology. DSDM takes a fundamentally different perspective on project control. Rather than viewing requirements as fixed and attempting to match resources to the project, DSDM fixes resources for the project, fixes the time available and then sets out to deliver only what can be achieved within these constraints.

DSDM is based upon nine underlying principles (Stapleton, 1998):

- Active user involvement is imperative. Many other approaches effectively restrict user involvement to requirements acquisition at the beginning of the project and acceptance testing at the end of the project. In DSDM users are members of the project team and are known as 'Ambassador' users.

- DSDM teams are empowered to make decisions. It can make decisions that refine the requirements and possibly even change them without the direct involvement of higher management.

- The focus is on frequent product delivery. A team is geared to delivering products in an agreed time period and it selects the most appropriate approach to achieve this. The time periods are known as timeboxes and are normally kept short (2 to 6 weeks). This helps team members to decide in advance what is feasible. Products can include analysis and design artefacts as well as working systems.

- The essential criterion for acceptance of a deliverable is fitness for business purpose. DSDM is geared to delivering the essential functionality at the specified time.

- Iterative and incremental development is necessary to converge on an accurate business solution. Incremental development allows user feedback to inform the development of later increments. The delivery of partial solutions is considered acceptable if they satisfy an immediate and urgent user need. These solutions can be refined and further developed later.

- All changes during development are reversible. If the iterative development follows an inappropriate development path then it is necessary to return to the last point in the development cycle that was considered appropriate. Changes are limited within a particular increment.

- Requirements are initially agreed at a high level. Once requirements are fixed at a high level they provide the objectives for prototyping. The requirements can then be investigated in detail by the DSDM teams to determine the best way to achieve them. Normally the scope of the high level requirements is not changed significantly.

- Testing is integrated throughout the life cycle. Since a partially complete system may be delivered it must be tested during development, rather than after completion. Each software component is tested by the developers for technical compliance and by user team members for functional appropriateness.

- A collaborative and co-operative approach between all stakeholders is essential. The emphasis here is on the inclusion of *all* stakeholders in a collaborative development process. Stakeholders not only include team members, but others such as resource managers and the QA team.

21.4.1 The DSDM life cycle

The DSDM life cycle has these phases:
- feasibility study;
- business study;
- functional model iteration;
- design and build iteration;
- implementation.

The relationships between the phases are shown graphically in Figure 21.7, and each is described in more detail below. Note that the last three are actually iterative processes, but for the sake of clarity this is not shown explicitly in Figure 21.7.

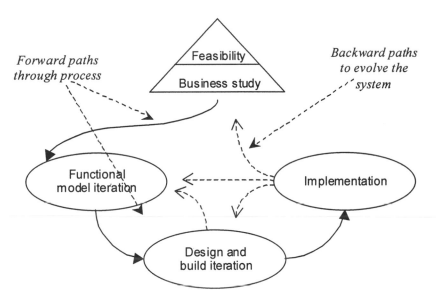

Figure 21.7 *Simplified DSDM life cycle (adapted from Stapleton, 1997).*

The *feasibility study* phase determines whether the project is suitable for a DSDM approach. It typically lasts only weeks, whereas the feasibility stage can last months on a traditionally run project. The study should also answer questions such as the following:
- Is the computerized information system technically possible?
- Will the benefit of the system be outweighed by its costs?
- Will the information system operate acceptably within the organization?

The *business study* phase identifies the overall scope of the project and results in agreed high level functional and non-functional requirements. Maintainability objectives are set at this stage and these determine the quality control activities for the remainder of the project. There are three levels of maintainability:
- maintainable from initial operation;

- not necessarily maintainable when first installed but this can be addressed later;
- short life-span system that will not be subject to maintenance.

Where the third level of maintenance is chosen care should be taken to ensure that the system is discontinued at the end of its allotted time, otherwise it may become subject to maintenance requests that are difficult to service.

The *functional model iteration* phase is concerned with the development of prototypes to elicit detailed requirements. The intention of DSDM is to develop prototypes that can ultimately be delivered as operational systems, so these must be built to be sufficiently robust for operational use and also to satisfy any non-functional requirements such as performance. When completed the functional model comprises high level analysis models and documentation together with prototypes that are concerned with detailed functionality and usability. During the functional model iteration the following activities are undertaken:

- the functional prototype is identified;
- a schedule is agreed;
- the functional prototype is created;
- the functional prototype is reviewed.

The *design and build iteration* phase is concerned with developing the prototypes to the point where they can be used operationally. The distinction between the functional model iteration and the design and build iteration is not clear-cut and both phases can run concurrently. The activities for the design and build iteration phase are very similar to those described above for the functional model iteration phase.

The *implementation* phase deals with the installation of the latest increment including user training. At this point it is important to review the extent to which the requirements have been met. If they have been fully satisfied the project is complete. If some non-functional requirements have yet to be addressed the project may return to the design and build iteration phase. If some element of functionality was omitted due to time constraints the project may return to the functional model iteration phase. If a new functional area is identified the project may return to the business study phase. The return flows of control are shown with dashed arrows in Figure 21.7. Implementation comprises the following iterative activities:

- user guidelines and user approval.
- training users.
- implementing the system.
- review of the business requirements.

21.4.2 Timeboxing

Timeboxing is an approach for fixing the resource allocation for a project or a part of a project. It limits the time available for the refinement of requirements, design, construction and implementation as appropriate. A RAD project has a fixed completion date that defines an overall timebox for the project. A DSDM approach to project management will then identify smaller timeboxes within this, each with a set of prioritized objectives. Each timebox produces one or more deliverables that allow progress and quality to be assessed. Within a timebox the team have three major concerns. They must first carry out any investigation needed to determine the direction that should be taken for that part of the project. They must then develop and refine the specified deliverables. Finally they must consolidate their work prior to the final deadline.

It is sometimes difficult to prioritize the requirements that will be actioned during a timebox. One way of doing this is to apply the set of rules that are known as the *MoSCoW* rules (for Must ... Should ... Could ... Want).

Must have requirements are crucial. If these are omitted the system will not operate. In DSDM the set of *Must have* requirements are known as the minimum usable subset.

Should have requirements are important but if necessary the system can operate usefully without them.

Could have requirements are less important and provide less benefit to the user.

Want to have but will not have this time around requirements can reasonably be left for development in a later increment.

All of these requirements are important for the final system but not to the same extent. If the full set cannot be addressed within a timebox, the MoSCoW categorization can be used to focus the requirements in an appropriate way.

Review Questions

21.1   How is the slack time for an activity calculated?
21.2   How can you decide whether a milestone is on the critical path?
21.3   What are the advantages of a Gantt chart when compared to a CPA chart?
21.4   What are the advantages of a CPA chart when compared to a Gantt Chart?
21.5   What are the main life cycle stages for DSDM?
21.6   What are the MoSCoW rules and how do they help in the management of a timeboxed activity?

21.5 Software Metrics

Planning and managing a software development project requires the estimation of the resources required for each of its constituent activities. A resource estimate for an activity can be based upon subjective perceptions of the activity or it can be based upon measurements of size and complexity, either of the activity itself or of the artefact that is produced. DeMarco's (1982) frequently quoted aphorism sums it up:

> You can't control what you can't measure.

We are used to applying measures in many forms of human endeavour. For example, the productivity of a car factory can be measured quite accurately in terms of the time taken to construct a car. The use of metrics in software development is still rather limited. The term software engineering was coined in the 1960s as part of an attempt to introduce the greater degree of rigour that was seen in the management of other types of engineering practice. It is a feature of most engineering disciplines that careful measurement is used to assess the efficiency and effectiveness of artefacts and of processes. A *software metric* is a measure of some aspect of software development, either at project level—usually its cost or its duration—or at the level of the application—typically its size or its complexity.

Software metrics can be divided broadly into two categories: *process metrics* that measure some aspect of the development process and *product metrics* that measure some aspect of the software product. (We use the term software product broadly to include anything that is produced during a software development project, including for example analysis models and test plans as well as program code). Two examples of process metrics are the project cost to date and the amount of time spent so far on the project. Product metrics relate to the information system that is under development. One of the simplest product metrics is the number of classes in an analysis class diagram.

Software metrics can also be categorized as *result* or *predictor* metrics, which are used respectively to measure outcomes and to quantify estimates. The current cost of a project is a result metric (even though this is only an interim outcome that will probably be modified tomorrow). A measure of class size (a crude measure might be a simple count of attributes and operations) would be a predictor metric, so called because it can be used as a basis for predicting the time that it will take to produce program code for that class. Result metrics are also known as control metrics (Sommerville, 1992) since they are used to determine how management control should be exercised. For example, a measurement of the current level of progress in the project is used to decide whether action is necessary to bring the project back onto schedule. The term 'predictor metric' is generally applied only to a measure of some aspect of a software product that is used to predict some other aspect of the product or of the project progress. Predictor metrics are not used solely for estimation. The results obtained from their application to a project may indicate, for example, that the system will be difficult to maintain or that it may offer very low levels of reuse. Since neither outcome is desirable, managers may attempt to change the design for the system or the process for its development in order to improve the system.

The validity of predictor metrics is based upon three assumptions, often made only tacitly by managers (Sommerville, 1992):

- there is some aspect of a software product that can be accurately measured;
- there is a relationship between the measurable aspect and some other relevant characteristics of the product;
- this relationship has been validated and can be expressed in a model or a formula.

The last of these assumptions suggests that a significant volume of historical data must be collected so that an appropriate statistical analysis can validate the relationship. In practice this is only feasible if the data collection is automated (for example, it is done by a CASE tool). Generally speaking, software developers do not regard the capture of metrics data as an important part of their role. Their attention is inevitably focused on the delivery of the product on time. Another factor that limits the uptake of metrics is the possibility of using them to monitor the performance of the developers themselves. This may give rise to concern among the developers about how data on their performance might be used (this is of course an ethical issue as well as a management one).

A number of metrics have been identified for use with structured analysis and design approaches. For example, De Marco (1982) developed a complexity metric known as the Bang Metric for use on structured analysis projects. Other metrics that focused on the degree of coupling and cohesion between program modules have also been suggested for use with a structured design approach. Most of these metrics were used little despite the introduction of automated CASE support.

A number of authors have identified metrics for object-oriented systems development (Lorenz, 1994; de Champeaux, 1997; Graham, 1995). De Champeaux suggests a list of desirable features as part of a general description of a useful metric:

- either it is elementary and focuses on a single well-defined aspect, or it is an aggregation of elementary metrics;
- it is suitable for automated evaluation;
- gathering the metric data is not too costly;

- it is intuitive;
- application to a composite is equivalent to application to the components and summation of the individual results;
- the metric can be measured numerically and arithmetic operations are meaningful.

de Champeaux (1997) lists a series of quality metrics, for example, a dependency metric that provides a measure of the stability of a system. When applied to a package or a sub-system this measures the degree of inter-package or inter-sub-system coupling. It is calculated by the formula:

$$I \quad = \quad \frac{Ce}{(Ca + Ce)}$$

where I is the instability of the system, Ca is the level of afferent coupling (the number of classes outside the package that depend on classes within the package), and Ce is the level of efferent coupling (the number of classes outside the package upon which classes within the package depend) (adapted from de Champeaux, 1997).

Where I is zero the package is maximally stable and has no dependencies on classes in other packages. Where I is 1 the package is maximally unstable and has dependencies only on classes outside itself.

The ability of a package to absorb change is reflected (in part at least) by the ratio of abstract classes to all classes within the package. Where this is zero then the package consists solely of concrete classes and is difficult to change. A ratio of one indicates the presence of no concrete classes at all and it is easier to change.

Lorenz and Kidd (1994) suggest a wide range of metrics including metrics for application size and class size. Application size is essentially determined from the number of use cases and the number of domain classes[2] together with multiplying factors that reflect the complexity of the user interface. The size of a class is determined by the number of attributes and operations it has and the size of the operations. Size metrics such as this can be used to estimate the resource requirement for a project providing that appropriate historical data is available to derive and validate the relationship.

21.6 Process Patterns

We have already discussed design patterns in Chapter 14. Coplien (1995) has defined a pattern language that is focused on the development process. The pattern language comprises both organization and process patterns. As with design patterns they capture elements of experience as problem solution pairs. The patterns address issues such as team selection, organizational size, team structure, the roles of team members and so on. Conway's Law (pattern 14) discusses how the architecture of the system comes to reflect the organizational structure or vice versa. Mercenary Analyst (pattern 23) is concerned with producing project documentation successfully. This pattern suggests that it is frequently more effective to hire a technical writer who can focus solely on the documentation.

---

[2] Lorenz and Kidd use the terminology scenario scripts and key classes.

21.7 Legacy Systems

There are many definitions of the term legacy system. We take it to mean any computerized information system that has been in use for some time, that was built with older technologies (perhaps using a different development approach at different times) and, most importantly, that continues to deliver benefit to the organization. Most computerized information systems interact with other computerized information systems. They may share data, the output from one may be an input to another and so on. Any new information system is likely to need to interact with older legacy systems that have not been built using the same technologies. Redeveloping legacy systems so that they interact appropriately with new systems is likely to be prohibitively expensive and probably involves too much risk. These legacy systems may be critical to the operation of the organization. The problem is one of integrating new object-oriented systems with non-object-oriented systems.

One strategy that enables the interoperation of old and new is the use of an *object wrapper* (Graham, 1995). An object wrapper functions as an interface that surrounds a non-object-oriented system so that it presents an interface suitable for use with new object-oriented systems. Essentially the old system appears to be object-oriented. The form of the wrapper depends on the nature of the legacy system. Where the old system uses text or form-based screen interfaces the wrapper may involve program code that reads data from the screen and writes data to the screen (sometimes known as screen scrappers) using some form of virtual terminal.

When an organization embarks upon object-oriented software development there may be an intention to migrate all existing software to the new technologies. The cost and risk involved may constrain this but where it is feasible to migrate systems it is important to manage the process carefully. Wrappers may be used initially to provide an object-oriented interface to the system. Then, once the system has been wrapped, it can be redeveloped (perhaps one sub-system at a time) without affecting its interface to other systems. A further variation on this approach is to use the Facade pattern (see Chapter 19) to wrap sub-systems so that they can be migrated incrementally.

21.8 Introducing Object Technology

The introduction of any new way of working requires careful planning. Introducing object technology is a very significant change as not only is the approach to systems development new but the development technology is also new. Staff must be trained in the principles of object-orientation, in new analysis and design techniques, in one or more object-oriented programming languages, in the use of new development environments and perhaps also in how to use a particular object-oriented database management system. Almost every aspect of the software development activity will change. This is not just a matter of training the developers; they also must gain experience in applying the new techniques and using the new technology. The safest way to achieve this is to apply object-orientation within a pilot project in the first instance (as suggested for FoodCo in Box B1.2). The move to object-orientation must be carefully planned and controlled and should ideally follow the steps shown below:
- identify a suitable pilot project;
- train the relevant staff;
- monitor the project carefully when it is under way;

- review the project implementation;
- identify the lessons learned and migrate this experience to other suitable projects.

Identification of a suitable pilot project. A pilot project should ideally not be subject to very tight timescales. The team is likely to require additional time to become familiar with the concepts and notation. The project should be amenable to an object-oriented approach in terms of its application domain and implementation technology.

Training. The move to object-orientation requires a different perspective on the problem domain. Training should be planned so that developers are trained in the new techniques immediately before they need to use them. It is important that a developer should understand the conceptual basis of object-orientation before embarking upon learning a new modelling notation (for example UML) or a new programming language (for example Java). Training and familiarization with the selected CASE toolset should be included. Users who will interact with the project should also be trained so that they can participate effectively in user reviews.

Monitoring the project. To gain maximum benefit from the move to object-orientation it is important to monitor its use within the organization. This provides feedback on the effectiveness of the training, the most appropriate way of applying the techniques and the project management requirements.

Review project implementation. A post-implementation review provides an opportunity to determine whether or not the expected benefits of the approach have been realized. It is important not to place unreasonable expectations upon a pilot project since unfulfilled aspirations may then over-shadow its successes. For instance, the potential benefits of reuse are unlikely to be achieved immediately after an organization has migrated to object-orientation. Even with careful management, significant levels of reuse may not appear for a further two or three years.

Migrate experience to other suitable projects. The experience of the pilot project will suggest adjustments that can be made to improve the way that object-orientation is to be used in the organization. Members of the initial project team may be used to seed other projects with their experience of the use of object-oriented development methods.

21.9 Summary

Systems development projects are similar to any other project in their need for sound management to ensure that they are completed within budget and on time. Some standard project management techniques used for this purpose include Critical Path Analysis and Gantt Charts, both of which can be used for project planning, to monitor resource utilization and to monitor project progress.

Object-oriented projects generally use a rapid development approach, and DSDM has emerged as the predominant framework for RAD project management. An object-oriented approach works particularly well with DSDM.

The successful management of systems development relies on the quantification of various measures of effort, progress and complexity. A number of software metrics have been developed for this purpose including some that are intended specifically for use within object-oriented development.

The introduction of object-orientation to an organization should be carefully planned and managed. It is also important to ensure that existing computerized information systems can continue to operate alongside object-oriented system, for example by the use of object wrappers.

The success of an information systems development project depends upon many factors, but in the final analysis the most important factor by far is the skill of the development staff. Effective project management enables developers to apply their knowledge and expertise in the most productive way.

When used by suitably skilled developers and managed by appropriately qualified managers, object-orientated development offers a means for capturing, modelling and building complex information systems that fully meet the needs of their users.

Review Questions

21.7 How do object wrappers help integrate new object-oriented systems with existing non-object-oriented systems?

21.8 What factors should be considered when migrating to object technology?

Case Study Work, Exercises and Projects

21.A Prepare a CPA chart, a Gantt chart and a staffing profile for the project activities shown in the following table.

Activity	Description	Mile-stone	Preceding activities	Expected duration	Staffing
A	Interview users	2		4	3
B	Prepare use cases	3		3	2
C	Review use cases	4	A, B	3	3
D	Draft screen layouts	5	C	4	2
E	Review screens	6	D	2	2
F	Identify classes	7	E	2	3
G	CRC analysis	8	F	3	2
H	Prepare draft class diagram	9	F	4	3
I	Review documentation	10	G, H	4	4

21.B FoodCo had never used object-oriented development methods until the start of the Production Costing System project. Identify other projects at FoodCo that would be alternative candidates for a pilot project, when considered as the first step in a company-wide migration process. Compare these in terms of their appropriateness.

21.C Critically evaluate a project management software package that is available to you. What changes to its functionality and non-functional characteristics would be required for it to be integrated with a CASE tool that supports UML?

Further Reading

De Marco (1982) provides a good introduction to the issues of managing software projects although the development techniques discussed are not object-oriented.

Stapleton (1997) is a comprehensive and readable introduction to DSDM. Information is also available at www.dsdm.org.

Graham (1995) examines the migration to object technology and is well worth reading.

Lorenz and Kidd (1994), Graham (1995) and de Champeaux (1997) address the use of object-oriented metrics.

Melton (1995) is a collection of research articles on the subject, the first chapter being an interesting history of the development of software metrics.

Coplien's pattern language (Coplien, 1995) describes many interesting patterns that will be familiar to project managers, albeit in many different guises.

Appendix A

Notation Summary

Use Case Diagram

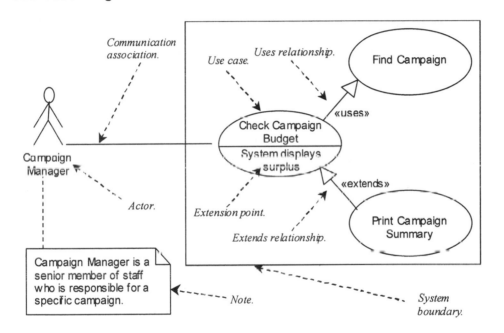

Static Structure Diagrams

Object instance notation

Class notation

Associations

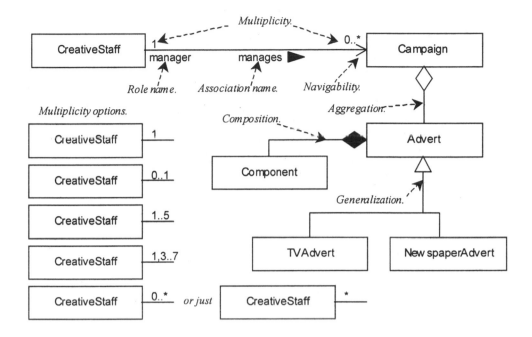

Behaviour Diagrams

Statoohart diagram

Activity diagram

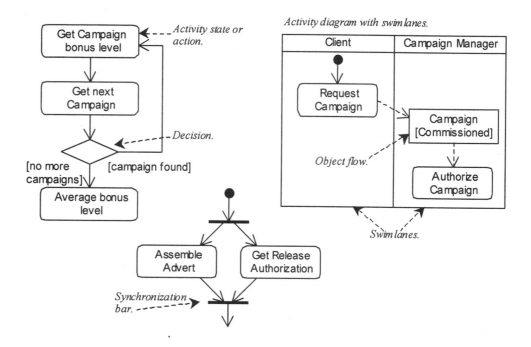

Interaction Diagrams

Sequence diagram

Collaboration diagram

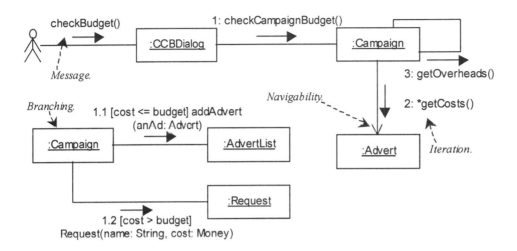

Implementation Diagrams

Component diagram

Deployment diagram

Appendix B

Selected Solutions and Answer Pointers

In this section we give solutions to a selection of the review questions in the chapters, and also pointers on possible approaches to some of the end-of-chapter case study work, exercises and projects.

Answers to Selected Review Questions

1.3   Even if something isn't a system (or might not be one) thinking of it as one still gives useful insights.

1.5   Feedback is sampling one or more outputs of a system for comparison to a control value. Feed-forward is sampling a system input, usually before it enters the system. The control value may be an output, input or an internal measure of system performance.

1.7   A management support system provides information that helps managers to make decisions. Most use feedback or feed-forward to monitor the performance of that part of the organization for which the manager is responsible.

1.9   Business goals and strategy are defined first. An IS strategy identifies applications that can help to meet business goals, and an IT strategy identifies IT needed to develop and run the applications. Each informs its predecessor about what can realistically be achieved. The process is iterative.

1.10   One of the simplest definitions says that information is data with a structure and a meaning derived from the context in which it is used.

2.1   They differ in their view of the problems because their view of the meaning and purpose of IS development also differs (see also Review Question 1.6).

2.3   The simplest definition is 'fitness for purpose'. Since it can be hard to identify the purpose, a better alternative is 'meeting all user requirements, both stated and implied'.

2.6   The system may address irrelevant problems. It may not fit the way that people work. It may be unsuitable for its environment. It may be out of date before delivery. Political difficulties may lead to delay or cancellation.

2.7   A stakeholder has an interest in a project because they are (or will be) affected by its progress or by its results.

3.1   Advantages of the traditional waterfall life cycle include:
   • teams with specialized skills can be assigned to tasks in particular phases,
   • progress can be evaluated as the end of each phase,
   • attendant risk can be controlled and managed.

3.2   Disadvantages of the traditional waterfall life cycle include:
- real projects rarely follow a simple sequential life cycle,
- iterations are almost inevitable,
- the lapsed time between inception and delivery is frequently too long,
- it is unresponsive to changes in the technology or requirements.

3.5   Prototyping in not necessarily concerned with delivery a working system whereas an incremental approach delivers a working system in successive increments.

3.7   Syntactic correctness is concerned with using the notation (e.g. UML) correctly, consistency relates to producing a set of models or diagrams that are consistent with each other and completeness refers to producing models that are completely defined.

3.8   The term requirements traceability refers to the capability of tracking each requirement to all the systems development deliverables, from analysis models to program code, that relate to it.

3.9   A diagram may be syntactically correct, consistent with other diagrams and models, and complete but it may not relate accurately or completely to the user requirements, the most important criterion for any diagram or model.

4.2   Semantics is the study of meaning. In object-oriented development it is used to denote the meaning that an element has for the user. The semantics of an object include its purpose, description, relationships and behaviour seen from a user perspective.

4.3   Other parts of a system only see an object's interface (services it can perform and operation signatures). Internal details including data are hidden and can only be accessed by a message that contains a valid signature.

4.4   Polymorphism means that when one message is sent to objects of different types, each has an appropriate, but different, implementation for its response. The object that sends the message need not know which type of object is addressed. One way of implementing polymorphism is through inheritance and overriding.

4.6   A subclass inherits all characteristics of its superclass and other ancestors (some may be overridden, but are still technically inherited). Each subclass is different from its ancestors in at least one way.

5.1   Examples of functional requirements are: the need for a process to be run that allocates staff to lines based on their skills and experience, and holidays and sick leave; printing out an allocation list; amending the allocation list. Examples of non-funtional requirements include: printing the allocation list by 12.00 noon; the need to handle 200 operatives' details.

5.4   Use cases are produced to model the functionality of the system from the users' point of view and to show which users will communicate with the system. They show the scope of the system.

5.6   An essential use case documents the interaction between user and system in a way that is free of technological and implementation details, and a real use case describes the concrete detail of a use case in terms of its design.

6.2   A purely graphical model does not have the precision to cover every aspect of system behaviour and structure. A purely textual model would be too large, too complex and extremely difficult to understand and to maintain.

6.3   An attribute is a characteristic of a class (every person has a height). An attribute value is a characteristic of an instance (this author is 1.75m tall).

6.4   An element's stability is the relative infrequency of change in its description.

6.7 Multiplicity denotes the number of objects that can be linked to a single object by a specific association. It is a constraint because it limits the behaviour of a system. If a client can have only one staff contact, it should be impossible to link a second.

6.11 A link is a connection between two objects. 'Changing' a link (say by substituting another object at one end) is equivalent to destroying the link and creating a new one. (Think about two objects tied with a single length of string. In a substitution, there is a moment when neither one nor the other is connected—unless you tie on the second before untying the first, but an object link cannot do this).

7.1 Use of a component saves time and work. A friend of one of the authors once said 'Have you ever wondered how much it would cost to make your own light bulb?'

7.3 Objects are well encapsulated, and object structures can be designed this way. The hierarchic nature of generalization abstracts out the more general features of a class. Hierarchic organization of models helps the developer to find components easily when they are needed. Composition encapsulates whole structures within a composite object.

7.4 A component of a composition cannot be shared with another composition. The component has a coincident lifetime with the composition (although a component can be explicitly detached before the composition is destroyed).

7.5 This is a basis for polymorphism. The superclass operation defines the signature, but each subclass has a different method that implements the behaviour (see Chapter 9).

7.6 An abstract class has no instances and exists only as a superclass in a hierarchy. It provides a generalized basis for concrete subclasses that do have instances.

8.1 Collaboration diagrams discourage both using a large number of messages between two objects and having too many parameters for each message as these are clumsy to represent on the diagram.

8.2 Small self-contained classes are easier to develop, test and maintain.

8.4 Sequence diagrams have a time dimension (normally vertically down the page) while collaboration diagrams do not. Collaborations show the links between objects which are not shown on sequence diagrams.

8.6 An object lifeline represents the existence of an object during an interaction represented in a sequence diagram.

8.7 The focus of control indicates which operation is executing at a particular stage in an interaction represented in a sequence diagram.

8.9 Sequence numbers are written in a nested style in a collaboration diagram to represent nested procedural calls.

9.1 An algorithm defines the step-by-step behaviour of an operation. A non-algorithmic approach defines only inputs and results.

9.2 Non-algorithmic methods of operation specification emphasize encapsulation.

9.5 Usually to record doubts and uncertainties about the model, or to include information that cannot be expressed in the formal notation of the diagram, such as invariants.

9.7 Decision tables are particularly suited to representing decisions with complex multiple input conditions and complex multiple outcomes, where the precise sequence of steps is either not significant or is not known.

9.10 They confirm the user's view of a model. They specify what the designer and programmer must produce to meet the users' requirements.

10.2 A guard condition is evaluated when a particular event occurs and only if the condition is true does the associated transition fire.

10.3 All the guard conditions from a state should be mutually exclusive so that for each set of circumstances there is only one valid transition from a state. If they are not mutually exclusive more than one transition may be valid and the behaviour of the statechart is indeterminate.

10.4 An object can be in concurrent states if it has a complex composite state with concurrent substates.

10.5 Nested states are substates of a composite state and describe the life cycle of an object in detail. Nested states need not be concurrent but may be. Concurrent nested states are substates that an object may occupy at the same time.

10.6 An action is considered to be instantaneous (its duration is actually determined by the processing environment) while an activity persists for the duration of a state, for example (its duration is dependent upon the occurrence of events that may cause a state change).

10.7 The parent of a state chart may be a class or an activity within a statechart.

11.2 User wants report etc.—analysis. Selection of business objects etc.—logical design. Size of paper etc.—physical design.

11.4 Seamlessness means that the same model (class diagram) is used and successively refined throughout the project.

11.7 We add user interface, data management and application control classes. The class diagram is also updated to show how associations are designed (see Chapter 13).

11.9 Functional, efficient, economical, reliable, secure, flexible, general, buildable, manageable, maintainable, usable, reusable.

12.1 Open layered architectures are more difficult to maintain because each layer may communicate with all lower layers hence increasing the degree of coupling in the architecture. A change to one layer may ripple to many layers.

12.2 A closed layer architecture many require more processing as messages have to be passed through intervening layers.

12.4 The main differences between the MVC and the layered architecture include the update propagation mechanism and the separation of the presentation layer into the View and Controller components in the MVC.

12.6 A broker decouples sub-systems by acting as an intermediate messaging-passing component through which all messages are passed. As a result a sub-system is aware of the broker and not directly in communication with the other sub-systems. This makes it easier to move the sub-systems to distributed computers.

12.7 User requirements may dictate concurrent behaviour that cannot be supported using facilities such as multi-tasking but requires the use of custom-built scheduler component.

13.1 Private, public or protected visibility.

13.2 Attributes should be designated private to enforce encapsulation.

13.3 The application of coupling and cohesion produce coherent, focused classes that are loosely coupled and more amenable to reuse.

13.5 Collection classes can be used to hold the object identifiers of the linked objects at the many end of an association. Collection classes provide collection specific behaviour for manipulating the collection.

13.6 A collection class of object identifiers should be included in a class if it is not used by another class and it does not increase the complexity of the class unduly.

13.8   Normalization is useful when decomposing complex objects or when implementing the system with a relational database management system.

14.1   A pattern captures proven solutions to commonly occurring problems. An antipattern documents unsuccessful attempts at solving problems and suggests how the failed solution may be adapted to solve the problem successfully.

14.2   The Transaction–Transaction Line Item pattern might be applied as Deposit–Deposit Line Item, Sale–Sale Line Item and Withdrawal–Withdrawal Line Item.

14.3   The Singleton pattern ensures that only one instance of a class exists and provides system wide access to that instance.

14.5   A pattern catalogue is group of largely unrelated patterns, which may be used together or independently. A pattern language is a group of related patterns that relate to a particular problem domain.

15.3   Dialogue metaphor describes interaction in terms of conversation between user and system involving different kinds of communication. Direct manipulation metaphor represents objects of interest to the user as objects on the screen that they can manipulate through the use of the mouse. Dialogue follows sequence determined by system. Direct manipulation is event-driven, and user can determine sequence of events.

15.6   User may hit Return key without thinking and delete the Client in error.

15.9   Possible advantages: structured—aid management of projects, apply standards that aid communication, force consideration of all aspects of HCI design; ethnographic—analyst gets detailed understanding of context of system, active user involvement, social and political factors taken into account; scenario-based—help to think through possible alternative routes in use cases, can be used to justify design decsisions, valuable for testing programs. Possible disadvantages: structured—can be bureaucratic; ethnographic—can be time-consuming; scenario-based—generates large volume of documentation.

16.2   A horizontal prototype deals with only one layer of the system architecture, usually the user interface. A vertical prototype takes one sub-system and develops it through each layer.

16.5   We use statechart diagrams to model the lifetime of instances of business classes.

16.8   Java `EventListener` only handles changes to interface objects. MVC deals with changes to Model objects. The Java `Observer` and `Observable` interfaces provide MVC mechanisms.

17.3   Tagged data, with a tag for the class of each object and the attribute of each value within each object. This way, it is possible to reconstruct any object from the data in the file without having to hard code the structure of every possible complex object. (This is the approach used by the Java `ObjectOutputStream` and `ObjectInputStream` classes.)

17.6   1. Remove repeating groups. Ensure all row column intersections contain only atomic values.
2. Make sure every attribute is dependent on the whole primary key. Create a separate table for part-key dependencies.
3. Ensure every attribute is dependent on the primary key and not on another non-key attribute. Create a separate table for non-key dependencies.

17.8   Object Manipulation Language and Object Definition Language.

18.3 Package diagrams show the logical grouping of classes in a system, whereas component diagrams show the physical components of a system. During implementation, package diagrams can be used to show the grouping of physical components into sub-systems; component diagrams can be combined with deployment diagrams to show the physical location of components of the system.

18.6 Possible tests would be to test validation of date, test validation of start time, check validation of job number etc. More detailed validation of time could be to check it is within a certain amount of time of current time—if not, then a warning should be displayed.

19.1 Review of cost benefit analysis. Summary of functional requirements met and amended. Review of achievement of non-functional requirements. Assessment of user satisfaction. Problems and issues with the system. Extract of quantitative data for future planning. Identification of candidate components for reuse. Possible future developments. Actions required. (See Section 19.2.1 for the detail.)

19.5 An executable unit of code that provides physical black-box encapsulation of related services. Its services can only be accessed through a consistent published interface that includes an interaction standard. A component must be capable of being connected to other components (through a communications interface) to form a larger group. (Allen and Frost, 1998)

20.1 A methodology is a set of principles. A method is an instantiation of the principles in a given situation.

20.2 A task is something you do in a particular project. Tasks have products. A technique specifies how to carry out a task. A task might be 'Construct high level model of user requirements'. The technique for doing this would be the UML use case.

20.5 An abstract use case is one that is carried out by an actor role, rather than a real actor.

20.8 Hard systems methodologies assume that system objectives can be clearly identified and defined. This can lead to solving the wrong problem or developing a wonderful system that nobody needs. Or the project may sink in quicksand and end by being cancelled.

20.10 The problems already mentioned for Review Question 20.8 apply here too. Also disproportionate cost and delay if the methodology is too complex; quality problems if its techniques do not provide adequate models of the application; user rejection if it does not encourage sufficient participation.

21.1 The slack time for an activity is the difference between its latest and earliest start times.

21.2 A milestone is on the critical path if its earliest start time and its latest start time are the same.

21.3 The Gantt chart can be used to show the partial overlap of activities, it is a useful aid for resource smoothing and it can represent progress on the project activities.

21.7 Object wrappers provide existing non-object-oriented systems with an object-oriented style of interface hence making integration easier.

Answer Pointers for Selected Case Study Work, Exercises and Projects

1.B   Some main sub-systems are: on-line sales, retail shops, supplies, deliveries, systems support and accounts, and more. Some of the control mechanisms involve supplier re-orders, the product catalogue, network performance and security. Most have some human activities and automated support. One example of feedback includes on-line shoppers, watching the progress of their orders. The market researcher uses feed-forward (what attracts customers to webpages).

1.C   Main business aim: say, 'To establish FoodCo as an independent branded name supplying a range of high quality food products to consumers'. Subsidiary aims: diversification of customer base; achievement of international recognition and sales. Each will be translated into measurable objectives, for example as a basis for the selection of information systems development projects.

2.C   Stakeholders should include many of the following. Good if you listed them all, excellent if you thought of some not shown below (provided you can justify their inclusion).

Patients and potential patients, patients' relatives, ambulance drivers, paramedics, control room operators, accident and emergency staff, supervisors of professional stakeholders, managers who control affected budgets, policy level managers, taxpayers (or purchasers of medical insurance policies if this is how the system is funded), general medical practice staff, politicians (particularly if the service is publicly funded), members of the general public (who make the emergency calls), other road users.

3.A   An incremental development can be justified for the following reasons:
- Useful increments can be delivered quite quickly. For example, staff management and material tracking could be implemented initially.
- Users can gain experience with the systems.
- Risk is minimized.

Requirements can be refined in light of the initial increments.

4.A   The human activity system referred to is the application domain for the proposed IS. Other systems might include the project team, the analyst's department, the business planning system and the wider (political and cultural) system of the organization. Formal and informal structures of communication and relationships are the interfaces. Other installed software and hardware systems may be important (consider the discussion in Box B1.2).

4.D   You should have equivalents for most of the following. The names are not significant at this stage, nor is an exact match in the way that you have grouped concepts together.

`Factory, Product, ProductRange, PackedProduct, SaladPack, VegetablePack, CookedProduct, Sauce, Pickle, SandwichTopping, Ingredient, Customer, Supermarket, Brand, Farm, Supplier, Employee, Consumer.`

5.B   Here are some of the use cases that should be in the diagram with the actor in brackets. `Check Staff Availability (Production Planner)`, `Enter Details of Staff Illness (Production Planner)`, `Print Availability Lists (Production Planner)`. There is a need for some means of entering details of staff holidays. The decision about who does this will affect the scope of the system. It could be done by the staff themselves and authorized on-line by the factory manager, or this process could be done on paper and only authorized holidays entered by the production planners.

6.A   The following are sample descriptions for two of the use cases.

Record Employee Leaving The Line

Normally employees are recorded as leaving the line when they clock off at the end of a working shift. Although there are breaks in the operation of the line during a shift these are not normally recorded as employees leaving the line. Date, time and location are recorded.

Stop Run

When the production line stops for a routine reason, e.g. for a break, to restock or to reload equipment, the time the run stopped is recorded and a reason is recorded. The line supervisor or chargehand can do this.

6.B   You should probably have the following classes (or reasonable equivalents to them), although other solutions are possible depending on the assumptions that are made.
Employee, Supervisor, ProductionLine, ProductionLineRun, EmployeeAbsence, LineProblem, LineRunInterruption.

7.C   An obvious superclass is Vehicle. Specializing operations might include land, takeOff, attachTrailer and graze. Attributes might include typeOfFuel, numberOfWheels, canTowTrailer, numberOfEngines, takeOffWeight and wingType.

7.E   Some generalization and composition can be justified by the inclusion of the following classes, together with those already suggested for Exercise 6.B.
Operative, RoutineBreak, AbsenceRecord.

8.A   Possible classes include Supervisor, ProductionLine, LineRun and Product. ProductionLine would be the co-ordinating class.

8.B   An alternative allocation of responsibility could be the class Product as the co-ordinating class.

9.A   One of the more complex (and therefore one that is well worth trying) would be ProductionLineRun.start(). Preconditions for this operation should be suggested by your use case description.

9.C   Most decision tables can be converted easily into Structured English with either case or nested-if. For very simple tables (two outcomes) if-then-else may be enough.

10.A   The events that affect ProductionLine include start run, end run, detect problem, pause run. The possible states for ProductionLine include Idle, Running, ProblemInterrupted and Paused.

10.E   The advantages of swimlanes include a clear delineation of the responsibility for each activity showing how behaviour is distributed over parts of the system.

11.A   Examples include sequence of entry, branching points (where the user has a choice), repetition of entries (can the user enter more than one holiday at the same time?), commands that the user might need to use while entering data (but not whether they use a menu, function keys or control keys).

11.D   In Windows there are many standards, for example: the use of function keys, particularly in combination with Alt and Ctrl keys; the standards for the appearance of menus, for example, menu entries followed by dots ('...') when the menu entry leads to a dialogue box; the positioning of certain buttons in dialogue boxes ('OK' and 'Cancel').

12.A The procedure should include:
- determine criteria for sub-division,
- partition system according to the these criteria,
- review sub-systems to minimize message passing and maximize potential for reuse,
- specify interfaces of each sub-system.

12.B The FoodCo Production Control System could contain the sub-systems Employee Management, Product and Production Line Management.

13.A All attributes should designated private and operations public. Choose data types that reflect the domains from which the attribute values are selected.

13.B The one-way associations are Line–LineFault. The two-way associations are Supervisor–ProductionLine and Line–LineRun.

14.B The ProductionLine class could use the state pattern with the state subclasses Idle, Running, ProblemInterrupted, Waiting. This use of the state pattern reduces the complexity of the ProductionLine class but may increase the storage and processing requirements for the application as a whole.

15.A See answer pointer to 11.D.

15.C Something along these lines ...

First, Rik runs off the three availability lists to show who is available for work the following week. He then starts with operatives who are available all week. For each operative, he views their record on screen, looking at their skills and experience, the line they are currently working on, and how long they have been on that line. He allocates each operative in turn to a line and a session in one of the factories.

This does not provide any detail of actual interaction with the system.

16.A If you are expecting to develop for a windowing environment, you will need a dialogue box as a minimum. Depending on how you handle looking up information, for example a list of valid reasons for stopping the line, you may need separate windows in which to display these look-ups. However, in the factory environment you may want to use a simpler device with an LCD screen for display and a sealed keyboard with dedicated keys. A PC with a mouse may be unnecessary and unsuitable in a messy environment. However, we are assuming a windowing environment for the other exercises in this chapter.

17.B You should end up with the following tables (or similar names): SalesOrder, OrderLine, Customer and Product. See the book web site for a more detailed solution.

17.E This will be similar to Figure 17.23, with a ClientBroker class to handle the operation to find each Client.

18.B Many libraries now use a web browser to access catalogue services. If this is the case, then your deployment diagram will include the client machines (PCs, Apple Macs or workstations), the web server, and probably another machine running the library software. Library staff may access the system from simple terminals for use cases to issue and return books. They will connect directly to the machine running the software, not via the web server. (The actual configuration will depend on your particular system.)

18.E   Issues to consider are as follows. Is the manual organized around the tasks a user carries out? Has it got an index? Can you find the terms in the index that you, the user, know, or does it use computer jargon? Does it show screen shots? Are they the same as actual screens or windows in the version you are using? (You should be able to think of other criteria.)

19.A   Possible inclusions for bug reports. User name, telephone no., building, room, address etc. Date and time bug occurred. Type of machine on which bug occurred. Operating system of machine. Software in which bug occurred. Other software running at the same time. Program/window/function being used at time of bug. Any error messages displayed for the user. What the user expected to happen. What actually happened. What the user did (key strokes, mouse clicks on buttons or menus etc.) immediately beforehand.

19.D   You could either include the encryption package within the security package or have it as a separate package. The Security Core Classes will need its services and will have a dependency on it. Will it need any kind of user interface classes, for example, if it requires setting up with some kind of parameters? If it does, it will presumably need somewhere to store these parameters, in which case it will also need data storage services.

21.B   Possible pilot systems include Sales Order Processing or Bar Code Production. Although these are important systems their current operation is sufficient for the company and hence new replacement systems although useful, are not critical. Of the two, the Bar Code Production system is preferable as the pilot project as it is smaller.

Glossary

Abstract class: a class that has no instances; a superclass that acts only as a generalized template for its instantiated subclasses.

Abstract operation: an operation that is not implemented in the class in which it appears (usually an abstract superclass), but will be implemented in a subclass.

Abstraction: a simplified representation that contains only those features that are relevant for a particular task; the act of separating out the general or reusable parts of an element of a system from its particular implementation.

Activation: the execution of an operation, represented in interaction sequence diagrams as a long thin rectangle.

Activity: an activity is some behaviour that may persist for the duration of a state.

Activity diagram: a variation of a statechart diagram that focuses on a flow of activity driven by internal processing within an object rather than by events that are external to it. In an activity diagram most (or all) states are action states (also called *activities*), each of which represents the execution of an operation.

Actor: an actor is an external entity of any form that interacts with the system. Actors may be physical devices, humans or information systems.

Aggregation: a whole–part association between two or more objects, where one represents the whole and the others parts of that whole.

Antipattern: documents unsuccessful attempts at providing solutions to certain recurring problems but includes reworked solutions that are effective.

Association: a logical connection, usually between different classes although in some circumstances a class can have an association with itself. An association describes possible links between objects, and may correspond either to logical relationships in the application domain or to message paths in software.

Association class: a class that is modelled in order to provide a location for attributes or operations that properly belong to an association between other classes.

Attribute: an element of the data structure that, together with operations, defines a class. Describes some property of instances of the class.

Capta: data that has been selected for processing due to its relevance to a particular purpose.

Class: a descriptor for a collection of objects that are logically similar in terms of their behaviour and the structure of their data.

Class diagram: a UML diagram that shows classes with their attributes and operations, together with the associations between classes.

Class Responsibility Collaboration (CRC): CRC cards provide a technique for exploring the possible ways of allocating responsibilities to classes and the collaborations that are necessary to fulfil the responsibilities

Class-scope: a class-scope attribute occurs only once and is attached to the class not to any individual object. A class-scope operation is accessed through the class (i.e. prefixed with the class name) not through an object. Model elements that are of class scope are underlined in class diagrams.

Cohesion: a measure of the degree to which an element of a model contributes to a single purpose.

Collaboration: describes how a group of objects interact to perform some required function. A collaboration is seen as containing a set of modelling elements that interact within a given context for a particular purpose

Collaboration diagram: a collaboration diagram shows an interaction between objects and the context of the interaction in terms of the links between the objects.

Collection class: provides collection-specific behaviour to maintain a collection. Used when designing associations with a many multiplicity to hold collections of object identifiers.

Common Object Request Broker Architecture (CORBA): a mechanism to support the construction of systems in which objects, possibly written in different languages, reside on different machines and are able to interact by message passing.

Component: an executable software module with a well-defined interface and identity.

Component diagram: a diagram that shows the organization of and dependencies among components. One of two UML implementation diagrams (qv.).

Composition: a strong form of aggregation with a lifetime dependency between each part and the whole. No part can belong to more than one composition at a time, and if the composite whole is deleted its parts are deleted with it.

Concrete class: a class that may have instances.

Concurrent states: if an object may be in two or more states at the same time, then these states are concurrent states.

Constructor operation: an operation that creates new instances of a class.

Contract: a black box description of a service (of a class or sub-system) that specifies the results of the service and the conditions under which it will be provided.

Coupling: relates to the degree of interconnectedness between design components and is reflected by the number of links and the degree of interaction an object has with other objects.

Critical Path Analysis (CPA): a diagrammatic technique for analysing the dependencies between project tasks and determining those tasks that must be completed on time if the project itself is to be completed on time.

Data: raw facts, not yet identified as relevant to any particular purpose.

Deployment diagram. A diagram that shows the run-time configuration of processing nodes (qv.) and the components, processes and objects that are located on them. One of two UML implementation diagrams (qv.).

Design constraint: a constraint that limits the design options that may be used. Common design constraints include cost and data storage requirements.

Destructor operation: an operation that destroys instances of a class.

Encapsulation: hiding internal details of a sub-system (typically a class) from the view of other sub-systems, so that each can be maintained or modified without affecting the operation of other parts of the system.

Enterprise (or business) rule: a statement that expresses business constraints on the multiplicity of an association; for example, an order is placed by exactly one customer.

Event: an occurrence that is of significance to the information system.

Exception: a mechanism for handling errors in object-oriented languages.

Extends relationship: a relationship between use cases where one use case extends or adds new actions to another. Written as a stereotype: «extends».

Functional requirement: a requirement that specifies a part of the functionality required by the user.

Generalization: the abstraction of common features among elements (for example, classes) by the creation of a hierarchy of more general elements (for example, superclasses) that encapsulate the common features.

Guard condition: a Boolean expression associated with a transition that is evaluated at the time the event fires. The transition only takes place if the condition is true. A guard condition is a function that may involve parameters of the triggering event and also attributes and links of the object that owns the statechart.

Implementation diagram: a generic term for the two UML diagrams used in modelling the implementation of a system.

Incremental development: involves some initial analysis to scope the problem and identify the major requirements. The requirements are then reviewed and those that deliver most benefit to the client become the focus of the first increment of development and delivery. The installation of first increment provides valuable feedback to the development team and informs the development of the second increment and so on.

Information: facts that have been selected as relevant to a purpose and then organized or processed in such a way that they have meaning for that purpose.

Inheritance: the mechanism by which object-oriented programming languages implement a relationship of generalization and specialization between classes. A subclass automatically acquires features of its superclasses.

Instance: a single object, usually called an instance in the context of its membership of a particular class or type (also object instance).

Instance diagram: a UML diagram similar in form to a class diagram, but that contains object instances instead of classes, links instead of asociations and may show attribute values.

Instance value (of an attribute): the value of an attribute that is taken by a particular object at a particular time.

Integrity constraint: a constraint that has to be enforced to ensure that the information system holds data that is mutually consistent and is manipulated correctly. Referential integrity ensures that an object identifier in one object actually refers to an object that exists. Dependency constraints ensure that attribute dependencies, where one attribute may be calculated from other attributes, are maintained consistently. Domain integrity ensures that attributes only hold permissible values.

Interface: that part of the boundary between two interacting systems across which they communicate; the set of all signatures for the public operations of a class or package.

Interface class: a system interacts with its actors via its interface or boundary classes

Invariant: an aspect of a UML model expressed as a formal statement that must always remain true. For example, the value of a derived attribute `totalCost` may need always to be equal to the total of all `cost` attribute values.

Knowledge: a complex structure of information, usually one that allows its possessor to decide how to behave in particular situations.

Legacy system: any computerized information system, that has probably been used for some time, that was built with older technologies (maybe using different development approaches at different times) and that, most importantly, continues to deliver benefit to the organization.

Life cycle (of a project): the phases through which a development project passes from the inception of the idea to completion of the product and its eventual decommissioning.

Lifeline: a lifeline is a vertical dashed line that represents the existence of an object on an interaction sequence diagrams. An object symbol containing the object's name is placed at the top of a lifeline.

Link: a connection between objects; an instance of an association.

Message: a request to an object that it provide some specified service, either an action that it can carry out or some information that it can provide.

Message passing: a metaphor for the way that objects interact in an object-oriented system by sending each other messages that request services, or request or supply information. Since objects interact only through the messages they exchange, their internal details can remain hidden from each other.

Method: the implementation of an operation.

Methodology: comprises an approach to software development (e.g. object-orientation), a series of techniques and notations (e.g. the Unified Modelling Language—UML) that support the approach, a life cycle model (e.g. spiral incremental) to structure the development process, and a unifying set of procedures and philosophy.

Modular construction: an approach that aims to build systems that are easy to maintain, modify or extend. Modular construction relies on modules that are essentially decoupled sub-systems, with their internal details encapsulated.

Multiplicity: a constraint on an association that specifies the number of objects at one end of the association that are permitted to form links with a specific object at the other end of the association.

Node: A physical computational resource used by a system at run-time, typically having processing capability and memory.

Non-functional requirement: a requirement that relates to system features such as performance, maintainability and portability.

Normalization: a technique that groups attributes based upon functional dependencies according to several rules to produce normalized data structures that are largely redundancy free.

Object: a single thing or concept, either in a model of an application domain or in a software system, that can be represented as an encapsulation of state, behaviour and identity; a member of a class that defines a set of similar objects.

Operation: an aspect of the behaviour that defines a class; an element of the services that are provided by a class; a specification of an element of system functionality that will be implemented as a method of an object.

Operation signature: determined by the operation's name, the number and type of its parameters and the type of the return value if any. Polymorphically redefined operations have the same signature.

Package: a mechanism for grouping UML elements, usually classes, into groups. Packages can be nested within other packages.

Pattern: a pattern is an abstract solution to a commonly occurring problem in a given context.

Polymorphism: the ability of different methods to implement the same operation, and thus to respond to the same message in different ways that are appropriate to their class. For example, objects of different subclasses in an inheritance hierarchy may respond differently to the same message, yet with a common meaning to their responses.

Post-condition: part of an operation specification; those conditions that must be true before the operation can execute.

Pre-condition: part of an operation specification; those conditions that must be true after the operation has executed—in other words the valid results of the operation.

Primary operation: an operation to create or destroy an instance of an class, or to get or set the value of an attribute.

Probe: a probe is an element of the notation of sequence diagrams (not UML) that represents a link between one sequence diagram and another via a «uses» or «extends» relationship (qv.).

Prototype: a prototype is a system or partially complete system that is built quickly to explore some aspect of the system requirements. It is not intended as the final working system.

Query operation: an operation that returns information but causes no change of state within a model or a software system.

Realizes relationship: a relationship between two classes where one class offers the interface of the other but does not necessarily have the same structure of the other. Commonly used to show that a class supports an interface. Written as a stereotype: «realizes».

Relation: a group of related data items organized in columns and rows, also known as a table.

Responsibility: a high level description of behaviour a class exhibits. It reflects the knowledge or information that is available to that class, either stored within its own attributes or requested via collaboration with other classes, and also the services that it can offer to other objects.

Sequence diagram: or interaction sequence diagram, shows an interaction between objects arranged in a time sequence. Sequence diagrams can be drawn at different levels of detail and also to meet different purposes at several stages in the development life cycle.

Service: a useful function (or set of functionality) that is carried out by an object (or a sub-system) when requested to do so by another object.

Software architecture: describes the sub-systems and components of a software system and the relationships between the components.

Specialization: the other face of generalization; an element (for example, a class) is said to be specialized when it has a set of characteristics that uniquely distinguish it from other elements. Distinguishes subclasses from their superclass.

Stakeholders: anyone who is affected by the information system. Stakeholders not only include users and development team members, but also resource managers and the quality assurance team, for example.

State: the state of an object is determined by values of some of its attributes and the presence or absence of certain links with other objects. It reflects a particular condition for the object and persists for a period of time until a transition to another state is triggered by an event.

Stereotype: a specialized UML modelling element. The stereotype name is contained within matched guillemets «...». For example, an interface package is a stereotype of a package.

Subclass: a specialized class that acquires general features from its ancestor superclasses in a generalization hierarchy, but that also adds one or more specialized characteristics of its own.

Sub-system: a part of a system that can be regarded as a system in its own right.

Superclass: a generalized class that is an abstraction of the common characteristics of its subclasses in a generalization hierarchy.

Synchronizing operation: an operation that ensures that attribute values which are dependent upon each other (e.g. may be calculated from each other) have consistent values.

System: an abstraction of a complex interacting set of elements, for which it is possible to identify a boundary, an environment, inputs and outputs, a control mechanism and some process or transformation that the system achieves.

Table: group of related data items organized in columns and rows. Used to store data in relational databases.

Task: a specific activity or step in a project.

Technique: a method for carrying out a project task.

Transaction: an elementary exchange, say of an item of capta (qv.) or of a unit of value.

Transition: the movement from one state to another, triggered by an event. A transition may start and end at the same state.

Type: a stereotype of class that is distinct from an implementation class; a type is defined by its attributes and operations but, unlike an implementation class, may not contain any methods. Classes that represent the concepts of the application domain are in fact types. An object may change its type dynamically during system execution, and may thus appear at different times to belong to different classes.

Usability requirement: user requirement that describes criteria by which the ease of use of the system can be judged.

Use case: describes, from a user's perspective, a behaviourally related set of transactions that are normally performed together to produce some value for the user. Use cases can be represented graphically in a use case diagram, each use case being described in the data dictionary. Use cases may be modelled at varying degrees of abstraction, essential use cases, the most abstract, are technologically and implementation independent whereas real use cases describe how the use case actually operates in a particular environment.

User requirement: something that users require a software system to do (functional requirement); alternatively, a standard for the performance of a system (non-functional requirement).

Uses relationship: a relationship between use cases where one use case uses the actions described in another use case. Written as a stereotype: «uses».

Visibility: UML modelling elements (e.g. attributes or operations) may be designated with different levels of accessibility or visibility. Public visibility means that the element is directly accessible by any class; private visibility means that the element may only be used by the class that includes it; and protected visibility means that the element may only be used by either the class that includes it or a subclass of that class.

Wrapper: or object wrapper, used to integrate object-oriented and non-object-oriented systems by encapsulating the non-object-oriented system with an object-oriented style of interface.

Bibliography

Adams, D., *The Restaurant At The End Of The Universe*, London: Pan Books, 1980.

Alexander, C., Ishikawa, S., Silverstein, M., Jacobson, M., Fiksdahl-King, I. and Angel, S., *A Pattern Language: Towns, Buildings, Construction*, New York: Oxford University Press, 1977.

Allen, C. P., *Effective Structured Techniques: From Strategy to CASE*, Hemel Hempstead, UK: Prentice-Hall, 1991.

Allen, P. and Frost, S., *Component-Based Development for Enterprise Systems: Applying the SELECT Perspective™*, Cambridge: Cambridge University Press; SIGS Books, 1998.

ANSI, *American National Standard for Information Systems: Database Language SQL*, ANSI X3,135-1986. New York: ANSI, 1986.

Apple Computer Inc., *Macintosh Human Interface Guidelines*, Reading, MA: Addison-Wesley, 1996.

Avison, D., and Fitzgerald, G., *Information Systems Development*, Oxford: Blackwell Scientific, 1988.

Barker, C., 'London Ambulance Service gets IT right', *Computing*, 25 June, 1998.

Beck, K. and Cunningham, W., 'A laboratory for teaching object-oriented thinking', *Proceedings of OOPSLA '89*, pp.1–6, 1989.

Bell, D., *The Coming Of Post-Industrial Society*, New York: Basic Books, 1973. Cited in Webster (1995).

Bellin, D. and Simone, S. S., *The CRC Card Book*, Reading, MA: Addison-Wesley, 1997.

Beveridge, T., 'Java Documentation' in *Java Unleashed*, Indianapolis: Sams.net Publishing, 1996.

Blaha, M. and Premerlani, W., *Object-Oriented Modelling and Design for Database Applications*, Upper Saddle River, NJ: Prentice-Hall, 1998.

Boehm, B. W., *Software Engineering Economics*, Englewood Cliffs, NJ: Prentice-Hall, 1981.

Boehm, B. W., 'A Spiral Model of Software Development and Enhancement', in Thayer, R. H. (Ed.), *Tutorial:Software Engineering Project Management*, Los Alamitos, CA: IEEE Computer Society Press, 1988.

Booch, G., *Object-Oriented Analysis and Design with Applications* (2nd Ed.), Menlo Park, CA: Benjamin/Cummings, 1994.

Booch, G., Rumbaugh, J. and Jacobson, I., *The Unified Modeling Language User Guide*, Reading, MA: Addison-Wesley; ACM Press, 1999.

Booth, P., *An Introduction to Human-Computer Interaction*, Hove: Lawrence Erlbaum Associates, 1989.

Brown, K. and Whitenack, B. G., 'Crossing Chasms: A Pattern Language for Object-RDBMS Integration' in Vlissides, J. M., Coplien, J. O. and Kerth, N. L. (Eds.) *Pattern Languages of Programme Design 2*, pp.227–238, Reading, MA: Addison-Wesley, 1996.

Brown, W. J., Malveau, R. C., McCormick, H. W. and Mowbray, T. J., *AntiPatterns: Refactoring Software, Architectures and Projects in Crisis*, New York, NY: John Wiley, 1998.

Browne, D., *STUDIO: Structured User-interface Design for Interaction Optimization*, Hemel Hempstead: Prentice-Hall, 1994.

Budgen, D., *Software Design*, Reading, MA: Addison-Wesley, 1994.

Buschmann, F., Meunier, R., Rohnert, H., Sommerlad, P. and Stal, M., *Pattern Oriented Software Architecture*, Chichester: John Wiley, 1996.

Capers Jones, T., *Programming Productivity*, New York, NY: McGraw-Hill, 1986. Cited in Yourdon, 1989.

Carlow International Inc., *Human-Computer Interface Guidelines*, prepared for the Goddard Space Flight Center, Falls Church, VA: Carlow International Inc., 1992.

Carroll, J. M. (Ed.), *Scenario-Based Design: Envisioning Work and Technology in System Development*, New York: John Wiley, 1995.

Cattell, R. G. G. and Barry, D. K. (Eds.) *The Object Database Standard: ODMG 2.0*, San Francisco: Morgan Kaufmann, 1997.

Chambers Twentieth Century Dictionary, (Revised edition), Edinburgh: W & R Chambers, 1977.

Checkland, P., *Systems Thinking, Systems Practice*, Chichester: John Wiley, 1981.

Checkland, P., 'Information Systems and Systems Thinking: Time To Unite?', *International Journal of Information Management*, **8**, pp.234–243, 1988.

Checkland, P., unpublished presentation given at *Systems for Sustainability: People, Organisations and Environments*, 5th International Conference of the United Kingdom Systems Society, Milton Keynes: The Open University, July 1997.

Checkland, P. and Holwell, S., *Information, Systems and Information Systems: Making Sense Of The Field*, Chichester: John Wiley, 1998.

Checkland, P. and Scholes, J., *Soft Systems Methodology In Action*, Chichester: John Wiley, 1990.

Coad, P. and Yourdon, E., *Object-Oriented Analysis* (2nd Ed.), Englewood Cliffs, NJ: Yourdon Press; Prentice-Hall, 1990.

Coad, P. and Yourdon, E., *Object-Oriented Design*, Englewood Cliffs, NJ: Yourdon Press; Prentice-Hall, 1991.

Coad, P. with North, D. and Mayfield, M., *Object Models: Strategies, Patterns and Applications* (2nd Ed.), Upper Saddle River, NJ: Yourdon Press; Prentice-Hall, 1997.

Codd, E. J., 'A Relational Model for Large Shared Data Banks', *Communications of the ACM*, **13**, (6), pp.377–387, 1970. (Also **26**, (1), pp.64–69, 1983)

Coleman, D., Arnold, P., Bodoff, S., Dollin, C., Gilchrist, H., Hayes, F. and Jeremaes, P., *Object-Oriented Development: The Fusion Method*, Englewood Cliffs, New Jersey: Prentice-Hall International, 1994.

Collins, T., 'Banking's Big Brother option', *Computer Weekly*, 19 November, 1998a.

Collins, T., 'MPs lambast NHS for double project fiasco', *Computer Weekly*, 12 December, 1998b.

Connor, D., *Information System Specification And Design Road Map*, Englewood Cliffs, NJ: Prentice-Hall, 1985.

Constantine, L., 'The Case for Essential Use Cases', *Object Magazine*, May 1997.

Cook, S. and Daniels, J., *Designing Object Systems: Object-Oriented Modelling With Syntropy*, Hemel Hempstead: Prentice-Hall, 1994.

Coplien J. O., *Advanced C++: Programming Styles and Idioms*, Reading, MA: Addison-Wesley, 1992.

Coplien, J., O., 'A Generative Development Process Pattern Language' in Coplien, J. O. and Schmidt, D. C., (Eds.), *Pattern Languages of Program Design*. Reading MA: Addison-Wesley, 1995.

Coplien, J. O. and Schmidt, D. C. (Eds.), *Pattern Languages of Program Design*, Reading, MA: Addison-Wesley, 1995.

Coplien, J. O., *Software Patterns*, New York: SIGS Books, 1996.

Cunningham, W., 'The CHECKS Pattern Language of Information Integrity' in Coplien, J. O. and Schmidt, D. C., (eds.), *Pattern Languages of Program Design*. Reading, MA: Addison-Wesley, 1995.

Daniels, J., unpublished presentation given at *Migrating to Object Technology*, TattOO '95, Leicester: De Montfort University, January 1995.

Date, C. J., *An Introduction to Database Systems* (6th Ed.), Reading, MA: Addison-Wesley, 1995.

Date, C. J. and Darwen, H., *Foundation for Object/Relational Databases: The Third Manifesto*, Reading, MA: Addison-Wesley, 1998.

de Champeaux, D., *Object-Oriented Development Process and Metrics*, Upper Saddle River, NJ: Prentice Hall, 1997.

Deitel, H. M. and Deitel, P. J., *Java: How to Program*, Upper Saddle River, NJ: Prentice-Hall, 1997.

DeMarco, T., *Structured Analysis and System Specification*, Upper Saddle River, NJ: Yourdon Press; Prentice-Hall, 1979.

DeMarco, T., *Controlling Software Projects*, Englewood Cliffs, NJ: Yourdon Press, 1982.

Dix, A., Finlay, J., Abowd, G. and Beale, R., *Human–Computer Interaction*, Hemel Hempstead: Prentice-Hall, 1998.

Drummond, H., 'The politics of risk: trials and tribulations of the Taurus project', *Journal of Information Technology*, **11**, pp.347–357, 1996.

Eaglestone, B. and Ridley, M., *Object Databases: An Introduction*, Maidenhead: McGraw-Hill, 1998.

Ellwood, W., 'Seduced by Technology', *New Internationalist*, December 1996.

Eriksson, H-E. and Penker, M., *UML Toolkit*, NewYork: John Wiley, 1998.

Fidler, C. and Rogerson, S., *Strategic Management Support Systems*, London: Pitman, 1996.

Finkelstein, C., *An Introduction to Information Engineering*, Wokingham: Addison-Wesley, 1989.

Flowers, S., 'IS Project Risk—The Role of Management in Project Failure', *Proceedings of BIT '97*, Manchester Metropolitan University, November 1997.

Flynn, D., *Information Systems Requirements: Determination And Analysis*, (2nd Ed.), Maidenhead: McGraw-Hill, 1998.

Fowler, M. with Scott, K., *UML Distilled: Applying The Standard Object Modeling Language*, Reading, MA: Addison-Wesley, 1997a.

Fowler, M., *Analysis Patterns: Reusable Object Models*, Reading, MA: Addison-Wesley, 1997b.

Gabriel, R., *Patterns of Software: Tales from the Software Community*, New York: Oxford University Press, 1996.

Gaines, B. R. and Shaw, M. L. G., 'Dialogue engineering' in Sime, M. E. and Coombs, M. J. (Eds.), *Designing for human-computer communication*, London: Academic Press, 1983.

Gamma, E., Helm, R., Johnson, R. and Vlissides, J.M., *Design Patterns: Elements of Reusable Object-Oriented Software*, Reading, MA: Addison-Wesley, 1995.

Gane, C. and Sarson, T., *Structured Systems Analysis: Tools and Techniques,* Englewood Cliffs, NJ: Prentice-Hall, 1978.

Gilb, T., *Principles of Software Engineering Management*, Wokingham: Addison-Wesley, 1988.

Goldsmith, S., *A Practical Guide to Real-Time Systems Development*, Hemel Hempstead: Prentice Hall, 1993.

Goodland, M. with Slater, C., *SSADM Version 4: A Practical Approach*, Maidenhead: McGraw-Hill, 1995.

Goss, E.P., Buelow, V. and Ramchandani, H., 'The Real Impact of a Point-of-Care Computer System on ICU Nurses', *Journal of Systems Management*, pp.43–47, January/February 1995.

Graham, I., Henderson-Sellers, B. and Younessi, H., *The OPEN Process Specification*, Harlow: Addison-Wesley, 1998.

Graham, I., *Object Oriented Methods,* Wokingham: Addison-Wesley, 1993.

Graham, I., *Migrating to Object Technology*, Wokingham: Addison-Wesley, 1995.

Greenbaum, J. and Kyng, M., *Design at Work: Cooperative Design of Computer Systems*, Hillsdale, NJ: Lawrence Erlbaum Associates, 1991.

Hammersley, M. and Atkinson, P., *Ethnography: Principles in Practice*, (2nd Ed.), London: Routledge, 1995.

Harel, D., 'Statecharts: a visual formalism for complex systems', *Science of Computer Programming*, **8**, pp.231–274, 1987.

Harel, D., 'On visual formalisms', *Communications of the ACM*, **31**, (5), pp.514–530, 1988.

Hart, A., *Knowledge Acquisition for Expert Systems*, London: Kogan Page, 1989. (New edition 1997 published by Chapman & Hall.)

Hatley, D. J. and Pirbhai, I. A., *Strategies for Real-Time Systems Design*, New York: Dorset House, 1987.

Hay, D., *Data Model Patterns Conventions of Thought*, New York: Dorset House, 1996.

Hicks, M., J., *Problem Solving in Business and Management,* London: Chapman & Hall, 1991.

Hopkins, T. and Horan, B., *Smalltalk: an introduction to application development using VisualWorks*, Hemel Hempstead: Prentice-Hall, 1995.

Howe, D. R., *Data Analysis for Data Base Design*, London: Edward Arnold, 1989.

Humphrey, W. S., *Managing The Software Process*, Wokingham: Addison-Wesley, 1989.

IBM, *IBM San Francisco Base*, Rochester, MN: IBM, 1998.

Ince, D., *ISO 9001 And Software Quality Assurance*, Maidenhead: McGraw-Hill, 1994.

Jackson, M., *Principles of Program Design*, London: Academic Press, 1975.

Jacobson, I., Booch, G., and Rumbaugh, J., *The Unified Software Development Process,* Reading, MA: Addison-Wesley; ACM Press, 1999.

Jacobson, I., Christerson, M., Jonsson, P. and Övergaard, G., *Object-Oriented Software Engineering: A Use Case Driven Approach,* Wokingham: Addison-Wesley, 1992.

Jayaratna, N., *Understanding and Evaluating Methodologies; NIMSAD: A Systemic Framework*, Maidenhead: McGraw-Hill, 1994.

Koestler, A., *The Ghost In The Machine*, London: Hutchinson, 1967.

Larman, C., *Applying UML and Patterns: An Introduction to Object-Oriented Analysis and Design*, Upper Saddle River, NJ: Prentice-Hall, 1998.

Loomis, M. E. S., *Object Databases: The Essentials*, Reading, MA: Addison-Wesley, 1995.

Lorenz, M. and Kidd, J., *Object-Oriented Software Metrics: A Practical Guide*, Englewood Cliffs, NJ: Prentice-Hall, 1994.

Maguire, M., *RESPECT User Requirements Framework Handbook*, European Usability Support Centres, 1997.

Martin, D., Riehle, D. and Buschmann, F. (Eds.), *Pattern Languages of Program Design 3*, Reading, MA: Addison-Wesley, 1998.

McBride, N., 'Business Use Of The Internet: Strategic Decision Or Another Bandwagon?', *European Management Journal*, **15**, pp.58–67, 1997.

Melton, A., (Ed.), *Software Measurement*, London: Thompson Computer Press, 1995.

Meyer, B., *Object Success: A Manager's Guide to Object Orientation, its Impact on the Corporation, and its Use for Reengineering the Software Process,* Hemel Hempstead: Prentice-Hall, 1995.

Meyer, B., *Object-Oriented Software Construction* (2nd Ed.), Upper Saddle River, NJ: Prentice-Hall PTR, 1997.

Microsoft Corporation, *The Windows Interface Guidelines for Software Design*, Redmond, WA: Microsoft Press, 1997.

Muller, P-A., *Instant UML*, Birmingham: Wrox Press, 1997.

Mumford, E., *Effective Systems Design And Requirements Analysis*, Basingstoke: Macmillan, 1995.

Mumford, E., *Systems Design: Ethical Tools for Ethical Change*, Basingstoke: Macmillan, 1996.

OMG, *The Common Object Request Broker Architecture and Specification; Revision 2.0*, Framingham, MA: Object Management Group Inc, 1995.

Orfali, R. and Harkey, D., *Client/Server Programming with JAVA and CORBA* (2nd Ed.), New York, NY: John Wiley, 1998.

Page-Jones, M., *The Practical Guide to Structured Systems Design* (2nd Ed.), Englewood Cliffs, NJ: Prentice-Hall, 1988.

Peters, T., *Thriving on Chaos: Handbook for a Management Revolution*, London: Macmillan, 1988.

Philips, T., 'A Slicker System For The City', *The Guardian On-Line*, 3 April, 1997.

Porter, M., *Competitive Strategy*, New York: Free Press, 1985.

Preece, J. with Rogers, Y., Sharp, H., Benyon, D., Holland, S. and Carey, T., *Human–Computer Interaction*, Wokingham: Addison-Wesley, 1994.

Pressman, R., *Software Engineering: A Practitioner's Approach*, European Edition, London: McGraw-Hill, 1997.

Rational Software Corporation, *OCL Specification—Unified Modeling Language v1.1*, http://www.rational.com/uml/1.1/, 1997.

Rational Software Corporation, *UML Notation Guide—Unified Modeling Language v1.1*, http://www.rational.com/uml/1.1/, 1997.

Rational Software Corporation, *UML Semantics—Unified Modeling Language v1.1*, http://www.rational.com/uml/1.1/, 1997.

Rational Software Corporation, *UML Summary—Unified Modeling Language v1.1*, http://www.rational.com/uml/1.1/, 1997.

Riehle, D. and Zullighoven, H., 'Understanding and Using Patterns in Software Development', *Theory and Practice of Object Systems*, **2**, (1), pp.3-13, 1996.

Rifkin, J., *The End Of Work*, New York: Putnam Publishers, 1995. Cited in Ellwood, W., 'Seduced by Technology', *New Internationalist*, December 1996.

Roberts, G., *The Users' Role in Systems Development*, Chichester: John Wiley; Central Computer and Telecommunications Agency, 1989.

Robinson, B. and Prior, M., *Systems Analysis Techniques*, London: International Thomson Publishing, 1995

Rosson, M. B. and Carroll, J. M., 'Narrowing the Specification-Implementation Gap in Scenario-Based Design', in Carroll, J. M. (Ed.), *Scenario-Based Design: Envisioning Work and Technology in System Development*, New York: John Wiley, 1995.

Rumbaugh, J., 'Models through the development process', *Journal of Object-Oriented Programming*, May 1997.

Rumbaugh, J., Blaha M., Premerlani, W., Eddy, F. and Lorensen, W., *Object-Oriented Modeling and Design*, Englewood Cliffs, NJ: Prentice-Hall International, 1991.

Rumbaugh, J., Jacobson, I. and Booch, G., *The Unified Modeling Language Reference Manual*, Reading, MA: Addison-Wesley; ACM Press, 1999.

Sachs, P., 'Transforming Work: Collaboration, Learning and Design', *Communications of the ACM*, **38**, (9), pp.36–44, 1995.

Sanders, J. and Curran, E., *Software quality: a framework for success in software development and support*, Wokingham: Addison-Wesley, 1994.

Sauer, C., *Why Information Systems Fail: A Case Study Approach*, Maidenhead: McGraw-Hill, 1993.

Schneider, K., 'Bug delays £25m court case system', *Computer Weekly*, 22 December, 1997.

Senn, J. A., *Analysis & Design of Information Systems*, (2nd Ed.), New York: McGraw-Hill, 1989.

Shackel, B., 'Human factors and usability', in Preece, J. and Keller, L. (Eds.), *Human-Computer Interaction: Selected Readings*, Hemel Hempstead: Prentice-Hall, 1990.

Shlaer, S. and Mellor, S., *Object-Oriented Systems Analysis: Modeling the World in Data*, Englewood Cliffs, NJ: Prentice-Hall, 1988.

Shlaer, S. and Mellor, S., *Object Lifecycles: Modeling the World in States*, Englewood Cliffs, NJ: Prentice-Hall, 1992.

Shneiderman, B., *Designing the User Interface* (3rd Ed.), Reading, MA: Addison-Wesley, 1997.

Silberschatz, A., Korth, H. F. and Sudarshan, S., *Database System Concepts* (3rd Ed.), New York: McGraw-Hill, 1996.

Skidmore, S., *Introducing Systems Analysis* (2nd Ed.), Oxford: NCC Blackwell, 1994.

Skidmore, S., Mills, G. and Farmer, R., *SSADM Models & Methods*, Manchester: NCC Blackwell, 1994.

Sommerville, I., *Software Engineering* (4th Ed.), Reading, MA: Addison-Wesley, 1992.

Soni, D., Nord, R. and Hofmeister, C., 'Software Architecture in Industrial Applications', in *Proceedings of the 17th International Conference on Software Engineering*, pp.196–207, Seattle, WA: ACM Press, 1995.

Standish Group, *Chaos Report*, Standish Consulting Group, 1995. Cited in Flowers, S., 'IS Project Risk—The Role of Management in Project Failure', *Proceedings of BIT '97*, Manchester Metropolitan University, November 1997.

Stapleton, J., *Dynamic Systems Development Method*, Harlow: Addison-Wesley, 1997.

Symons, V., 'Evaluation of information systems: IS development in the processing company', *Journal of Information Technology*, **5**, pp.194–204, 1990.

Tanenbaum, A. S., *Modern Operating Systems*, Englewood Cliffs, NJ: Prentice Hall, 1992.

Texel, P., P., and Williams, C., B., *Use Case Combined with Booch/OMT/UML: processes and products*, Upper Saddle River, NJ: Prentice Hall, 1997.

The Concise Oxford Dictionary (9th Ed.), Oxford: OUP, 1995.

Vlissides, J.M., Coplien, J. O. and Kerth, N., (eds.), *Pattern Languages of Program Design 2*, Reading, MA: Addison-Wesley, 1996.

Wall Street Journal, 22 September, 1998, reprinted in *The Guardian Editor*, 26 September, 1998.

Ward, P. and Mellor, S., *Structured Development for Real-Time Systems Vol 1: Introduction & Tools*, Englewood Cliffs, NJ: Yourdon Computing Press, 1985.

Ward, P. and Mellor, S., *Structured Development for Real-Time Systems Vol 2: Essential Modelling Techniques*, Englewood Cliffs, NJ: Yourdon Computing Press, 1985.

Ward, P. and Mellor, S., *Structured Development for Real-Time Systems Vol 3: Implementation Modelling Techniques*, Englewood Cliffs, NJ: Yourdon Computing Press, 1986.

Webster, F., *Theories Of The Information Society*, London: Routledge, 1995.

Weir, C. and Daniels, J., 'Software Architecture Document', in *Proceedings of Object Technology 98*, Oxford, 1998.

Whiteside, J., Bennett, J. and Holtzblatt, K., 'Usability engineering: our experience and evolution', in Helander, M. (Ed.), *Handbook of Human-Computer Interaction,* Amsterdam: North-Holland, 1988.

Whitten, J., Bentley, L. and Barlow, V., *Systems Analysis and Design Methods* (3rd Ed.), Irwin, 1994.

Willcocks, L. and Lester, S., 'Any Way Out Of The Labyrinth? Information Technology Productivity Revisited', *The Future Of Information Systems: UK Academy For Information Systems—First Conference*, Cranfield University, April 1996.

Wirfs-Brock, R., Wilkerson, B. and Wiener, L., *Designing Object-Oriented Software,* Englewood Cliffs, NJ: Prentice-Hall International, 1990.

Yourdon, E. and Constantine, L. L., *Structured Design: Fundamentals of a Discipline of Computer Program and Systems Design*, Englewood Cliffs, NJ, USA: Yourdon Press; Prentice-Hall, 1979

Yourdon, E., *Modern Structure Analysis*, Englewood Cliffs, NJ: Prentice-Hall, 1989.

Yourdon, E., *Object-Oriented Systems Design: An Integrated Approach*, Englewood Cliffs, NJ: Prentice-Hall International, 1994.

Yourdon, E., *Structured Walkthroughs*, Englewood Cliffs, NJ: Yourdon Press, 1985

Zuboff, S., *In The Age Of The Smart Machine: The Future Of Work And Power*, Oxford: Heinemann, 1988.

Index